Computational Biomedicine

COMPUTATIONAL
BIOMEDICINE

MODELLING THE HUMAN BODY

Edited by

Peter V. Coveney, University College London

Vanessa Díaz-Zuccarini, University College London

Peter Hunter, University of Auckland and University of Oxford

Marco Viceconti, University of Sheffield

OXFORD
UNIVERSITY PRESS

Great Clarendon Street, Oxford OX2 6DP,
United Kingdom

Oxford University Press is a department of the University of Oxford.
It furthers the University's objective of excellence in research, scholarship,
and education by publishing worldwide. Oxford is a registered trade mark of
Oxford University Press in the UK and in certain other countries

Published in the United States of America by Oxford University Press
198 Madison Avenue, New York, NY 10016, United States of America

British Library Cataloguing in Publication Data

Data available

Library of Congress Control Number: 2013956438

ISBN 978–0–19–965818–3

Printed in Great Britain by
Bell & Bain Ltd, Glasgow

PREFACE

At the present time, medicine is on the verge of a radical transformation driven by the inexorably increasing power of information technology. We live in what is sometimes called the digital era, wherein information is widely accessible in electronic form, made available over the Internet by the worldwide web. Medical information is no different, and today there is a relentless push to digitize healthcare data, for example in the form of electronic health records, making it more facile for patients, doctors, and other healthcare workers to track the records of individual patients, to improve treatments, and to monitor outcomes of clinical care.

Of course, medicine and human health are intensely personal matters, as well as being subject to scientific understanding, and so accessing such data is surrounded by necessary considerations of privacy and confidentiality. The security of personal healthcare data is subject to rules and regulations in information governance which frequently threaten to undermine the entire programme of digital medicine, when the balance between access to such records is countered by arguments about the need to maintain patient privacy. At all times it is essential to remember that access to such data has the potential to enable medical cures and improve people's lives, while overly zealous attention to protection of personal privacy denies this.

As and when access to human healthcare data is permitted, modern science can progress rapidly. The immediate wins in basic and clinical medicine centre on the analysis of large collections of such data, using a variety of methods collectively referred to today as machine learning. These methods are inherently statistical in nature, providing inferences about medical conditions based on correlations between 'input' and 'output' variables. Their computational cost is relatively low and many medical discoveries of this nature are possible by the mining of vast quantities of data. One of the most promising of such approaches is stratification, the clustering of subsets of the population into groups which are found to be susceptible to a particular disease and, more importantly, a specific form of medical treatment. With increasing quantities of data, particularly whole human genomes being acquired with astonishing speed from so-called next-generation sequencing technology, there is an expectation that we will discover well-defined clustering of human populations, along with the aspiration that we shall be able to target these groups with different treatments (thus, for example, drugs, regimens, and so on can be tailor-made to stratified groups). At this point, medicine for the masses (a kind of average way of treating everyone) becomes more about medicine for well-defined groups of patients.

Thus digital technology points to a new era of stratified medicine, in which specific electronic information on a patient allows a doctor to perform a treatment better tailored to the individual concerned. We are all different, however, and such statistical approaches, being based on data acquired on populations of many different patients, inevitably smear out individual differences. In the case of rare diseases, for example, there may not even be a sufficient number of other similar conditions to make reliable inferences. As we discover more about the intricacies of other diseases—cancer is one case in point—it is becoming clear that the mechanisms behind a disease can vary significantly from one patient to another.

To go beyond the power of inference-based methods, and to develop a truly personalized form of medicine, a different approach is required, one in which a deeper understanding of the *mechanisms* of disease are taken into account. These mechanisms may vary to a greater or lesser extent between individuals, and understanding these differences opens the door to more scientific ways of treating the disease and curing it in a convincing manner. Full control of disease cases will emerge when we understand these mechanisms in a quantitative manner, and are able to reliably *predict* individual outcomes before applying treatments. This depends on our ability to mathematically represent and thus model the human body and its pathologies. The human body is a complex system, and its mathematical description necessitates the running of models on computers, a medical form of computer simulation. This too is heavily dependent on information technology, including an ability to access powerful computers as well as large-scale data, the aim being to produce 'high-fidelity' descriptions of medical conditions and the effects of proposed treatments, before these are performed. It is this new approach to medical science that forms the basis of this textbook.

We look at the many levels of biological organization in the human body, from the molecular through cellular and tissue to organ systems, and how one can successfully begin to model these using principles taken from the physical and engineering sciences. These approaches always combine some patient-specific data—it might be from a genomic analysis, one or more imaging modalities, or combinations of these—with a mathematically based mechanistic model. In some cases, such models are already gaining a reasonable degree of personal human fidelity and are being used in research contexts to address clinical conditions.

Because the various levels of physiological organization are not independent but in reality influence one another to varying degrees, the modelling challenge we face is compounded by the need to consider how processes occurring on these levels, often on very different space and time scales, interact. This is the heart of systems biology, although it might better be referred to as 'systems medicine' in the present context. The use of 'systems' here implies that we are addressing a multi-scale problem upon which the overall physiological or pathological behaviour depends. And so we look at how multi-scale modelling and simulation are performed today too.

Beyond that, there are the practical issues of how to perform many complex and demanding human body simulations, in a timely and reliable manner, through the use of computer automation, by exploiting computational workflows and distributed high-performance computing environments.

Underpinning all of these emerging capabilities is the management of healthcare data collection and provision with all the attendant privacy issues, which is linked closely to the legal frameworks that pertain within individual countries and for international collaboration. We describe the architecture of the informatics platforms necessary to manage access to patient data in a suitably anonymized format, while permitting the reverse linkage to patient identity in cases where this is needed, in conformance with legal and ethical requirements.

For the future deployment of such models and simulations in clinical decision making it is vital that they are fully verified and validated. This issue, central for building confidence in the medical community, is also in fact a key issue in all forms of computational science today, and receives attention in the final chapter of the book.

To whom is our textbook addressed? Plainly the scope of the domains covered is dauntingly broad and we cannot reasonably expect all readers to be able to comprehend everything in detail. We have primarily focused on providing a comprehensible account for advanced undergraduates and beginning postgraduate students who have a background in any of mathematics, physical sciences, engineering, and computer science. That is, we have tried to present the required biology and medicine to readers who are familiar with mathematics (including discrete and continuous dynamical systems, and ordinary and partial differential equations), numerical analysis, and programming, but who need more explanation about the human phenomena that they will need to model.

We hope that this textbook will provide a firm foundation for future generations to build on the initial progress now being made in predictive mechanistic modelling of the human body using computers. These are the very first steps along the path to virtual humans which we hope will one day assist in enhancing health and wellbeing for all the inhabitants of this planet.

Finally, we wish to thank the many friends and colleagues who have contributed chapters to this book (they are all listed at the start of this book and at the end of the chapters they have composed), along with numerous others who provided critical and constructive feedback to us at all stages along the way to the published book. We are especially grateful to Dr Clare Sansom who has acted as Assistant Editor to us in the labour of taking all of the original content and transforming it into a homogeneous textbook. In all matters pertaining to the production of the book we are indebted to Jonathan Crowe, our commissioning editor at Oxford University Press, who has taken a very close interest in this work from inception to completion.

Peter V. Coveney
Vanessa Díaz-Zuccarini
Peter Hunter
Marco Viceconti

25 October 2013

ABOUT THE EDITORS

Peter Coveney holds a Chair in Physical Chemistry, is Director of the Centre for Computational Science, and is an Honorary Professor in Computer Science at University College London. He is also Professor Adjunct within the Medical School at Yale University. Coveney is active in a broad area of interdisciplinary theoretical research including condensed matter physics and chemistry, materials science, life and medical sciences. He has published over 350 scientific papers and texts including two best-selling popular science books, *The Arrow of Time* and *Frontiers of Complexity*.

Vanessa Díaz-Zuccarini is a lecturer in Bioengineering at University College London. Working at the interface of systems biology and engineering, she is the founder of the Multiscale Cardiovascular Engineering Group in UCL, which she now leads. Her research aims to apply multi-physics and multi-scale computational techniques to tackle challenges in cardiovascular science from a patient-specific point of view. She focuses on the modelling and simulation of cardiovascular physiology and pathology by using a combination of systems biology and information and communication technologies for health and biomechanics. She has a strong interest in bringing research results to the clinic and has developed close links with UCL-affiliated hospitals, particularly the vascular unit at University College Hospital and the Institute of Child Health.

Peter Hunter FRS is Professor of Engineering Science and Director of the Bioengineering Institute at the University of Auckland, New Zealand, and co-Director of Computational Physiology at the University of Oxford, UK. As a recent Chair of the Physiome Committee of the International Union of Physiological Sciences he has helped to lead the international Physiome Project, which is using computational methods to understand the integrated physiological function of the body in terms of the structure and function of tissues, cells, and proteins. He has been involved with developing many of the standards necessary to facilitate reproducible multi-scale modelling.

Marco Viceconti holds the chair of Biomechanics within the Department of Mechanical Engineering of the University of Sheffield and serves as Scientific Director of the Insigneo Institute for *in silico* medicine there. His main research interests are in the development and validation of medical technology for neuromusculoskeletal diseases. Viceconti is a central figure in the emerging Virtual Physiological Human (VPH) community: co-ordinator of the VPH research roadmap and of the VHPOP osteoporosis project, and 'VPH ambassador' for the VPH Network of Excellence. He is the current chair of the Board of Directors of the VPH Institute, an international non-profit organization.

CONTENTS

ONLINE RESOURCE CENTRE

The Online Resource Centre to accompany *Computational Biomedicine*, at http://www.oxfordtextbooks.co.uk/orc/coveney/, features the following materials to support teaching and learning:

- Figures from the book in electronic format, for use in lecture slides [for registered adopters of the book only];
- Additional materials to augment topics discussed in certain chapters.

CONTRIBUTING AUTHORS

Assistant Editor: Clare Sansom, Birkbeck, University of London, London, UK

Chapter 1

Peter V. Coveney, University College London, UK

Denis Noble, University of Oxford, Oxford, UK

Chapter 2

Imad Abugessaisa, The Unit of Computational Medicine, Department of Medicine, Center for Molecular Medicine, Karolinska Institutet, Karolinska University Hospital, Stockholm, Sweden

David Gomez-Cabrero, The Unit of Computational Medicine, Department of Medicine, Center for Molecular Medicine, Karolinska Institutet, Karolinska University Hospital, Stockholm, Sweden

Jesper Tegnér, The Unit of Computational Medicine, Department of Medicine, Center for Molecular Medicine, Karolinska Institutet, Karolinska University Hospital, Stockholm, Sweden

Chapter 3

Jon Olav Vik, Centre for Integrative Genetics, Department of Animal and Aquacultural Sciences, Norwegian University of Life Sciences, Ås, Norway

Arne B. Gjuvsland, Centre for Integrative Genetics, Department of Animal and Aquacultural Sciences, Norwegian University of Life Sciences, Ås, Norway

Bernard De Bono, Auckland Bioengineering Institute (ABI), University of Auckland, New Zealand; Centre for Health Informatics & Multiprofessional Education (CHIME), University College London, 3rd Floor, Wolfson House, Stephenson Way, London NW1 3HE, UK

Stig W. Omholt, Department of Circulation and Medical Imaging, Norwegian University of Science and Technology, Trondheim, Norway

Chapter 4

Alejandro Frangi, Center for Computational Imaging & Simulation Technologies in Biomedicine, Information & Communication Technologies Department, Universitat Pompeu Fabra, Barcelona, Spain and Department of Mechanical Engineering, University of Sheffield, Sheffield, UK

Denis Friboulet, Université de Lyon, CREATIS, CNRS UMR5220, Inserm U1044, INSA-Lyon, Université Lyon 1, France

Nicholas Ayache, Inria Asclepios Team, Sophia Antipolis, France

Hervé Delingette, Inria Asclepios Team, Sophia Antipolis, France

Tristan Glatard, Université de Lyon, CREATIS, CNRS UMR5220, Inserm U1044, INSA-Lyon, Université Lyon 1, France

Corné Hoogendoorn, Center for Computational Imaging & Simulation Technologies in Biomedicine, Information & Communication Technologies Department, Universitat Pompeu Fabra, Barcelona, Spain

Ludovic Humbert, Center for Computational Imaging & Simulation Technologies in Biomedicine, Information & Communication Technologies Department, Universitat Pompeu Fabra, Barcelona, Spain

Karim Lekadir, Center for Computational Imaging & Simulation Technologies in Biomedicine, Information & Communication Technologies Department, Universitat Pompeu Fabra, Barcelona, Spain

Ignacio Larrabide, Center for Computational Imaging & Simulation Technologies in Biomedicine, Information & Communication Technologies Department, Universitat Pompeu Fabra, Barcelona, Spain

Yves Martelli, Center for Computational Imaging & Simulation Technologies in Biomedicine, Information & Communication Technologies Department, Universitat Pompeu Fabra, Barcelona, Spain

Françoise Peyrin, Université de Lyon, CREATIS, CNRS UMR5220, Inserm U1044, INSA-Lyon, Université Lyon 1, France

Xavier Planes, Center for Computational Imaging & Simulation Technologies in Biomedicine, Information & Communication Technologies Department, Universitat Pompeu Fabra, Barcelona, Spain

Maxime Sermesant, Inria Asclepios Team, Sophia Antipolis, France

Maria-Cruz Villa-Uriol, Department of Mechanical Engineering, University of Sheffield, Sheffield, UK

Tristan Whitmarsh, Center for Computational Imaging & Simulation Technologies in Biomedicine, Information & Communication Technologies Department, Universitat Pompeu Fabra, Barcelona, Spain

David Atkinson, UCL Centre for Medical Imaging, 250 Euston Road, London NW1 2PG, UK

Chapter 5

James B. Bassingthwaighte, Department of Bioengineeering, University of Washington, Seattle, USA

Colin Boyle, Department of Mechanical Engineering, University College London, London, UK

Cesar Pichardo-Almarza, InScilico Ltd, London, UK

Vanessa Díaz-Zuccarini, Department of Mechanical Engineering, University College London, London, UK

Chapter 6

S. Randall Thomas, IR4M UMR8081 CNRS, Orsay & Villejuif, France and Université Paris-Sud XI, Orsay, France

Rod Smallwood, University of Sheffield, Sheffield, UK

Nic Smith, King's College, London, UK

Cesar Pichardo-Almarza, InScilico Ltd, London, UK

Chapter 7

Alfons Hoekstra, Computational Science, Faculty of Science, University of Amsterdam, The Netherlands

Bastien Chopard, Computer Science, University of Geneva, Switzerland

Pat Lawford, Medical Physics Group, Department of Cardiovascular Science, University of Sheffield, UK

Chapter 8

Daniel A. Silva Soto, University of Sheffield, Sheffield, UK

Steven Wood, Sheffield Teaching Hospitals, Sheffield, UK

Susheel Varma, University of Sheffield, Sheffield, UK

Rodney Hose, University of Sheffield, Sheffield, UK

Chapter 9

Stefan J. Zasada, Centre for Computational Science, University College London, London, UK

Peter V. Coveney, Centre for Computational Science, University College London, London, UK

Chapter 10

Ali Nasrat Haidar, Centre for Computational Science, University College London, 20 Gordon Street, London, WC1H 0AJ, UK

Nikolaus Forgó, Institut für Rechtsinformatik/Institute for Legal Informatics, Leibniz Universität, Hannover, Germany

Hartwig Gerhartinger, Institut für Rechtsinformatik/ Institute for Legal Informatics, Leibniz Universität, Hannover, Germany

Chapter 11

Marco Viceconti, University of Sheffield, Sheffield, UK

Norbert Graf, Saarland University, Saarbrücken, Germany

Appendix

Poul Nielsen, Auckland Bioengineering Institute (ABI), University of Auckland, New Zealand

Bernard De Bono, Auckland Bioengineering Institute (ABI), University of Auckland, New Zealand ; Centre for Health Informatics & Multiprofessional Education (CHIME), University College London, 3rd Floor, Wolfson House, Stephenson Way, London NW1 3HE, UK

Peter Hunter, Auckland Bioengineering Institute (ABI), University of Auckland, New Zealand and University of Oxford, Oxford, UK

With additional thanks to Benjamin Bhattacharya-Ghosh for his assistance with the production of a number of the figures.

1 Introduction

Learning Objectives

After reading this chapter, you should:

- understand the concept of and the philosophy behind human systems biology and its application in translational medicine;
- be able to describe the philosophy behind major international and multidisciplinary initiatives in this field such as the European Commission's Virtual Physiological Human programme;
- have an overview of topics that will be introduced throughout this book.

1.1 Introduction

When our children and grandchildren visit their doctors in the latter part of this century, what will they find? They may encounter computer-based avatars of themselves, programmed with their individual genetic makeup and physiological conditions. Doctors will be able to use these to refine and test tailor-made, personalized treatment programmes. 'One size fits all' medicine would then truly be part of the past.

If this vision of the future of medicine is to come to pass it will owe much to the new discipline of computational biomedicine. This represents an important paradigm shift in the biomedical sciences and clinical medicine in the early twenty-first century, and it is the topic of this book. Computational biomedicine itself owes much to the last revolution in biomedicine: human genomics [1]. The human genome sequence was first published in 2003 as the culmination of an international project that had taken about 18 years and cost an estimated US$2.7 billion (see **http://www.genome.gov**). Yet the pace of technological development has advanced so rapidly in the 10 years or so since then that the cost of sequencing a single human genome is already well below $10 000 and may well drop to the tantalizing figure of $100 or even less within a few years. By then, the much heralded era of personalized genomics—but not yet that of fully personalized medicine—will have arrived. Already, direct-to-consumer businesses are being set up to profile genomes and feed information back to individuals, providing advice and counselling where needed. The first of these to be set up, such as California-based 23andMe[1], work with profiles based on 0.5–2 million 'interesting' base positions out of the 3.2 billion in the human genome, but similar services based on profiles of the complete genome are not far behind. Furthermore, 23andMe engages its users in research into the genetic causes of disease by encouraging them to take part in surveys and linking user-reported disease incidence to known genotypes.

The growing influence of molecular biology on medicine during the 60-odd years since the discovery

---

[1] https://www.23andme.com/

of the structure of DNA cannot be doubted. However, it is becoming increasingly clear even to professional geneticists that knowing an individual's genome sequence cannot answer all the questions about their risk of and predisposition to disease and that genomics by itself will never be able to answer all research questions in biomedicine. After all, we humans have (approximately) a mere 23 000 genes in our genomes, fewer than many more primitive organisms. This apparent paradox can be at least partly resolved by considering the complexities that arise from non-linear interactions between genes and their protein products. As yet, we understand only a small part of the function and mechanism of gene–gene, gene–protein, and protein–protein networks and interactions.

Yet even this is not sufficient to explain the full complexity of human physiology. These complex regulatory networks are themselves influenced by events that occur at longer length and timescales than molecular ones, at the level of cells, tissues, organs, individual organisms, and their environment [2,3]. The ways in which interactions and events that occur on the level of an organism's phenotype can act as determinants of genetic behaviour—so-called downward causation as opposed to the upward causation that is the equally important influence of genetic interactions on physiology—are described by Denis Noble in *The Music of Life* [4].

1.2 Systems Biology

The study of living organisms as systems built from different spatial and temporal levels, and the interactions between them, is termed systems biology. Due to its intrinsic complexity, this type of study necessarily involves computer-based modelling and simulation. The techniques that are used can be generically termed computational biomedicine, and this is the title that has been given to our book.

Any given biological phenomenon, whether associated with normal human (or animal) physiology or with disease, is modelled, in Nobel laureate Sydney Brenner's famous phrase, 'from the middle out' [5]. This involves constructing a computational description of the phenomenon based on the most appropriate level of complexity while building in

connections from levels both above and below it in the hierarchy [6]. For example, a model of a single cardiac cell might start with equations for currents passing across its membrane through ion channels, whereas a model of the whole heart would start with equations that model its electrical activation and muscle contraction on a much larger scale. This approach is interdisciplinary and relies on scientists trained in disciplines as disparate as mathematics and computation, physics, chemistry, molecular and cell biology, physiology, and medicine collaborating together. It also involves engineers, at least where the modelling of tissues and organs is involved, and, wherever possible, clinicians.

Systems biology is often thought of as being intimately linked to another computational discipline: bioinformatics. This term is used to describe the computational analysis of biological data: generally, but not always, molecular-scale data such as gene sequences and protein structures. Yet these two complementary disciplines are essentially based on completely different philosophical approaches. Systems biology can be thought of as a scientific implementation of the ideas of philosopher Sir Karl Popper (1902–1994), in which models based on aspects of reality are used to make predictions that are tested through observation. The models are judged on their ability to correctly predict natural phenomena. These models will always be approximate and provisional, as adjustments will always be necessary following observations of the external system being modelled: much more rarely, a model will be completely redesigned after the paradigm on which it is constructed is rethought. This cycling from model to observation and back again (or hypothesis testing) is in tune with the way that many biologists think.

Bioinformatics, in contrast, is based on an 'inductivist' view of science derived from the thinking of philosopher Sir Francis Bacon (1561–1626), a pioneer of the scientific method. Put simply, this states that data (and look-up tables) are all that are needed for a complete scientific understanding of phenomena. There is no need for models; should discrepancies between predicted and actual phenomena be observed, it will always be possible to resolve these once sufficient data have been collected. These two approaches remain complementary, however. We still know remarkably little about the fundamental processes that are involved in human health and disease, and both predictive modelling and data mining will remain equally important in translational research into the foreseeable future.

1.3 Initiatives for Modelling Human Physiology

Computational physiology on a large scale is necessarily an international and collaborative endeavour. The European Union has for decades taken a more interdisciplinary and applied approach to scientific research than many of its constituent national governments. It is, therefore, not surprising that it has been at the forefront of funding developments in human systems biology.

One example of a programme that aims to study human systems biology on a large scale is the Virtual Physiological Human (VPH) initiative, which was part of the European Commission's Framework Programme Seven (2007–2013). The aim of initiatives like this one, and their successors, is the eventual representation of the human body as 'a single, coherent and dynamical system'. It encompasses and combines models at a variety of scales, from single genes and proteins to the individual and, beyond that, into population studies. This European Commission initiative has evolved from previous international endeavours that sought to integrate a computational and systems-based approach into physiology and medicine. The first of these was the grass-roots International Physiome Project, set up under the auspices of the International Union of Physiological Sciences [7].

One important feature of this programme, as of all serious initiatives within the sphere of computational biomedicine, is a very strong emphasis on the development of applications in medical and clinical practice. The ideas behind systems physiology imply that computational and experimental scientists should, wherever possible, collaborate with clinicians as integral partners and work towards, if not produce, models and tools that will be of practical use in the clinic.

Researchers and clinicians can find themselves equally challenged by the demands of this collaboration. Scientists need to think about how their models can be applied to clinical practice by, for example, taking care to develop tools that are easy for busy, stressed, non-specialist clinicians to use on a day-to-day basis. Clinicians need to learn, or perhaps to remember, about the rigour of research methods and to explain their requirements to non-medically trained colleagues clearly and without jargon. Computational biomedicine is bound to influence future medical training, and we would urge medical students who wish to understand how their discipline is likely to develop during the coming century to read this book.

1.4 Book Synopsis

This book comprises a complete overview of the tools and techniques of computational biomedicine and its applications to human physiology and clinical medicine. Its contents fall naturally into two main sections, and its many examples and case studies have a particular focus on the translation of models into clinical practice, both now and in the future. These case studies are illustrative of the general approach.

The first section (Chapters 2–7) covers the different types of modelling that are applicable to the spatial and temporal scales in the human body, and gives examples of how these are being combined and applied in a translational context. Chapters 2 and 3 describe how research at the omics level—the genome, the proteome, and the transcriptome—relates to and can be applied in current medical practice. Genomic and proteomic information is already being used in the diagnosis and treatment of disease on an individual level in a number of ways. It is perhaps most advanced in cancer, where patients are often selected for clinical trials based on the presence or absence of mutations in particular genes in their tumour genomes, and where new genetic aberrations are being identified through the comparison of tumour and equivalent normal genomes (or, often, the coding regions of those genomes, termed the exomes) [8].

The next three chapters each describe types of modelling that are being applied at different scales within the human body, moving roughly from smaller to larger scales. These cover, in order, image- and signals-based modelling, cell modelling, and tissue and organ modelling. We should eventually be able to combine these into a model of a complete organ or system, perhaps even the virtual human envisaged in the term VPH.

The last chapter in this first section explains how models on various temporal and spatial scales can be combined, before discussing some examples of multi-scale modelling that have, or that are expected to have, practical applications in the clinic. The examples selected give some idea of the breadth, depth, and applicability of current research in computational biomedicine, and also highlight the importance that is placed on ensuring that all models are properly verified and validated.

The large proportion of our selected examples that concern the modelling of the heart and cardiovascular system simply reflects the fact that cardiovascular modelling is more mature as a subdiscipline than that of any other organ system and that, in general, models of the heart are the closest to having substantial impact in clinical practice [9]. Other important physiological or pathological domains in which significant progress is being made, and which are covered here, include the pulmonary system and the processes of cancer growth and development.

In the second section (comprising Chapters 8–11) we turn from specific types, levels, and applications of modelling to the infrastructure that is needed for such computationally intensive modelling to take place and for models to be integrated together and, eventually, implemented in a clinical environment. Many (although not all) models of human physiological systems are large in scale, three-dimensional, and often, indeed ideally, patient-specific, and these require access to substantial computational resources. They are typically too large to be accessed from any one location, so grid- or cloud-based systems have become the norm. Often, enormous volumes of distributed and highly heterogeneous data must be made readily and quickly accessible by all the partners in a large, international project. This, in turn, can make significant technical demands on the managers of a project, and can also raise ethical questions; for example, under which jurisdiction's law should issues related to privacy, data protection, etc., be considered? There are many cases where it is impossible or it may not be clinically desirable to completely anonymize patient data, and this has implications for privacy and patient consent.

The first two chapters of this part of the book concern workflows, which are the series of automated steps through which a complex computational simulation and analysis is managed. We start by discussing the basic principles of workflows in human systems biology and their clinical applications, and go on to describe the implementation of workflows on high-performance computational (HPC) grids. In Chapter 10 we discuss some of the important technical and clinical issues that must be assessed before computational models of human physiology can be applied in practice; and the associated data-management issues. All models and simulations are based on data and rely on more data for modification and final validation. Managing and maintaining secure access to data from disparate sources, some of which is patient-specific, is a complex process [10].

Our concluding chapter describes how computational biomedicine is beginning to impact clinical practice, explaining, for instance, how computational models are treated by regulatory agencies and how they can be evaluated through clinical trials. Finally we touch on a few of the future developments that will be needed before the virtual patient can become a useful part of routine medical care.

Throughout the book we highlight the importance of common standards in multidisciplinary research. These are touched on in many places and summarized in more detail in the Appendix. Where data and models from different groups and often from different countries need to be combined, as is undoubtedly the case with large-scale systems biology projects, standardization becomes crucially important. Initiatives such as the well-known Gene Ontology project [11] are increasingly necessary to ensure that concepts and terms are used consistently and systematically by all researchers involved in such a project. Markup languages based on XML, the 'universal' Extensible Markup Language—including CellML for cellular models [12] and Systems Biology Markup Language (SBML) for models of metabolism and cell signalling—have been developed to enable models built by different groups to be combined. These are also touched on wherever relevant and expanded on in more depth in the Appendix.

References

1. Collins, F.S. (2010) *The Language of Life: DNA & the Revolution in Personalized Medicine*. New York: Harper Collins.

2. Kohl, P. and Noble, D. (2009) Systems biology and the virtual physiological human. *Molecular Systems Biology* 292, 1–6.

3. Kohl, P., Crampin, E.J. et al. (2010) Systems biology: an approach. *Clinical Pharmacology & Therapeutics* 88, 25–33.

4. Noble, D. (2006) *The Music of Life*. Oxford: Oxford University Press.

5. Brenner, S., Noble, D. et al. (2001) Understanding complex systems: top-down, bottom-up or middle-out? In *Complexity in Biological Information Processing*, vol. 239, Novartis Foundation Symposium, Bock, G. and Goode, J. (eds), pp. 150–159. Chichester: Wiley.

6. Coveney, P.V. and Fowler, P.W. (2005) Modelling biological complexity: a physical scientist's perspective. *Journal of the Royal Society Interface* 2, 267–280.

7. Hunter, P.J. and Borg, T.K. (2003) Integration from proteins to organs: the Physiome Project. *Nature Reviews Molecular and Cellular Biology* 4, 237–243.

8. Puente, X.S., Pinyol, M. et al. (2011) Whole genome sequencing identifies recurrent mutations in chronic lymphocytic leukaemia. *Nature* 475(7354), 101–105.

9. Bassingthwaighte, J., Hunter, P. and Noble, D. (2009) The Cardiac Physiome: perspectives for the future. *Experimental Physiology* 94(5), 597–605.

10. Heenay, C., Hawkins, N. et al. (2010) Assessing the privacy risks of data sharing in genomics. *Public Health Genomics* 14, 17–25.

11. Ashburner, M., Ball, C.A. et al. for the Gene Ontology Consortium (2000) Gene ontology: tool for the unification of biology. *Nature Genetics* 25(1), 25–29.

12. Miller, A.K., Marsh, J. et al. (2010) An overview of the CellML API and its implementation. *BMC Bioinformatics* 11, 178.

Chapter written by
Peter V. Coveney, University College London, UK;
Denis Noble, University of Oxford, Oxford, UK.

2 Molecular Foundations of Computational Bioscience

Learning Objectives

After reading this chapter, you should:

- understand the diversity, formats, origin, and structures of different nucleic acid, protein, and metabolite (omics) datasets and databases;
- know current procedures for and challenges with accessing and managing phenotypic (clinical) data;
- realize that there are both practical and conceptual challenges in developing multi-scale models incorporating omics data that are of real clinical relevance;
- understand that computational modelling has the potential to bridge between omics and the clinic;
- be prepared to begin developing your own computational models, incorporating both omics and clinical data.

2.1 Introduction

As a reader of this book, it is important for you to understand that computational modelling and analytical tools can be useful, if not necessary, for understanding complex natural processes, particularly those of human physiology. Hitherto, most computational models produced and applied to biomedicine have been physiological in nature. Thus, they capture and model the input/output properties of a process within a tissue or an organ.

One example of this is the model of gas exchange in the lung developed over recent decades by Peter Wagner and his colleagues [1]. This model captures the essential input/output processes of gas exchange in the lung using a system of five differential equations. This simple but useful approach is easily understood and can be widely applied. Yet there are enormous challenges surrounding how this type of model can

be applied in the clinic if patient-specific (phenotypic) information such as lifestyle, medication, disease history, and clinical records is to be included. There is also now a vast amount of molecular omics data that has been generated from basic research in molecular biology and that can be incorporated into models, enabling personal genomic information, for example, to be included. This molecular information is generally stored in publicly accessible databases. It is important to remember, however, that data from clinical studies—including omics information; for example, about an individual patient's genomic variation—are less accessible as a rule.

Huge research efforts are being invested in the medical community in interpreting such molecular data and correlating them with selected parts of a clinical phenotype profile. For example, during the last decade genome-wide association studies have revealed many associations between genetic variation and different disease phenotypes [2] (see also Chapter 3). However, as a rule these effects are small and it is very difficult to understand how each genomic locus is involved in a disease and how the genomic data should be linked to differential diagnosis, predicting the progression of the disease, or selecting the correct treatment.

Moving from simply listing the associations between genetic variants and disease and developing a mechanistic understanding of each subgroup of patients presents one of the largest challenges for today's medical and pharmaceutical industries. Leroy Hood of the Institute of Systems Biology, Seattle, WA, USA, and his colleague Stephen Friend have established an idea of medicine that is 'personalized, predictive, participatory, and preventive' (so-called P4 medicine [3]; see also **http://www.p4mi.org**). This visionary paradigm requires an ability to bridge molecular information (omics data) and the clinical phenotypes of a disease that have been stratified into relevant subgroups.

It is difficult to conceive how to tackle this 'grand challenge' in medicine without resorting to computational models. And if this modelling is to be successful, models should incorporate and integrate many different types of omics and physiological data at many different spatial and temporal scales, from the molecular (nanometres, nanoseconds) to the clinical (metres, days, and years). The challenge for those involved in computational modelling is to bridge effectively between molecular information and clinical phenotypes. Molecular-level models analysed using bioinformatics and biostatistics must be combined seamlessly with information generated in the clinic. The culture of clinical research is very different from that of basic biomolecular research, and it will first be necessary to make all the information available in data formats and using languages and tools that can be clearly understood by researchers in both communities.

If novice researchers in computational biomedicine are to make a contribution to the models and tools that will be incorporated into P4 medicine they need to develop an understanding of the different types of data that are involved and how they can be accessed. To this end, this chapter describes the most important types of molecular data that can be yielded from biological systems and how these data can be linked to physiological models at different scales. Chapter 3 then explains in detail the linkage between genetic data and clinical phenotypes.

This chapter is not designed to be a primer for modern molecular biology. You will, however, need to have at least a basic understanding of cell and molecular biology if you are to develop models that are relevant to biomedicine. We recommend that if you need to understand any area of molecular biology in detail you should consult a recent edition of a comprehensive textbook such as those listed as Recommended Reading at the end of this chapter.

It is easy to forget that omics data do not just relate to the information contained in the DNA sequence. Instead, we can obtain valuable data from messenger and non-coding RNA, proteins, metabolites, and different physical attributes of proteins and nucleic acids, such as histone modifications and DNA methylation.

2.1.1 Introduction to Biological and Clinical Data

Each of the types of molecular data that we discuss in this chapter requires specific techniques and technologies for its profiling and analysis. The techniques that are used to generate these data types are generally high throughput in nature; that is, they generate an enormous amount of data in a short time. The bottleneck is generally in the analysis, and each combination of data type and technology needs specific bioinformatics pipelines and methodologies for this analysis.

The use of high-throughput techniques has exploded during the last decade or so and now allows the rapid and relatively cheap generation of large-scale molecular datasets that are an almost essential prerequisite for constructing computational models that are relevant to biomedicine. However, this effort requires a thorough understanding of the richness of the molecular data types that are important now and will be even more so in the future. This is particularly relevant when modellers come to consider how to fit together models operating on a molecular scale with models at larger scales. Two of the high-throughput techniques that are currently in regular use are DNA sequencing and microarrays for generating RNA transcript data. These will be described briefly in Sections 2.2.2 and 2.3.2 respectively.

Databases—even clinical databases—that are created by research consortia are generally accessible from

Table 2.1 Some examples of publicly accessible omics databases and analysis tools

Resource name	Description	URL
Genomic Basic Local Alignment Search Tool (BLAST) databases	BLAST finds regions of local similarity between sequences The program compares nucleotide or protein sequences to sequence databases and calculates the statistical significance of matches	http://blast.ncbi.nlm.nih.gov
GenBank	The National Institutes of Health (NIH) genetic sequence database; an annotated collection of all publicly available DNA sequences	http://www.ncbi.nlm.nih.gov/genbank/
Entrez Genome, Protein, Nucleotide, and Structure databases	A cross-database search engine linking a wide range of omics and related data types	http://www.ncbi.nlm.nih.gov/sites/gquery

within the partners' institutes. These databases provide an arena for scientific collaboration and foster the improvement of the quality and effectiveness of medical research. There are now more than a thousand curated biomedical databases, a fact that indicates the level of fragmentation of this knowledge and the importance of methods and tools to enhance the integration of these different types of data. The journal *Nucleic Acids Research* (**http://nar.oxfordjournals.org/**) publishes a special issue each January devoted to biomedical databases (see Recommended Reading).

Table 2.1 lists a small number of the most important and widely used biomedical databases and analysis tools. Some of these (and many more) are provided by the National Center for Biotechnology Information (NCBI), which is based in Bethesda, MD, USA, and which also publishes the widely used database of the biomedical literature, MEDLINE.

In the next sections of this chapter we briefly describe the structure and main functions of each type of molecule in turn, and then describe some of the formats used to hold those types of molecular data and the databases in which they are stored.

2.2 DNA and its Data Formats

2.2.1 The Structure and Function of DNA

Modern molecular biology and biomedicine would have been impossible without a comprehensive knowledge and understanding of deoxyribonucleic acid (DNA). There are many aspects of the study of this crucially important macromolecule: its physical and biochemical properties, the effects of DNA variation within populations, the study of the DNA configurations, and the mechanisms involved in gene transcription are only a few.

The basic structure of DNA is very well known. It consists of two long chains or strands of polynucleotides held together by hydrogen bonds between the nucleotide bases and is generally coiled into the iconic double-helix structure. There are four types of nucleotide subunit in DNA, differing only in the chemical structure of the base. Polynucleotide chains are formed by sequences of the four nucleotides: adenine (A), cytosine (C), guanine (G), and thymine (T). A always forms hydrogen bonds with T, and C with G (Figure 2.1). This genetic code allows DNA both to replicate itself precisely and to generate the messenger RNA (mRNA) molecules that are translated into protein sequences.

DNA sequences can be very long; the total size of the human genome has been estimated to be 3.2 gigabases (Gb), or 3.2×10^9 bases. Humans and all other eukaryotic organisms package almost all their DNA into a number of chromosomes in the cell nucleus; in contrast, the genetic information of prokaryotes (bacteria and archaea), which lack nuclei, is all contained in the cytoplasm.

The DNA Sequence

The sequencing of the human genome, completed at the turn of the millennium, was one of the most

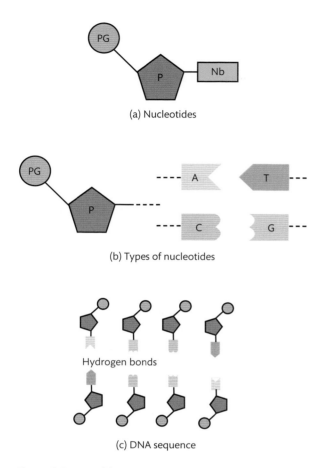

(a) Nucleotides

(b) Types of nucleotides

Hydrogen bonds

(c) DNA sequence

Figure 2.1 DNA. (a) A nucleotide is composed of a nitrogenous base (green), a pentose sugar (red), and a phosphate group (PG; blue). (b) Four types of nucleotide have been found in DNA. The base can be any of the four types shown; A always binds with T, and C binds with G to form the double helix. (c) A DNA molecule comprises two nucleotide strands; one strand is a 'negative copy' of the other.

genes, often by similarity (homology) with other known gene sequences, and by the location of sites that mark the start of the transcription of coding DNA sequences into RNA.

DNA Mutations and Single Nucleotide Polymorphisms

The DNA sequences of two individual, unrelated humans are approximately 99.9% identical. Therefore, if we are to understand the phenotypic variation between individuals, and differences in susceptibility and response to disease in particular, it is very important to understand the remaining 0.1%: the sequence variation between individual human genomes.

A DNA mutation is defined as a change in the sequence of a genome as compared to a reference sequence. Human sequences are often compared to a sequence known as the Genome Reference Consortium human genome, which was compiled using DNA obtained from 13 anonymous volunteers from Buffalo, New York, USA. Mutations can be classified according to how they affect the structure and/or function of the DNA. Single nucleotide polymorphisms (SNPs) are the most widely studied type of mutation. An SNP is defined as a difference between a sequence of interest and the reference sequence at a single nucleotide position (for example, where an A is replaced by a C, G, or T) that is observed in at least 1% of the population.

As a rule, SNPs have a small impact upon the physical structure of the DNA. Yet such variants can have functional consequences: a different gene sequence may alter the composition of the encoded protein, or it may alter the extent to which a regulatory protein can bind, influencing the level of its expression. SNPs that have no biological effect are referred to as silent mutations. However, some point mutations (many much rarer than SNPs) are responsible for single-gene (Mendelian) disorders such as haemophilia. In other words, a single point mutation (that is, a single base change) is enough to cause a disease state. Our ability to identify SNPs has increased dramatically since the completion of the first human genome sequence. The Thousand Genomes Project[1] was one of the first of many projects that aim to provide a global picture of the genetic variability of humankind.

important scientific achievements of the twentieth century [4,5]. However, it has been known for much longer that this genome consists of 22 pairs of homologous chromosomes and one further pair of sex chromosomes, with each female carrying two X chromosomes and each male one X and one Y.

Raw DNA sequences can be analysed in a number of ways, generating even more data. The percentage of a region of DNA sequence that comprises the nucleotides G and C (the so-called GC content of a region) may give insight into a region's stability or help distinguish between DNA regions that code for proteins and those that do not. A precise analysis of the DNA sequence can identify protein-coding

[1] http://www.1000genomes.org/

Copy-Number Variation

Initial analysis of human genome sequences suggested that SNPs were the main source of genetic variation and were therefore the main contributors to human phenotypic variation. However, genome-scanning technologies later uncovered a large amount of structural variation between the genomes of different individuals, including deletions and duplications of chromosome regions and large-scale variations in copy number. A copy number variation (CNV) refers to a situation in which a chromosome contains a segment of DNA that is 1 kilobase (kb) or larger in size and that has either been deleted or duplicated one or more times compared to a reference genome. A copy number polymorphism is a CNV that occurs in more than 1% of the population. We now know that 12% of the human genome has variable numbers of copies, and this copy-number-variable DNA includes many thousands of genes. This diversity may explain a significant proportion of normal phenotypic variation in the human population.

Other DNA Variation

SNPs and CNVs are not the only sources of DNA variation between individual humans. Alterations in the DNA sequence are considered *microscopic* if they are about 3 megabases (Mb) or larger in size. Examples of such microscopic variations are the presence of an abnormal number of chromosomes in a cell (aneuploidies), chromosomal rearrangements, and so-called heteromorphisms and fragile sites. *Submicroscopic* structural variants are changes that are between about 1 kb and 3 Mb in size, and are larger than those identified by early sequencing technologies (such as some inversions and deletions). There is strong evidence that both microscopic and submicroscopic variations can be related to markers that are visible in the phenotype.

Conformational Variation in DNA

Despite the apparent simplicity of the double-helical structure of DNA, conformational variations can modify the activity of the genome in terms of both RNA transcription (and therefore subsequent protein translation) and DNA replication. The term *conformation* refers to the shape of the DNA molecule, described in either two or three dimensions.

The study of DNA conformations includes an analysis of how DNA is packaged inside the cell and how the packing affects transcription by determining the access to genes (see Figure 2.2 and the following subsections on epigenetics and histone modification). Moreover, this analysis includes the study of how local conformation changes can affect the binding affinities of proteins to a given DNA sequence, thereby regulating the effect(s) of transcription factors on gene expression. Interestingly, no completely reliable general code of recognition between amino acids and nucleotides has thus far been uncovered that takes into account both the sequence and the conformation of the DNA; the discovery of this 'DNA code', if it in fact exists, could be described as a final goal for the conformational analysis of DNA.

Other conformational differences arise from the structures of the chromosomes themselves. The DNA that forms the chromosomes of eukaryotic cells is packaged with proteins to form a structure known as chromatin. In each chromosome, the chromatin is organized into *condensed* regions (known as heterochromatin) and more *open* regions known as euchromatin (Figure 2.2). These affect the extent to which the DNA at each point in the chromosome is able to be transcribed into RNA (is transcriptionally active), and this is subject to change.

Epigenetics: Regulating the Genome without Altering the DNA Sequence

Features of the structure and chemistry of DNA other than DNA sequence alterations that cause heritable changes in gene function are referred to as epigenetic modifications. The term *epi* comes from a Greek word for 'above', thus referring to changes above the level of the genome sequence. Epigenetic changes include changes to the packaging of DNA (Figure 2.2) and the addition of methyl groups to nucleotides in a process known as DNA methylation. A complementary definition of epigenetics that relates to the heritability of these changes is 'the study of mitotically and/or meiotically heritable changes in gene function that cannot be explained by changes in DNA sequence' [6].

DNA Methylation

DNA methylation, then, involves the addition of a methyl group to a nucleotide. This is the most widely

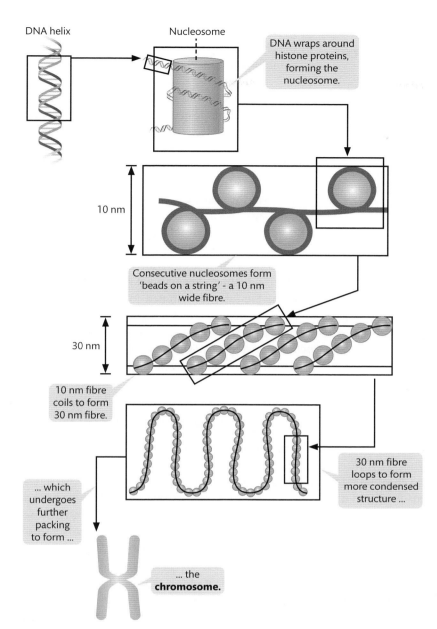

Figure 2.2 Packaging of DNA. The basic unit for DNA packaging in eukaryotes is the nucleosome. DNA is wrapped around histone proteins as a left-handed superhelix. The nucleosomes are arranged along the DNA as beads on a 10 nm-wide string, and this in turn is coiled to form a 30 nm-wide fibre. These fibres are further compacted by forming loops and finally packed into the structures that we know as metaphase chromosomes. Reproduced from *Chemistry for the Biosciences: The Essential Concepts*, 2nd edn by Jonathan Crowe and Tony Bradshaw (2010) by permission of Oxford University Press.

studied mechanism of epigenetic modification. The most common DNA methylation in mammals targets cytosine–guanine dinucleotides (CpG) at the 5′ end of the cytosine (Figure 2.3). CpG DNA methylation is considered to be regulatory. However, the precise locations of DNA methylations that are involved in regulation are not yet widely understood. The regions known to be involved include regions of DNA that are located near the start point for transcription of genes, and that bind proteins to initiate that transcription; these are known as the promoter regions of the genes concerned.

Recently, new advances in genomic technologies have enabled the initiation of large-scale epigenome-wide association studies of human diseases, similar to genome-wide association studies. These studies of DNA methylation across populations aim to characterize the key loci and biological

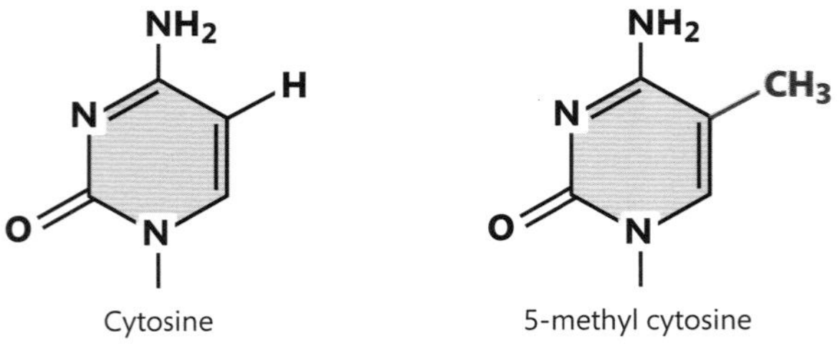

Figure 2.3 DNA base methylation. Methylation of mammalian DNA is most commonly via the conversion of cytosine (left) to 5-methyl cytosine (right), and this occurs most often when the methylated cytosine lies immediately after a guanine residue in the DNA sequence.

contexts in which DNA methylation is known to vary, such as:

- CpG sites that show differential methylation between individuals;
- multiple adjacent CpG sites that display differential methylation;
- allele-specific methylation in which the degree of methylation depends on the parent of origin or the presence of a variation in the DNA sequence (polymorphism).

Histone Modifications

The basic packaging unit of DNA into chromatin in eukaryotic chromosomes is the nucleosome (see Figure 2.2). A nucleosome core is formed by an octamer of proteins known as histones with 145–147 base pairs (bp) of DNA wrapped around it. The octamer consists of two copies of each histone: H2A, H2B, H3, and H4. Histones, which are highly conserved in the genome, were identified early on as regulatory elements of RNA synthesis. Histone regulation takes place through covalent chemical modifications to the histone proteins, including methylation, acetylation, and phosphorylation, with several modifications targeting the N-terminal regions of the histones in particular.

Histone modifications are able to:

1. modify the net charge of nucleosomes, which could weaken the interactions between the histones and the DNA;
2. be 'read' by other proteins that influence chromatin dynamics;

3. directly influence the higher-order structure of the chromatin; and
4. combine these mechanisms.

One example of a histone modification with functional associations is histone acetylation, which has become associated with active transcription. Acetylation of histones is catalysed by a group of enzymes (see Section 2.4) known as histone acetyltransferases (HATs), and the reverse process is catalysed by another group of enzymes, the histone deacetylases (HDACs). These types of enzymes have therefore become associated with activating and deactivating genes respectively. The histone code hypothesis [7] suggests that multiple modifications to histones can act in combination to specify distinct chromatin states. Histone modifications are often distributed in localized patterns close to the open reading frames (or ORFs) that are associated with protein-coding genes, and they are believed to regulate when each gene is transcribed into mRNA, as well as the amount of this RNA that is produced.

2.2.2 DNA Databases and Data Formats

The main source of genotypic variation is the changes in the DNA sequence of the gene. Some of these changes have no effect on the observable characteristics of the organism (its phenotype); others have a subtle effect; and still others can cause serious disease or even make the organism nonviable. As we have discussed in the previous section, the most common type of variation arises from changes in single bases; it is only when a variant of this type occurs in more than 1% of the population that it is correctly termed an SNP.

The output files from runs of DNA sequencing are used to build trace files and sequence files. The trace files represent the raw data; these are binary files, and their format depends on the type of sequencing technology that has been used. They are interpreted to give the sequence files, which are text files that generally show the DNA base sequence using the single-letter code (i.e. ACGT). Quantitative data concerning the quality of the trace obtained is also stored, and specialized software is needed to obtain the quality of the sequence from this. Sequence data obtained using modern methods (particularly the so-called next-generation sequencing methods

```
>sp|Q9UKB5|AJAP1_HUMAN
MWIQQLLGLSSMSIRWPGRPLGSHAWILIAMF
QLAVDLPACEALGPGPEFWLLPRSPPRPPRLW
SFRSGQPARVPAPVWSPRPPRVERIHGQMQMP
RARRAHRPRDQAAALVPKAGLAKPPAAAKSSP
SLASSSSSSSSAVAGGAPEQQALLRRGKRHLQ
GDGLSSFDSRGSRPTTETEFIAWGPTGDEEAL
ESNTFPGVYGPTTVSILQTRKTTVAATTTTTT
TATPMTLQTKGFTESLDPRRRIPGGVSTTEPS
TSPSNNGEVTQPPRILGEASGLAVHQIITITV
SLIMVIAALITTLVLKNCCAQSGNTRRNSHQR
KTNQQEESCQNLTDFPSARVPSSLDIFTAYNE
TLQCSHECVRASVPVYTDETLHSTTGEYKSTF
NGNRPSSSDRHLIPVAFVSEKWFEISC
```

Figure 2.4 Part of the human gene *AJAP1*, which is associated with obesity (see Section 2.7, Box 2.1), illustrating the FastA sequence format. The top line, starting with >, is the sequence title; all other lines give the sequence itself, using the single-letter DNA code.

developed during the last decade) is generally extremely precise [8].

Data obtained from initiatives like the Human Genome Project and many others are deposited in public databases and are accessible to the research community. Enormous quantities of data are generated in genomic studies, with the exact amount differing between sequence technology platforms.

Raw DNA sequences held in databases are commonly stored in a text format known as FastA, which consists of a title line starting with the greater-than symbol (>) followed by the nucleic acid sequence presented using the single-letter code. Figure 2.4 shows an example of a FastA sequence file representing part of the DNA sequence of a human gene.

One major focus of genomics research is the study of changes in the DNA base sequence. Alterations in the base sequence of coding regions of DNA self-evidently translates into changes in the mRNA sequence and, consequently, in the sequences and functionality of proteins. Moreover, cell metabolism is affected by changes in protein functionality, which in turn modifies the functionality of cells, tissues, and organs.

2.3 RNA and its Data Formats

Readers should be familiar with the way in which the DNA sequences of genes are transcribed into a

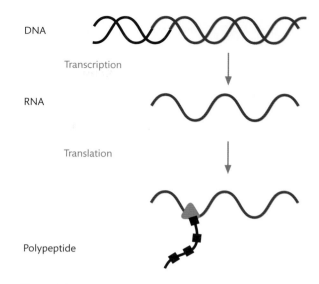

Figure 2.5 The Central Dogma of Molecular Biology. The dogma describes three classes of information transfer. First, DNA to DNA by replication. Second, DNA to RNA by transcription. The last is translation: RNA to polypeptide (protein).

type of ribonucleic acid (RNA) known as messenger RNA or mRNA, and these mRNA sequences (or, in eukaryotes, the coding parts of pre-mRNA sequences) are translated into functional proteins. This is referred to, along with DNA replication, as the Central Dogma of Molecular Biology, and it is shown in outline in Figure 2.5.

2.3.1 RNA Structure and Function

RNA molecules, like those of DNA, are made up of chains of nucleotides. There are, however, significant differences between DNA and RNA chains. The main differences between RNA and DNA are shown in Figure 2.6. Briefly, they are that:

1. RNA is generally a single-strand molecule;
2. RNA contains a ribose sugar, compared to the deoxyribose of DNA;
3. the DNA base thymine (T) is replaced by uracil (U) in RNA.

In addition, the secondary structure of RNA is very much more complex and variable than that of DNA, although it can be explained well enough simply from the thermodynamic properties of the molecules.

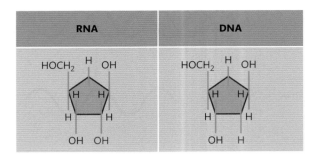

Figure 2.6 RNA and DNA. One of the main differences between ribose nucleic acid (RNA) and deoxyribose nucleic acid (DNA) is that the hydroxyl group on the 2-position of the sugar ring (shown in red) is replaced by a hydrogen atom. The other main difference is the replacement of the base thymine (T) with uracil (U); both these bases form hydrogen-bonded base pairs with adenine.

RNA Transcription

Transcription is the name given to the process of copying a portion of a DNA sequence into a nucleotide sequence of RNA. The name refers to the idea that both DNA and RNA are essentially written in the same language. The process begins with the opening and the unwinding of a portion of the double-stranded DNA. Next, one of the DNA strands acts as a template on which the RNA sequence is constructed by adding complementary nucleotides to this DNA strand, one at a time.

The initiation of RNA transcription is catalysed by an enzyme, RNA polymerase (or RNAp), which links successive nucleotides together. RNA polymerase differs from DNA polymerase, the similar enzyme that catalyses DNA replication, in that it catalyses the linkage of ribonucleotides rather than deoxyribonucleotides. The error rate of RNA transcription is higher than that of DNA replication, and also more variable. Very approximately, one DNA base in 10^{-4} is mis-transcribed into RNA; the error rate of eukaryotic DNA replication is more like one base in 10^{-8}. The biological effect of errors in RNA transcription is reduced by the degeneracy of the genetic code (see Section 2.4).

The process of transcription varies significantly between prokaryotes and eukaryotes, being more complex in eukaryotes. Eukaryotes cannot simply synthesize an mRNA molecule 'from scratch'; instead, transcription requires a *primer*, a short sequence of nucleotides from which the transcript is 'seeded'. In addition, the initiation of transcription in eukaryotes requires the assembly of a (sometimes large) set of proteins known as transcription factors on the initiation site of the genome, and also involves the unpacking of DNA from the nucleosomes. Transcription in prokaryotes is much simpler; the chromosome does not need to be unpacked and many fewer transcription factors are required.

Whereas bacteria have only one RNA polymerase, eukaryotes have three. Eukaryotic RNA polymerase II (PolII), which catalyses the synthesis of the mRNA molecules that encode proteins, is the most similar to the unique bacterial RNA polymerase. The structure and function of this polymerase, and its associated transcription factors, have been extensively studied. The transcription factor TFIID is now known to initiate transcription by binding to a consensus sequence in the promoter region known as a TATA box; this generates a distortion in the DNA that allows the assembly of the entire transcription-initiation complex. Another protein, known as TFIIH, then catalyses the phosphorylation of the last (tail) domain of the polymerase.

There is also evidence that the enzymes involved in histone acetylation and deacetylation (HATs and HDACs) are both targeted to transcribe regions of active genes by phosphorylated RNA PolII. Polymerase binding has also been observed to correlate with those histone modifications that are enriched in transcribed regions of genes. Many other regulatory features and mechanisms which are too complex to explain in detail here have been associated with the regulation of transcription factor binding and gene transcription.

Different Types of RNA

Apart from the chemical differences noted in earlier sections of this chapter, the most obvious difference between DNA and RNA molecules is length. Almost all RNA molecules are much shorter than the DNA molecules that form chromosomes. There are also many different types of RNA molecule with different functions. RNA molecules can be divided into two broad groups: firstly, the mRNA molecules that code for and are translated into protein sequences, and secondly, non-coding RNA (ncRNA). Many different types of ncRNA are known, and most of these have vital biological functions. However, several types are known to exert some regulatory influence on the genome by controlling which genes are transcribed in what quantities and in which cells.

The two best-known types of ncRNA are transfer RNA (tRNA) and ribosomal RNA (rRNA). A tRNA molecule recognizes and binds to a specific three-nucleotide sequence of mRNA while also binding to an amino acid. This sequence, termed a codon, determines which amino acid is added to the growing protein chain (Section 2.4.1). As such, tRNA facilitates the transfer of an amino acid to a growing polypeptide as it is synthesized by the ribosome, and forms the link between the amino acid and its corresponding codon. As its name implies, rRNA is part of that ribosome, the 'molecular machine' of protein synthesis. It interacts with tRNA to catalyse the synthesis of the new protein molecule.

mRNA is the RNA molecule that encodes the information for the coding and production of a protein. Figure 2.7 shows a summary of the steps involved in gene expression in eukaryotic cells, from RNA transcription to protein translation. In summary, these steps are as follows.

1. *Transcription*: the DNA sequence of a gene is transcribed into an RNA molecule that is termed a pre-mRNA.
2. *5′ Cap addition*: a modified guanine nucleotide is added to the 5′ end of the RNA molecule to protect it from degradation.
3. *Splicing*: the pre-RNA contains coding sequences of RNA (termed *exons*) interspersed with non-coding

segments (*introns*). During splicing, these introns are eliminated from the pre-RNA molecules.
4. *Editing*: the composition of the pre-RNA is edited if required.
5. *Polyadenylation*: a sequence of adenine nucleotides is added at the 3′ end of the molecule.
6. *Transport*: the RNA molecule (now just referred to as mRNA) is exported from the nucleus into the cytoplasm.
7. *Translation*: the RNA sequence is decoded by the ribosome to direct synthesis of a protein.

Knowledge of the mechanics of the transcription process has allowed us to identify certain important features within the base sequence of a mRNA molecule. These include the following.

1. *Introns*: these are the non-coding sequences of DNA described above, which are encoded as pre-mRNA but which will not be translated into protein sequences.
2. *Exons*: these form the coding sequence of the mRNA.
3. *UTRs*: the acronym stands for untranslated region, and refers to the sections situated at the 5′ and 3′ ends of the mRNA that are not translated but play an important role in the regulation of gene expression.
4. *Start codon*: the three-base RNA sequence that is recognized by the ribosome to begin translation.
5. *Stop codon*: the RNA sequence that is recognized by the ribosome to end translation.

Most studies of the links between nucleotide sequence and disease have been concerned with the

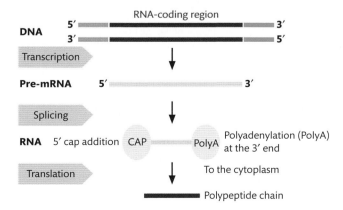

Figure 2.7 Steps in mRNA transcription.

protein-coding exons. However, more recently there has been interest in understanding putative relationships between ncRNA sequences, including introns, and disease. Furthermore, intron/exon boundaries have been explored in the context of alternative splicing mechanisms. Alternative splicing is a regulatory mechanism that allows the production in mammals of multiple mRNAs and, consequently, protein sequences, or isoforms, from the same gene. These sequences may have different or, in some cases, opposite functions.

Alternative splicing is common in higher eukaryotes, including humans [7]. For example, 92–94% of human genes are now known to undergo alternative splicing, and in 86% of these the minor isoform has a frequency of 15% or more. It is now believed that a large number of genes, encoding proteins with many different functions, are regulated by alternative splicing. In some cases, the differentially spliced proteins have distinctly different functions; in others they may, for example, be translated into protein in different cell types or different contexts.

ncRNA was initially thought to be involved only in the transcriptional machinery governing protein translation, in the forms of tRNA and rRNA; see Figure 2.8. This view, however, has had to be extended in recent decades as new types of ncRNA molecules with different functions have been discovered [9]. Both the number of types of ncRNA and the number of novel functions associated with these types are expanding. Some of the more important types of ncRNA and their functions are given in Table 2.2.

Since these ncRNA molecules can also exert regulatory functions, mutations within ncRNA can result in disease. For example, sequence variations in a number of microRNAs (see Table 2.2) have recently been linked to several types of cancer, including chronic lymphocytic leukaemia and colorectal cancer. However, most ncRNA mutations have a smaller phenotypic impact than mutations in coding DNA. In other words, their effects are more subtle and therefore less devastating, but these subtle phenotypes are often more difficult to identify.

Genomic DNA that codes for regulatory ncRNA sequences can be found on the opposite strand to

(a)

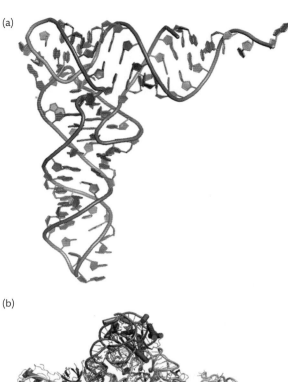

(b)

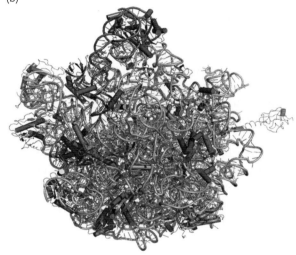

Figure 2.8 Non-coding RNAs. The two forms of ncRNA molecules with regular structures that are involved in protein translation. (a) A tRNA molecule, which forms the link between a specific three-base codon of mRNA and its associated amino acid. (b) The structure of an intact ribosome from the bacterium *Thermus thermophilus*. All ribosomes are complex structures formed from two subunits, each of which comprises many RNA and protein chains. Here the RNA backbone is shown in grey and the proteins in blue. Reproduced from *Molecular Biology: Principles of Genome Function* by Nancy Craig et al. (2010) by permission of Oxford University Press.

protein-coding genes (the antisense strand), in intergenic regions, and interleaved or overlapping with protein-coding genes. ncRNA is also associated with molecules that mediate epigenetic regulation

Table 2.2 Some of the most important types of small ncRNA molecules

RNA type	Cellular location	Function
MicroRNA	Cytoplasm	Negative control of gene expression
Small interfering RNA	Cytoplasm	RNA interference
Small nuclear RNA	Nucleus	RNA splicing
Small nucleolar RNA	Nucleolus	Chemical modification of RNA

such as DNA methyltransferases, histone-modifying enzymes, and the chromatin-remodelling complex.

2.3.2 RNA Transcript Databases and Data Formats

The technology that is most widely used to determine and compare the pattern of RNA expression in different cells is, rather confusingly, termed the DNA (or complementary DNA, or cDNA) microarray. This is based on the hybridization of oligonucleotides (cDNA) on glass supports and the measurement of binding between mRNA molecules expressed in the cells under test and the immobilized cDNA using fluorescence. The technology can be used to obtain information about the expression of a large group of genes or, increasingly, about the expression of each gene in a whole genome in a particular cell type or in response to a particular stressor. This latter type of experiment is termed global gene expression analysis. Microarray experiments typically generate an enormous quantity of data, and computational techniques—particularly statistical ones—are very widely used in interpreting the data that is derived from microarrays.

Microarray databases are widely used in a number of biomedical research fields, examples of which include the following.

1. *Basic research*. Microarrays generate quantitative patterns of gene expression that can be used to build an understanding of gene-regulation networks, metabolic pathways, cellular differentiation, and tissue development. In these experiments, researchers integrate expression data from various organisms with data from other molecular biology resources such as information on gene promoters and metabolic pathways and annotated genome sequences, some of which are described elsewhere in this chapter.

2. *Medical research and drug discovery*. Clinical researchers use microarrays to explore whether the over-expression of a set of genes is correlated with certain disease conditions, to find other conditions affecting the expression of a particular set of the genes, or identify groups of genes that have similar expression profiles. In the context of drug discovery, this helps to identify potential compounds that may alter the expression levels of genes that are associated with particular diseases.

3. *Comparative and diagnostic research*. Gene expression databases are also used in comparative studies to detect disease at an early stage. This involves, for example, the comparison of the gene expression profile for a sample tissue with profiles for similar tissue samples that have been associated with different diseases and that have already been deposited in a microarray expression database.

4. *Toxicology research*. Pharmaceutical companies often create and maintain databases that show how gene expression is affected by drugs in development, and other molecules in their databases and these are used to support drug discovery and toxicological research. A molecule that affects a class of genes in the same way as one known to produce a toxic response will be likely to produce a similar response. These profiles may be obtained from human tissue samples or from animal models; knowledge about the response of a gene to a toxin in an animal can allow researchers to make a better prediction of the response of the homologous (equivalent) human gene.

5. *Development of experimental techniques for profiling.* Depositing results derived from different profiling platforms into public databases facilitates the development of profiling technology since they help to avoid unnecessary repetition of the same experiments. Comparison of the results from laboratory samples with those from archived control samples is a common practice.

International efforts have been made to standardize the format of microarray gene expression data. For example, the Functional Genomics Data Society (FGDS) first developed the Minimum Information About a Microarray Experiment (MIAME) standards, a set of minimum information specifications for reporting data from functional genomics [10].

2.4 Proteins and their Data Formats

2.4.1 Protein Structure and Function

Proteomics, coined by analogy with genomics, is the study of the structure and function of the proteins coded by an organism's genome. Just as DNA and RNA are composed of sequences of nucleotides, proteins are widely known to be composed of sequences of amino acids. An amino acid is a molecule that contains an amino group ($-NH_2$), a carboxylic acid group (R-COOH), and a side chain with characteristic chemical features (Figure 2.9). When mRNA sequences are translated into proteins (see Figure 2.5, the Central Dogma of Molecular Biology), each three-nucleotide codon of mRNA encodes a particular amino acid. Thus the amino acid sequence of a protein is determined by the nucleotide sequence of the mRNA. There are enormous

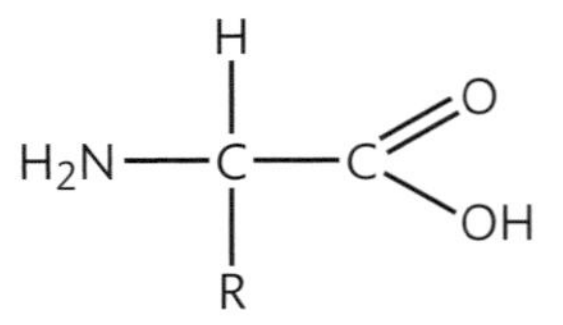

Figure 2.9 An amino acid. The basic chemical structure of an amino acid. The genetic code encodes 20 different amino acids, distinguished by their R groups. Further variety arises from chemical modifications to some of these groups after protein translation (post-translational modification).

numbers of amino acids, but only 20 different ones are encoded by the genetic code and incorporated into protein chains.

Proteins can be classified by their function into groups, and these include:

1. *enzymes*, which catalyse chemical reactions;
2. *structural or fibrous proteins*, which provide structural support to cells; these are particularly important in some cell types including muscle cells (actin and myosin) and tendons;
3. proteins that provide *communication signals*, such as those involved in signal transduction into and between cells;
4. *carriers* of (generally) small molecules round the body, such as the oxygen transporter haemoglobin.

Proteins are also characterized by their three-dimensional structures. The structure of a protein can be described in hierarchical terms, from primary to quaternary structure (Figure 2.10).

Briefly, the primary structure of a protein is simply another way of describing its amino acid sequence. Its secondary structure comprises repeated local structures that adopt characteristic three-dimensional shapes, such as α-helices or β-strands that are stabilized by a regular pattern of hydrogen bonds, such as turns or α-helices [11]. The tertiary structure describes the way the secondary structural elements experience spatial interactions (i.e. they fold) to give the overall three-dimensional shape of the protein chain. Finally, the term quaternary structure describes how two or more separate polypeptide chains may interact to form a complete, functional protein complex.

An understanding of protein conformation is often necessary to understand how proteins function incorrectly, or fail to function at all, in disease. There are many textbooks covering this important area, such as the recent one by Arthur Lesk listed in the Recommended Reading section of this chapter.

Post-Translational Protein Modifications

The human genome is now estimated to contain about 23 000 genes, which is only a few thousand more than organisms such as the fruit

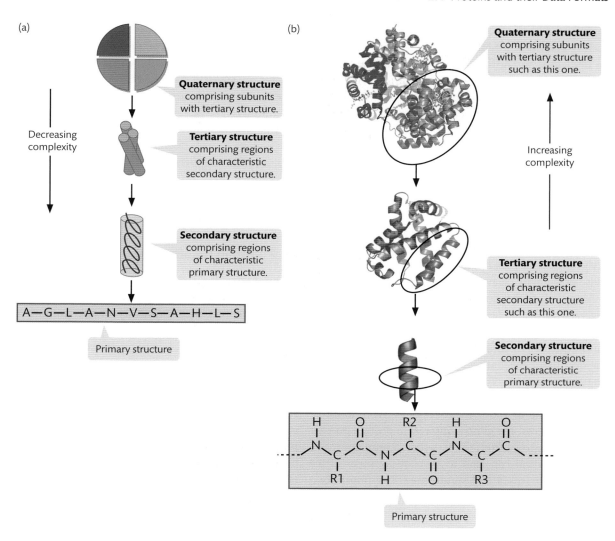

(a)

Quaternary structure
comprising subunits
with tertiary structure.

Decreasing
complexity

Tertiary structure
comprising regions
of characteristic
secondary structure.

Secondary structure
comprising regions
of characteristic
primary structure.

A—G—L—A—N—V—S—A—H—L—S

Primary structure

(b)

Quaternary structure
comprising subunits
with tertiary structure
such as this one.

Increasing
complexity

Tertiary structure
comprising regions
of characteristic
secondary structure
such as this one.

Secondary structure
comprising regions
of characteristic
primary structure.

Primary structure

Figure 2.10 Hierarchy of protein structure. (a) The general hierarchy of protein structure, from primary structure (the amino acid sequence) to quaternary structure (the arrangement of chains that makes up a functional protein). (b) These principles as applied to the oxygen-carrying protein haemoglobin. Reproduced from *Chemistry for the Biosciences: The Essential Concepts*, 2nd edn by Jonathan Crowe and Tony Bradshaw (2010) by permission of Oxford University Press.

fly. The differences between humans or other vertebrates and invertebrates, therefore, cannot arise from their simply having more genes. One possible explanation for this is the increased frequency of post-translational modifications of proteins in vertebrates. The size of the human proteome (the complete set of proteins encoded by the human genome) is considered to be two or three orders of magnitude larger than the size of the genome (i.e. there

are perhaps a million different molecular species of proteins).

There are two major mechanisms that effectively multiply the number of proteins. The first of these is alternative splicing, which we have already discussed. The second mechanism involves the post-transcriptional modifications of proteins, which occur after the newly translated proteins are released from the ribosome. The importance of these modifications is highlighted by the fact that about 5% of the protein-coding genomes of higher eukaryotes are known to be dedicated to coding for enzymes that carry out such modifications.

These modifications come in two types. The most abundant are the enzyme-catalysed covalent additions of chemical groups such as phosphates to amino acid side chains. More than 200 different types of covalent modification have been observed [12]. The second type of modification concerns the covalent cleavage of peptide backbones in proteins (proteolysis) either through the action of proteases or by autocatalytic cleavage. Some protein hormones, such as insulin, and some enzymes—particularly proteases—are active only after they have undergone one or more of these proteolytic cleavages.

Many of these covalent modifications can be reversed, which suggests that they can act as regulatory control parameters. The prototype of reversible modification is protein phosphorylation, which is the dominant mechanism for protein-based signalling in eukaryotes. There are enzymes that catalyse the removal of phosphate and acetate groups, and groups of one or more sugar residues, from amino acid side chains. The clinical interest in protein phosphorylation stems from the fact that deregulation of phosphorylation can lead to the development of a number of diseases and syndromes. The phosphorylation of histones (see Section 2.2.1) has also been widely studied.

Transcription Factors

A protein that regulates RNA transcription by binding to regulatory regions of DNA is referred to as a transcription factor (abbreviated to TF in protein names); the study of transcription factors has become a central research area within genomics because of the pre-eminent role that these proteins play in transcriptional regulation. Regulation is known to occur mainly at the level of transcription, and determines which genes will be transcribed into primary pre-RNA in any given cell type and at any given time. The remaining stages in transcription and translation generally occur more automatically. The complex area of transcription in eukaryotes is described in detail by Latchman (2005; see Recommended Reading).

All transcription factors have several structural domains (Figure 2.11):

1. a DNA-binding domain, which binds to promoters or enhancer regions;
2. a transactivation domain, which binds to other proteins to form complexes;
3. a signal-sensing domain, which responds to stimuli to activate or deactivate the protein's function and therefore regulate transcription.

The medical importance of transcription factors arises from their association with specific diseases and with their recently discovered potential as drug targets.

2.4.2 Protein Databases and Data Formats

Technologies for protein sequencing are by now very well developed. However, directly sequenced proteins represent the minority of the millions of protein sequences that are stored in publicly available databases. More are obtained by direct translation of the protein-coding regions of gene sequences stored in DNA databases. All protein sequences are often stored in a very richly annotated format, with each sequence linked to other resources describing its structure and function and to literature references; this is particularly true of the relatively small subsection of the UniProt database that is termed SwissProt (see Table 2.3). However, manipulation of protein

Figure 2.11 Domain structure of transcription factors. (a) The domain structure for a prototypical transcription factor. All transcription factors contain a DNA-binding domain (DB, blue) and a transactivation domain (TA, pink) which activates gene transcription, generally through the binding of further proteins. Most transcription factors also contain a signal-sensing domain (SS, green), which binds the ligand. These domains may be found in any order, and many transcription factors also contain other domains (peach). (b) The domain structure of the oestrogen receptor, a typical transcription factor. These domains are labelled as follows: ER, oestrogen receptor domain; Zn, zinc finger domain; Oest, ligand (oestrogen)-binding domain; CT, C-terminal domain (function unknown). The domains are colour-coded by function as in panel (a).

Table 2.3 Examples of public repositories for protein sequence and proteomics data

Database name	Description	Address
UniProt	Comprehensive, high-quality protein sequence database, divided into two sections: the smaller, more comprehensively (and manually) annotated section is termed SwissProt, and the larger automatically annotated one, TrEMBL	http://www.uniprot.org/
Human Protein Atlas	Genes with protein-expression profiles based on antibodies	http://www.proteinatlas.org/
GPM Proteomics Database (GPMDB)	Contains tens of thousands of datasets contributed by researchers	http://gpmdb.thegpm.org/
UniProtKnowledgebase (UniProtKB)	A central hub for the collection of sequence and functional information on proteins, with rich annotation	http://www.uniprot.org/help/uniprotkb
Peptide Atlas	Provides a method and a framework to accommodate proteome information coming from high-throughput proteomics technologies	http://www.peptideatlas.org/
Proteomics Identifications Database (PRIDE)	A standards-compliant data repository for proteomics data	http://www.ebi.ac.uk/pride/

sequences using bioinformatics tools often requires them to be represented using the same straightforward FastA sequence format as is often used for DNA sequences (Figure 2.4).

Proteins are the end products of DNA expression. They can be considered the fundamental building blocks of life as they are critical for almost all functions of living cells (see Section 2.4.1). The expression of proteins in a cell changes over time, resulting in changes to the physiological state of the cell. Furthermore, the majority of pharmacological compounds target proteins. For these reasons, the research community has a great interest in cataloguing changes in protein expression.

The development of proteomic technologies to characterize and measure the types and quantities of proteins present in a cell type plays a central role in functional genomics, illuminating the properties of biological systems via DNA and mRNA sequence analysis. Proteomics technologies are used to measure and quantify which protein products are present in a cell population and therefore which genes are over- or under-expressed in the cells.

This process, which is known as protein expression profiling, aims to understand the roles of functional molecules in biological systems. Protein expression profiling requires access to data, and involves:

- protein identification,
- protein quantification,
- determination of the functional activity of all proteins that are identified.

As with other omics technologies, proteomics generates a vast amount of data stored in complex datasets, which requires data-management methodology and tools to store, track, and access. Types of protein data other than basic sequence data that are stored in publicly accessible databases include protein expression patterns (involving protein separation and identification), protein–protein interactions, and protein structures.

Some popular protein databases are listed in Table 2.3.

2.5 Metabolism, Metabolites, and their Databases

Nucleic acids and proteins are not the only molecules that are enumerated and collected into collections of omics data. Just as the word *metabolism* refers to the set of chemical reactions, catalysed by enzymes, that occur in a living cell, the term *metabolite* is used to refer to the non-genetically encoded substrates, intermediates, and products of these reactions, while *metabolomics* refers to the study of these metabolites and the chemical processes that are associated with them. Humans (and any organisms) can be characterized by their metabolic profiles [13]—the quantities of metabolites that they produce, and the rates of those reactions—just as they can by their genome and/or proteome. This concept is particularly useful when considering the response of individuals to drugs and when analysing clinical samples.

Two major types of chemical process are involved in metabolism:

1. *catabolism*: the catabolic breakdown of large molecules, where the major function is to provide the system with energy or with raw material;
2. *anabolism*: the step-by-step synthesis of complex molecules from smaller molecules that are often the products of catabolic reactions.

Chemical reactions are not studied in isolation in the human body, as they are all involved in large highly interconnected cycles such as the Krebs cycle and the citric acid cycle (see [14] for a historical overview of several of these cycles). The key metabolites for any living system can be grouped into amino acids, lipids, carbohydrates, nucleotides, and minerals.

The chemical reactions and the metabolites that are synthesized and broken down during biochemical reactions in cells can be visualized as a set of linear and interrelated pathways. This has become the classical way to view metabolism. However, advancements in integrative system biology, mathematical modelling, and graph theory have set the stage for a recent shift towards the analysis of metabolism as a complete, complex network of inter-linked reactions. If this type of network analysis focuses on the substrates and products of the reactions it is known, again by analogy with genomics and proteomics, as metabolomics.

The most commonly used methods in metabolomics research are analytical tools based on nuclear magnetic resonance (NMR) spectroscopy and mass spectrometry. These technologies help researchers understand biochemical functionality and metabolic changes by profiling the range of metabolites found in cells. Metabolomic profiling aims to detect and measure the concentrations of the complete set of metabolites in a cell population, rather than using a predefined set of metabolites (i.e. there is no prior knowledge of the targets). In systems biology, metabolomic profiling facilitates the definition of biochemical interactions and the associated changes in individual metabolite concentrations. Tools such as System Biology Markup Language (SBML) are used to model and describe chemical reactions in metabolic pathways and networks. These are described in more detail elsewhere in this volume, particularly in the Appendix.

Metabolomics has a number of important uses in biomedicine, including in drug discovery and in identifying biochemical markers of disease (biomarkers). A number of databases holding information about metabolites and metabolomics have been developed and made publicly accessible. These can be divided into five types:

- metabolic pathway databases (e.g. KEGG[2], Cyc, and Reactome),
- compound-specific databases,
- spectroscopic databases,
- disease physiology databases,
- comprehensive, organism-specific metabolomics databases (e.g. Human Metabolome Database, HMDB[3]).

These databases have been designed to provide the research community with resources for the identification, quantification, and cataloguing of metabolites. The HMDB, for example, catalogues spectroscopic, quantitative, analytical, and molecular-scale data about human metabolism and metabolites.

2.6 Integrating Different Data Types and Sources

It is not sufficient to have secure and easy access to individual datasets: integration between databases is now a crucial requirement for biomedical research; particularly, perhaps, for research involving the modelling and simulation of the human body. Data integration is now a central activity in system biology, bioinformatics, and the development of personalized health systems. In addition, a prerequisite for data integration

[2] http://www.genome.jp/kegg/
[3] http://www.hmdb.ca/

is the identification of and access to a complete range of data sources, including, but not restricted to, the omics databases described in previous sections.

The ease with which the wider biomedical research community can access the data it needs is determined by a range of factors. These may be technical (such as how the data are stored and managed) or non-technical (such as the policies and regulations that govern their storage, which are particularly relevant for clinical data). In this section we will discuss the challenges that are involved in making these disparate data sources accessible to a degree that allows research involving the integration of multiple data types to take place (more details can be found in [15]). Research in these areas of systems biology and modelling is often supported with web-based workflow-management systems, which are used for the seamless integration of different datasets. These will be discussed in much more detail elsewhere in this volume, particularly in Chapters 8 and 9.

Biomedical data can essentially be classified into three different types:

1. information on individual patients obtained from clinical practice;
2. data from biological samples; and
3. the molecular (omics) data that have already been described.

Each of these sources has different rules and regulations that govern management of and access to the data. In general, clinical data sources are more tightly managed and access to them is more severely restricted than it is to research data. The latter are commonly shared through free-to-use public repositories or at the very least published in reputable scientific journals.

1. *Clinical data from individual patients.* Clinical data taken from individual patients or aggregated from, for example, participants in clinical trials are generally stored and managed in electronic medical records (or EMRs). Most of these are commercial products that are managed by their vendors. Software vendors usually do not disclose to third parties any information about the structure of their databases or the information about the data (known as metadata) that are stored there. However, international standards have been put in place to support data management and

the exchange of data between clinics. Standards such as the International Classification of Diseases (ICD) and the Systematized Nomenclature of Medicine (SNOMED) support the classification of diseases using agreed vocabularies. Clinical trial data that are stored in this way include the details of the protocol used for each study as well as information obtained from the individual patients in the study.

2. *Biological sample databases.* Biological data obtained from samples collected from patients and often also from relevant healthy individuals as controls are held in sample databases. These data are made available mainly in the context of specific studies of one or several diseases. For example, fluid found in the cavities of the flexible (synovial) joints will be collected and used in studies of rheumatoid arthritis, which affects these joints. In the best cases software systems are used to manage and track samples, to store details about them, and to relate this to information about the individuals from whom the samples were collected.

3. *Data from biological research sources.* Almost all databases that hold basic molecular biological data are open access and very easily accessible through the Internet and web resources. These contain (for example) different types of omics data, annotations of those data, and links between those data in the form of biochemical pathway analysis, and entries are often expressed using a common vocabulary. As biomedical research in general (and, arguably, computational biomedicine in particular) is necessarily multidisciplinary, it is essential for researchers from different backgrounds to use a standard terminology. Terms are often taken from one or more formal, hierarchical shared vocabularies known as ontologies. Many of these have been formulated in domains with relevance to biomedicine; the Gene Ontology, which impinges on the topics of both this chapter and Chapter 3, is one of the best known. Research databases usually support mechanisms for data retrieval and access via web services and application programming interfaces (APIs), and large databases may be used in the cloud or copies held offline for local use.

In addition to data retrieval and access mechanisms, public databases all provide some information about the individual datasets and the structure and schemas of the databases; this is collectively known as metadata (meaning 'data about data'). Researchers are obliged to publish their research data in full when submitting their manuscripts to journals. The editorial boards of an increasing number of journals are making sincere efforts to police these policies and to provide standard formats that researchers can use when submitting papers that involve the analysis of biological (or, less often, clinical) datasets.

2.6.1 Clinical Data Sources

Research in translational medicine, which seeks to understand the aetiology, molecular pathogenesis, prevention, and treatment of human disease, is critically dependent on the existence of quantitative and accurate information of the clinical characteristics of the diseases under investigation. A huge quantity of clinical data is stored in EMRs. Ideally, data stored in these records should include clinical characteristics of the disease, routine laboratory data, therapies and response to therapies, and comorbidities for each patient.

Clinical research has become increasingly data-driven and has come to rely on the analysis and correlation between different types of data. In particular, clinical researchers have made very wide use of advanced statistical techniques for many decades. However, the extent of mechanistic modelling using clinical data in combination with other data types is still in its infancy. The June 2011 issue of the journal *Interface Focus* (see Recommended Reading) forms a representative collection of papers illustrating the development of this emerging discipline.

Security and privacy are clearly central issues when dealing with data from medical records and with biological samples from traceable individuals. These issues are touched on in Section 2.8, with an emphasis on genetic data. They are dealt with at greater length in many other chapters of this book.

The EMR

Most clinical data, particularly data resulting from clinical trials, are now stored electronically in an EMR system. In clinical trials, special-purpose systems are deployed to capture and store both the trial protocol and the results and outcomes for each individual patient. An EMR must include a wide range of data types, including demographics, diagnoses, disease activity, and type of medication and dosage. Over time, we expect that fuller integration of this clinical information with all the omics data that are now available will improve our understanding of complex diseases and provide new insights into treatment methods and the efficacy and toxicity of drugs.

2.7 Management of Omics Data Types

The management of all the datasets generated from different technology platforms in basic biomedical research, clinical research, and healthcare is becoming an increasingly complex task. This is being driven by the heterogeneity of the data, the explosion in the sheer amount of data available, and the different cultural perspectives of the research communities involved. These complex management issues are not limited to data storage and access; there are also many challenges related to database integration and harmonization at different levels within the biological system. This section discusses current challenges of managing omics data in particular and presents some possible solutions.

2.7.1 Techniques for Data Integration and Management

The integration of data sources is a crucial and fundamental step in setting up any complex mathematical model, and in particular for biological and biomedical modelling—systems biology and bioinformatics—where many different complex data types are involved. Integration supports the process of hypothesis generation and thereby sets the stage for the discovery of novel quantitative relationships between different biological entities.

In this section we define the term *data integration* to mean 'the ability to search, query, and retrieve data from distributed and heterogeneous databases' [16], and we focus on how the necessity for integration reflects on the management of the data. Omics databases can be highly synergistic and the integration of newly generated experimental data with information from public databases will often lead to important insights into the meaning of the data.

The pressing need for data integration is not, however, because databases are dispersed across independent web systems: all public data are available on the open Internet. It is more because each of the data sources was built for a specific purpose; some of the systems have evolved from pre-web and even pre-Internet versions. The first versions of the Protein Data Bank (which holds information about protein structures), for example, were distributed on magnetic tape.

The data stored in each database contain a specific type of information and a typical research project will involve the use of a large number of these specialist databases. This is best illustrated using a simple example. The scenario described in Box 2.1 takes such an example from a specific biomedical research discipline—obesity—to demonstrate the need for querying and accessing multiple data sources [17]. The researcher in this scenario is

required to visit several databases and web resources including the University of California Santa Cruz (UCSC) Genome Browser, Ensembl, BLAST, Gene Ontology (GO), and several model organism databases. These resources are autonomous and highly diverse in terms of the type of data that is stored, the way the data are represented, the user interface, and the capabilities for querying the database. These databases are complementary and it is generally straightforward to link from one resource to another.

Vertical and Horizontal Data Integration

Horizontal data integration refers to the integration of complementary data sources at the same biological level of organization (cell to cell, or organ to organ, for example) whereas *vertical data integration* refers to integration of overlapping data sources spanning

---

Box 2.1 Example of the Use of Multiple Databases to Answer a Specific Research Question

Imagine a researcher in obesity who is trying to identify genetic factors associated with high body mass index, or 'obesity factors'. This researcher has localized such an obesity factor to a 5 Mb region on human chromosome 1. The finding and localization of this region will lead to many questions, such as, 'are there any genes in this region that are homologous, or evolutionarily related, to genes known to be involved in the regulation of lipid metabolism in any

experimental model system?' Data integration is required to identify genes involved in the regulation of lipid metabolism. The following algorithm [18] outlines one approach to this problem.

1. Identify all predicted genes in the 5 Mb region. This will requires access to a genome browser such as Ensembl or the University of California Santa Cruz (UCSC) Genome Browser.

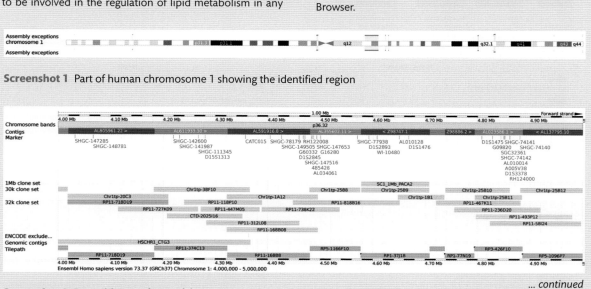

Screenshot 1 Part of human chromosome 1 showing the identified region

Screenshot 2 Ensembl view of part of that region

... continued

2. Find homologous genes in several model organisms. This requires access to a sequence database search tool such as BLAST at the NCBI.

Screenshot 3 List of genes in the region

3. Identify vocabulary terms that are stored in the Gene Ontology (GO) database and that are related to lipid metabolism.

4. Find out whether any of the homologous genes identified in step 2 are associated with any of the GO terms from step 3. This requires either:

a. querying each of the model-organism databases; or

b. downloading from GO and searching the entire list of genes related to lipid metabolism.

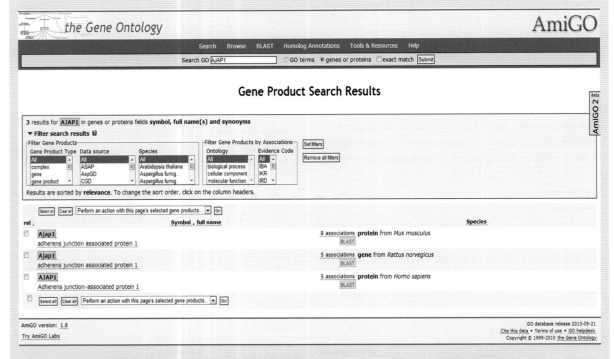

Screenshot 4 GO terms associated with the gene *AJAP1*

5. Finally, check whether one or more of the homologous genes from step 2 features on the gene list obtained from either substep 4a or substep 4b.

several of the different levels from the molecule to the organism. Both these types of integration must be central to research in systems biology and computational biomedicine if that research is to be of relevance to practical problems in the clinic. A research programme involving both these types of data integration will requires all relevant types of heterogeneous data to be accessible and searchable in a secure and interoperable manner.

Integration of Omics Databases

Improvements in algorithms and methods for information retrieval have led to new approaches to the integration of data within biological databases. Several different strategies for the integration of omics databases are enumerated below. Do not expect to understand these immediately; they will be described in much more detail elsewhere in this volume, particularly Chapter 10:

1. link-driven federation,
2. loosely and tightly coupled federation,
3. data warehouse integration,
4. database integration by ontologies,
5. database integration through web services.

Access and Usability of Omics Data

Even now, the methods used to access different omics databases are by no means uniform. For example, access to data collected in the clinic (e.g. individual patients' genotypes) is as a rule severely restricted within clinical settings, behind secure firewalls. Public data sources, however, make data available on the Internet. Some public sources permit online access only (i.e. data cannot be downloaded), while other public data sources (e.g. the genome browser Ensembl; **http://www.ensembl.org**) allow researchers to download entire datasets.

In general, researchers conduct database searches by submitting keywords or query parameters in the form of Boolean expressions. This can perhaps be explained best using a fairly complex example. The query set out below is used to retrieve a particular sample from a biobank that: contains more than 2×10^7 cells, has a given human leukocyte antigen (HLA) type, and has a particular base composition at a given SNP. Besides the biobank, this query uses two linked databases: a SNP database and a serology database that contains the HLA information.

```
Select
RS2064476,GTA_71.
Sample_Id
Amount_of_cells.
HLA_TYPE
CCP_SERUM
From
Serology database & SNP database & Biobank and
cells database
Where
HLA_TYPE = *01/*04','*03/*04','*04','*04/*04")
&
Amount_of_cells >= '20.10⁶ '
&
GTA_75.RS2064476 in ('AG','GG'))
```

Two challenges for those accessing biomedical data sources are the syntactic and semantic heterogeneity of data. Syntactic heterogeneity is said to occur when exactly the same entity, for example a gene, is given different names in different databases (for example, GenBank and the equivalent European Nucleotide Archive). Syntactic heterogeneity occurs when the same entity is stored in different databases with different meanings. In this example, one entry for human haemoglobin A in the structural Protein Data Bank is given the identifier 2DN1, and this is linked to the protein sequence with identifier HBA_HUMAN in UniProt and the genomic DNA sequence with identifier J00153 in GenBank.

Semantic heterogeneity exists because in practice terminologies are not defined using formal methods. This type of data heterogeneity is increasing because datasets are being generated and deposited by increasing numbers and types of platforms and groups.

The dynamic nature of biomedical data makes them unstable in the sense that the data frequently change in format and structure due to developments in technology and laboratory equipment; Chapters 10 and 11 present issues of accessibility, integration, and usability of data in much more detail.

Workflow Management and Access to Omics Databases

In computational research, scientific procedures and processes are generally described and implemented in a sequential and often automatic way; this is termed a workflow. Tasks within a computational workflow include subtasks such as running scripts, accessing and querying a database, interacting with high-performance computing (HPC) resources, and calling web services to reuse remote resources. Biomedical data sources should support effective workflow management by providing access to the latest sources and should provide facilities for distributing database queries across a range of processors. Workflow software-management systems play a central role in enabling researchers to share data and services. Issues related to workflow management are detailed in Chapter 8, and high-performance and distributed computing are discussed in Chapter 9.

Database Annotation

The usability and usefulness of omics databases depend critically on how the individual entities (e.g. proteins, genes, etc.) are annotated and on the quality and availability of the associated metadata. A lack of proper and complete annotation severely reduces and complicates the utilization of data generated from different platforms and between different experiments.

In addition to the omics data, the experimental design and protocols and different kinds of experimental model systems can be annotated.

The goal of gene annotation is to identify the segments of a genome that comprise the genes, and to define the coding sequences of these genes. It can therefore be described as a two-step process. The first step is the location of the genes and other elements in the genome, and the second the addition of a description and information to each element. For example, the gene will be annotated with the name and function of its protein product and the intron and exon sequences in that gene will also be marked. Many bioinformatics tools support automatic annotation. BLAST, for example, can be used to find similarities between genome elements and then to annotate genomes according to similarity. The most common types of genome annotation concern the function, and often the structure, of the proteins encoded by those genes.

Protein functional annotations can be performed manually, as in the smaller but better-annotated SwissProt section of the UniProt database. A curator selects and extracts annotation information from published articles and then attaches it to the record for a protein. Algorithmic approaches can accelerate the process of finding functionally related protein families and these are used in the annotation of most databases.

InterPro and InterProScan are examples of tools that can help curators with this annotation process.

Standards

The main rationale of providing and developing public omics databases is to store data produced by different researchers and to make the data accessible and sharable in standardized and interoperable formats. Standards form part of the infrastructure that is required for collecting, storing, retrieving, integrating, mining, and querying data generated by different researchers, laboratories, and platforms. Databases that apply strict standards and that make those standards publicly available are of greater value for the research community than less compliant ones.

These standards are usually developed by groups of individuals with expertise in specific areas (in this case, experts on the specific type of biological data concerned). Although they are voluntary, their wider adoption by a research community will increase reliability, reusability, and effectiveness for sharing and exchange of the product (i.e. dataset). Standards may be published informally on the web and/or as formal documents, for example in scientific journals.

Standards for data reporting, exchange, and sharing are particularly important in the omics field, because the data are so complex and so heterogeneous. In addition, the nature and amount of data generated by omics platforms[4] has greatly increased the need for the community to agree on guidelines, rules, and definitions of the various terminologies and formats that they need to use (e.g. genomes and genes and other data obtained from genome-sequencing platforms).

The practical implication of omics standards in translational and clinical research is that they give researchers confidence in the quality of the data and make it easier to use. This makes it more likely that clinicians, in particular, will be able to apply this approach [18]. In recent years, journals and publishers have required authors to comply with standards as a prerequisite to submitting supplementary data for publication. In general, standard-compliant published articles tend to be more visible and to gain larger numbers of citations and visibility.

Three types of essential data sharing standards have been identified and classified [19]:

4 Examples of omics technologies include microarrays, proteomics platforms, and high-throughput cell assays.

Table 2.4 Common standards for omics data and clinical phenotypes

Type of omics	Methodologies	Current standards	Consortium responsible
Genomics	Genotyping, CNV, comparative genomic hybridization (CGH), chromatin modification, ultra-high-throughput sequencing (UHTS), metagenomics	Minimum Information about a Genome/Metagenomic Sequence/Sample (MIGS/MIMS)	Genomic Standards Consortium
Transcriptomics	Gene expression via DNA microarrays or ultra-high-throughput sequencing, tiling, promoter binding (ChIP-chip, ChIP-seq), *in situ* hybridization studies of gene expression	Minimum Information about a Microarray Experiment (MIAME)	Functional Genomics Data Society
Proteomics	Mass spectrometry, gel electrophoresis, molecular interactions, protein modifications, sample processing	Minimum Information about a Proteomics Experiment (MIAPE)	Human Proteome Organization Proteomics Standards Initiative (HUPO-PSI)
Metabolomics	Biochemical network modelling, biochemical network perturbation analysis (environmental, genetic), network flux analysis	Core Information for Metabolomics Reporting (CIMR)	Metabolomics Standards Initiative (MSI)

1. *experiment description standards* (how the data were generated; e.g. descriptions of methods, protocols, and samples);
2. *data exchange standards* (e.g. file formats and types of structured and non-structured data representation);
3. *terminology standards* (e.g. the names of genes and gene products).

Providing standards for all these types of data will promote data sharing, improve researcher understanding of reported or published results, and help researchers to compare and reproduce published data and results.

In functional genomics there has been a continuous effort to create standards that have been agreed on by the community. BioStandard (**http://biostandards.info/wiki/Main_Page**) provides consolidated information about standards in omics and clinical databases. This resource classifies standards into four domains, which are listed in Table 2.4.

2.8 Software Systems: Security and Interoperability

We set out in this section what researchers need in a software solution if it is to be used in a biomedical setting, and most importantly if it is to be used with sensitive clinical data. We emphasize the central issue of data interoperability as required for data sharing and exchange in systems biology, bioinformatics, and computational modelling as applied to biomedical and, particularly, clinical research.

2.8.1 Software Solutions

Several technology platform solutions have become available for managing the access and use of biomedical data in translational medicine research. Many of these have been developed by research communities at universities and research institutes, and these are generally released as open source programs. One commonly used open-source platform is Informatics for Integrating Biology and the Bedside (i2b2)[5]. The i2b2 platform is funded by the US National Institutes of Health (NIH), uses ICD as a taxonomic standard to classify diseases, and facilitates the creation of formal ontologies to meet the specific requirements of different groups of users.

The design principle behind i2b2 was to provide software platforms and scalable solutions that facilitate repurposing of clinical data into the research setting and to secure the access and management of patient information for research purposes. This platform was

[5] https://www.i2b2.org/

implemented as a set of software modules communicating with each other via web service technology in a service-oriented architecture environment. This kind of architecture provides secure communication based on Simple Object Access Protocol messages. The principles behind the design of i2b2 paid attention to its query and data-retrieval performance. This design supported two predefined use cases:

1. to explore patient data to locate sets of patients that would be of interest for further research; and
2. to make use of detailed data provided by EMRs to identify different phenotypes within an identified set of patients in support of genomic, outcome, and environmental research.

The Stanford Translational Research Integrated Database Environment (STRIDE) is based on the HL7 data model and represents an integrated, standards-based platform for translational research informatics. STRIDE provides a number of the functionalities that are commonly required in translational research including data management for biobanks as well as data held in EMRs. STRIDE is similar to i2b2 in using ICD and other standards to build a semantic model to represent biomedical concepts and relationships between different types of biomedical data.

System and Data Interoperability

Interoperability has been defined generically as 'the ability of two or more systems or components to exchange information and to use the information that has been exchanged' [20]. A more specific definition has described the interoperability of web services as 'the sharing of data'. There is thus a strong relationship between interoperability, data exchange, and sharability. The more interoperable the system or corresponding software tools are, the easier it is to exchange data. Interoperability therefore reduces the demand for complex integration.

RICORDO[6] is a framework for building ontologies to support the semantic interoperability of data and model resources for the research community. It is a research and development project that provides the clinical communities in physiology and pharmacology research and in medical education with methods

[6] http://www.ricordo.eu/

and tools for ensuring that the same data and tools can be used and reused in a variety of contexts. To that end, RICORDO contributes to the development of community standards including reference ontologies and metadata. Both these types of standard have been implemented as a toolkit for representing complex biomedical concepts and supporting the sharing and annotation of data and model resources.

Security and Sharing of Clinical Data

Security and privacy issues pertaining to patient data is a critical concern. Rules and regulations for handling these issues have been introduced in many countries, but these are not always directly comparable.

Integrating omics data and clinical and biological research datasets into one data layer creates a particular security and authorization challenge to do with patient identity. Any patient data that may be used to identify an individual may not be used in a research setting. The process of anonymization of data from individual patients so that they may be used both legally and ethically in biomedical research is described in detail elsewhere in this volume.

Earlier in this chapter we saw something of how widely used genomic information is in both healthcare and clinical research. There is, however, a risk that it might be possible to identify an individual participant in a research or clinical study from his or her genotype at specific given loci. The challenge of protecting individuals' genomic information is discussed at some length elsewhere in this volume.

There is no technical solution that can absolutely prevent the re-identification of individuals from genetic data. The availability of summary-level allele frequencies for two matched sample groups and a more complete genome profile of one of the participants (i.e. the subject of interest) can make it relatively easy to re-identify individual subjects. There is therefore a current trend with annotators of genomics databases to hide some of the genotype/phenotype information to reduce the risk of re-identification.

2.9 Conclusions

In this chapter we have described a large number of different molecular omics data types which can be integrated and linked to different clinical phenotypes. These data can allow researchers to define novel disease subtypes corresponding to different groups of patients or healthy individuals within a population. You should now appreciate that there are several different types of DNA information and that a consideration of SNPs alone is not sufficient to define variation in human genomes (see Figure 2.12 for a schematic illustration of the complexity of interactions between DNA, RNA, and proteins in a cell).

In the genome era, constructing computational models of biomedical systems will very often require data on the genotypes of healthy or diseased individuals. Modellers will need to consider a wide range of different data types, including, but not restricted to genes; features of RNA, proteins, and metabolites; and levels of epigenetic regulation. It is also important to realize that molecular events occur in a context-dependent manner, and all multi-scale models that are relevant to disease will necessarily be specific to particular tissue and cell types.

To summarize, integrating clinical data with the rich universe of omics data is a prerequisite for systems biology and computational modelling if that research is to impact medical research. Moreover, this data- or model-driven integration will be central if modelling is to play its expected part in realizing the vision of P4 medicine (see Section 2.1), combating disease and nurturing wellness. However, there are still significant hurdles to be overcome to ensure that all researchers have access to the data they need, in relevant, data-rich, standardized formats, and with disparate types of data linked together across the complete range of spatial and temporal scales.

A purely statistical data-integration targeting prediction will not be sufficient to understand all the mechanisms that underlie human health and disease. A complete mechanistic understanding will be necessary, for example, to discover and target the most relevant specific molecular mechanisms for a drug-discovery programme. Hence, data integration has to be performed with the help of computational models. Many challenges remain, including developing systematic methodology for

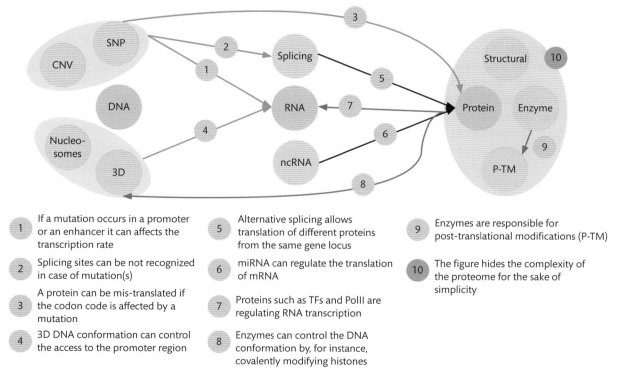

1. If a mutation occurs in a promoter or an enhancer it can affects the transcription rate

2. Splicing sites can be not recognized in case of mutation(s)

3. A protein can be mis-translated if the codon code is affected by a mutation

4. 3D DNA conformation can control the access to the promoter region

5. Alternative splicing allows translation of different proteins from the same gene locus

6. miRNA can regulate the translation of mRNA

7. Proteins such as TFs and PolII are regulating RNA transcription

8. Enzymes can control the DNA conformation by, for instance, covalently modifying histones

9. Enzymes are responsible for post-translational modifications (P-TM)

10. The figure hides the complexity of the proteome for the sake of simplicity

Figure 2.12 DNA, RNA, and protein interactions. Illustration of the complexity of the interactions between molecules in a cell. This figure depicts a reduced number of regulatory mechanisms through examples but it is not a complete overview of regulation; for instance, metabolites would also have interactions with all the molecules represented. miRNA, microRNA; TF, transcription factor.

merging two different but partially overlapping computational models into a single one; inferring new and improved computational models from a given set of data; and incorporating and making best use of information from a set of data given a particular computational model.

If we are to solve these fundamental challenges while keeping pace with the still increasing rate of molecular data production, we are bound to require innovative conceptual and algorithmic techniques. There is still an enormous amount of stimulating work for newcomers entering the field, and enormous opportunities for modellers to contribute to the development of predictive, preventive, personalized, and participatory medicine.

Recommended Reading

Alberts, B., Johnson, A. et al. (2007) *Molecular Biology of the Cell*, 5th edn. New York: Garland Science.

Berg, J.M. (2011) *Biochemistry*, 7th edn. New York: W.H. Freeman.

Interface Focus (2011) Themed issue: the virtual physiological human. *Interface Focus* 1(3), 281–473.

Latchman, D. (2005) *Gene Regulation: A Eukaryotic Perspective*, 5th edn. New York: Taylor & Francis.

Lesk, A. (2010) *Introduction to Protein Science: Architecture, Function & Genomics*. Oxford: Oxford University Press.

Nucleic Acids Research (2014) Databases issue, January 2014.

References

1. Hopkins, S.R., Barker, R.C. et al. (2000) Pulmonary gas exchange during exercise in women: effects of exercise type and work increment. *Journal of Applied Physiology* 89(2): 721–730.

2. McCarthy, M.I., Abecasis, G.R. et al. (2008) Genome-wide association studies for complex traits: consensus, uncertainty and challenges. *Nature Reviews Genetics* 9(5): 356–369.

3. Hood, L. and Friend, S.H. (2011) Predictive, personalized, preventive, participatory (P4) cancer medicine. *Nature Reviews Clinical Oncology* 8(3): 184–187.

4. Lander, E.S., Linton, L.M. et al. (2001) Initial sequencing and analysis of the human genome. *Nature* 409(6822): 860–921.

5. Venter, J.C., Adams, M.D. et al. (2001) The sequence of the human genome. *Science* 291(5507): 1304–1351.

6. Russo, V.E.A., Martienssen, R.A., and Riggs, A.D. (1996) *Epigenetic Mechanisms of Gene Regulation*. Woodbury: Cold Spring Harbor Laboratory Press.

7. Wang, E.T., Sandberg, R. et al. (2008). Alternative isoform regulation in human tissue transcriptomes. *Nature* 456(7221): 470–476.

8. Mardis, E.R. (2008) Next-generation DNA sequencing methods. *Annual Review of Genomics and Human Genetics* 9: 387–402.

9. Mattick, J.S. (2009) The genetic signatures of non-coding RNAs. *PLoS Genetics* 5(4): e1000459.

10. Brazma, A., Hingamp, P. et al. (2001) Minimum information about a microarray experiment (MIAME)-toward standards for microarray data. *Nature Genetics* 29(4): 365–371.

11. Pikaart, M. (2011). The turn of the screw: an exercise in protein secondary structure. *Biochemistry and Molecular Biology Education* 39(3): 221–225.

12. Walsh, C.T., Garneau-Tsodikova, S., and Gatto, Jr, G.J. (2005) Protein post-translational modifications: the chemistry of proteome diversifications. *Angewandte Chemie International Edition in English* 44(45): 7342–7372.

13. Gates, S.C. and Sweeley, C.C. (1978) Quantitative metabolic profiling based on gas chromatography. *Clinical Chemistry* 24(10): 1663–1673.

14. Kornberg, H. (2000) Krebs and his trinity of cycles. *Nature Reviews Molecular Cell Biology* 1(3): 225–228.

15. Schadt, E.E., Linderman, M.D. et al. (2010) Computational solutions to large-scale data management and analysis. *Nature Reviews Genetics* 11(9): 647–657.

16. Louie, B., Mork, P. et al. (2007) Data integration and genomic medicine. *Journal of Biomedical Informatics* 40(1): 5–16.

17. Stein, L.D. (2003) Integrating biological databases. *Nature Reviews Genetics* 4(5): 337–345.

18. Mayer, B. (2011) *Bioinformatics for Omics Data: Methods and Protocols*. Methods in Molecular Biology, vol. 719. Berlin: Springer.

19. Chervitz, S.A., Deutsch, E.W. et al. (2011) Data standards for omics data: the basis of data sharing and reuse. *Methods in Molecular Biology* 719: 31–69.

20. IEEE (1990) *IEEE Standard Computer Dictionary: A Compilation of IEEE Standard Computer Glossaries*. New York: IEEE.

Chapter written by

Imad Abugessaisa, David Gomez-Cabrero, and Jesper Tegnér, The Unit of Computational Medicine, Department of Medicine, Center for Molecular Medicine, Karolinska Institutet, Karolinska University Hospital, Stockholm, Sweden.

3 From Genotype to Phenotype

Learning Objectives

After reading this chapter, you should:

- be able to outline how genetics relates to computational modelling of biomedical systems;
- understand the complex relationship between genotypes and phenotypes, particularly in regard to quantitative traits;
- understand concepts from classical and quantitative genetics and apply them to problems in computational biomedicine;
- apply the phenotype concept to discussions of multilevel physiological simulations;
- understand what is meant by a causally cohesive genotype–phenotype study;
- know how to obtain genetic data and model resources that are relevant to these studies;
- understand the importance of ontological annotation and semantic interoperability for the integration and reuse of data and model resources in computational biomedicine.

3.1 Introduction

The science of genetics deals with heredity and variation in populations of living organisms. There is a distinction between *genotype* (an individual's particular makeup of hereditary material, generally DNA) and *phenotype* (its observable characteristics of interest, such as height, blood type, or sleep pattern). Early studies of the relation between genotype and phenotype were based on categorical phenotypes governed by a single gene. Later, *quantitative genetics* studied variation in continuous traits such as height, taking advantage of patterns of statistical similarity among individuals related within families.

Of course, phenotypes arise from physiology. However, genetics was developed decades before the discovery that DNA is the hereditary material, and before the advent of systems biology and computational physiology. Therefore, traditional genetics is rather limited in its scope. This chapter is about how computational physiology is finally beginning to help account for the steps from molecular genetics to physiology and phenotype.

To follow this chapter, and those on cell, tissue, and multilevel modelling in Chapters 5–7, you need a passing familiarity with some concepts from systems dynamics that we now briefly define. The term *system* denotes a set of interacting and interdependent components that form a unified whole. In particular, living beings are *dynamical* systems: their state changes over time due to the physiological processes of their component cells, tissues, and organs. The *rates* of the various processes depend on the body state and external influences. This is often formulated mathematically in *differential equations*:

$$dx/dt = \text{rate of change in state over time}$$
$$= \text{a function of state} = f(x)$$

You will see examples of this in Box 3.2 and Box 3.4, and in Chapter 5 on cell modelling. Such equations can be solved to give an explicit time course or trajectory as a function of time, $f(t)$. You could, of course, approximate this result by simple curve-fitting of whatever function of time you like. But this does not *explain* the dynamics in the same way that the equation above does. The notion that processes change the state, but also depend on it, is fundamental to the theory of dynamical systems. This echoes the distinction in Chapter 1 between bioinformatics and systems biology.

A *model* is a purposeful simplification of reality, designed to imitate certain phenomena or characteristics of a system while downplaying non-essential aspects. Its value lies in the ability to generalize insights from the model to a broader class of systems. Thus, a laboratory mouse can be a model representing mammals in general; an *in vitro* heart cell can represent the cells in an intact heart; and a set of differential equations can approximate the dynamic behaviour of a biological system.

Usually, a differential equation model such as the one above has *parameters*: these are the numbers that characterize the shape of the function $f(x)$, but remain constant over the timescale of the phenomenon being studied. Importantly, parameters may differ between instances of the system represented, such as different patients, species, or cell types.

In the rest of this chapter we briefly review classical and quantitative genetics before outlining how to incorporate systems biology and genetics into a common framework. Finally, we discuss how to link such models to data.

3.1.1 Genetics and its goals

Genetics is rare among scientific disciplines in that its goals were explicitly formulated at the outset. William Bateson, who coined the term genetics, defined the goals of the new discipline in this way, at the Third International Conference of Hybridization and Plant Breeding in 1906 [1] (cited in [2]): 'I suggest for the consideration of this conference the term *genetics*, which sufficiently indicates that our labours are devoted to the elucidation of the phenomena of heredity and variation: in other words, to the physiology of descent, with implied bearing on the theoretical problems of the evolutionist and the systematist, and application to the practical problems of breeders whether of animals or plants.'

The phrase 'physiology of descent' suggests that it is for genetics to explain how hereditary units actually cause phenotypic patterns in terms of physiological principles. However, while there have been tremendous advances in both the molecular understanding and the statistical modelling of heredity and variation, much less progress has been made in linking these insights together into a comprehensive theory.

It is likely that any such future comprehensive theory of genetics will have the following four characteristics:

1. its explanatory apparatus will be based on the regulatory principles observed in biological systems;
2. it will be capable of explaining population-level genetic phenomena with reference to these regulatory principles;
3. the mathematical language of non-linear system dynamics will be a core element;
4. the theoretical apparatuses of non-linear system dynamics and mathematical statistics will become highly integrated.

Thus, we predict that system dynamics will be at the core of any future mathematical structure that is capable of linking a genotype to a phenotype. This reflects the growing realization that statistics alone is not an adequate language to describe and analyse how complex dynamic phenomena result from the interactions of lower-level entities in a system. However, statistics is certainly useful in identifying patterns and generating hypotheses. We therefore now turn to a brief introduction to the theory of quantitative genetics.

3.2 Quantitative Genetics: A Brief Introduction

3.2.1 Genotypes and Haplotypes

At the heart of genetics is the distinction between a *phenotype*—that is, an observable trait of an organism—and a *genotype*. The genotype of an individual refers to that individual's specific genetic constitution at one or more *polymorphic loci* (singular: locus). We now know that these loci refer to locations within the genome where the nucleotide sequence of the DNA varies between individuals. However, the core of quantitative genetics was developed before the discovery that DNA is the hereditary material, and so this whole theory is built up around abstract loci that vary between individuals and contribute to phenotypic variation.

We discussed in Chapter 2 how the genomes of individual humans (or members of any species) differ in molecular terms, and described single-nucleotide polymorphisms (SNPs) and copy number variation (CNV). Millions of these polymorphic loci are catalogued in projects such as the International HapMap Project[1]. The variants found at a polymorphic locus, such as the two different bases observed at a SNP, are called *alleles*. Because humans are *diploid* organisms (i.e. each cell has two copies of each chromosome), the genotype at a locus is specified by *two alleles* at that position.

To illustrate some concepts relating to genotypes, we consider two loci, A and B, on the same chromosome with alleles A_1/A_2 and B_1/B_2. Figure 3.1 illustrates a segment of the genome of an individual who carries alleles A_1 and B_1 on the maternal strand and alleles A_2 and B_1 on the paternal strand. Thus, the full genotype of this individual is $A_1A_2B_1B_1$. It is described as *heterozygous* (i.e. containing different alleles) A_1A_2 at locus A and *homozygous* (i.e. containing the same allele) B_1B_1 at locus B. Furthermore, this diploid genotype consists of two *haplotypes* that can be named A_1B_1 and A_2B_1, referring to the maternal and paternal chromosomes respectively.

We next take this up to the level of the population, where analysis of genotypes focuses on the

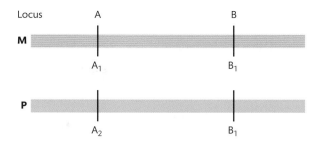

Figure 3.1 A segment of the genome of an individual who is heterozygous at one locus (labelled A) and homozygous at another (B). The maternal chromosome is labelled M and the paternal chromosome is labelled P.

frequencies of each allele at a locus (note that the sum of the frequencies of all alleles at a locus is 1). Now, consider a population where the frequencies of the four alleles A_1, A_2, B_1, and B_2 are given as $f(A_1) = p_{A1}$, $f(A_2) = p_{A2} = 1 - p_{A1}$, $f(B_1) = p_{B1}$ and $f(B_2) = p_{B2} = 1 - p_{B1}$. If haplotypes are formed by the independent sampling of alleles at the two loci, then the haplotype frequency of the haplotype p_{A1B1} would be given by the product $p_{A1}p_{B1}$ of the allele frequencies. For example, if allele A_1 occurs in 25% of the population, and B_1 in 10% of the population, then these alleles would be expected to occur together in (0.25 × 0.1) or 2.5% of individuals. If this is not the case—that is, if the haplotype A_1B_1 occurs more or less frequently than expected—the alleles are said to be in *linkage disequilibrium* (LD).

In populations of humans (and of many other species) genetic variation includes many blocks of SNPs that do not vary independently and thus are in very high LD. Scientists working for the HapMap Project are developing a map of LD patterns in human populations. SNPs that are physically close to each other on the same chromosome tend to be in high LD, and average LD tends to drop consistently with physical distance along a chromosome. This is generally understood in terms of balances between evolutionary forces. LD may be increased by mutation, selection, and genetic drift due to small population sizes. On the other hand, LD is reduced by the process of *genetic recombination*. This occurs during the formation of the gametes (egg and sperm cells) when DNA strands break and recombine to form chromosomes that are different from the maternal and paternal ones.

[1] http://hapmap.org

3.2.2 **From Genotype to Phenotype**

Recall that a phenotype is any observable trait of interest in an organism, though usually one that varies within a population. Phenotypes are often classified by whether their variation is caused by genetic variation at one, a few, or many genetic loci (mono-, oligo-, and polygenic traits). Furthermore, we can distinguish between phenotypes that vary continuously (quantitative phenotypes, e.g. height) and those that are divided into distinct categories (qualitative phenotypes, e.g. eye colour, albinism, or the presence or absence of a disease).

Single-gene, qualitative traits are also called Mendelian traits, after the monk Gregor Mendel, who discovered the basic inheritance patterns of qualitative traits by studying phenotypes such as colour in peas. The database Online Mendelian Inheritance in Man[2] contains over 7000 human phenotypes with Mendelian inheritance. In about half of these the molecular basis for the difference in phenotype is known. Well-known examples include cystic fibrosis, in which the defective gene encodes an ion-channel protein; haemophilia, which affects a blood-clotting factor; and sickle-cell anaemia, which affects the oxygen-carrying protein, haemoglobin.

Even though many human diseases are fully determined by a single polymorphic locus, most quantitative traits, and important diseases including diabetes, many types of heart disease, and many cancers, are complex traits that are affected by many genes as well as by the individual's lifestyle and environment. Much of the analysis of such complex traits is conceptually founded on single-locus genotype–phenotype maps similar to those that are used for Mendelian traits.

However, while Mendel distinguished between dominant and recessive traits (or alleles), there is a whole spectrum of possible types of *gene action* for loci underlying a quantitative trait. Figure 3.2 illustrates the genotypes of a single locus A underlying a quantitative trait (e.g. height) together with the mean phenotype for each genotype. Parameters a and m are chosen such that the mean phenotype of homozygous individuals with two copies of the allele A_1 is $m - a$, and the mean phenotype of those

[2] http://www.omim.org

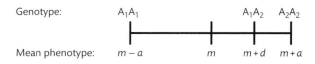

Figure 3.2 Genotypes and corresponding mean phenotypes for a single locus A. The parameters a and d are used to describe the gene action of the locus.

with two copies of the allele A_2 is $m + a$. The gene action of locus is determined by the mean phenotype value ($m + d$) of heterozygotes A_1A_2, relative to the two homozygotes. The locus A is said to show *additive gene action* if the heterozygote is mid-way between the homozygotes; that is, if $d=0$. Furthermore the locus is said to show *partial dominance* if $|d| < a$, *complete dominance* if $|d| = a$ (which implies that the heterozygote phenotype is identical to that of one of the homozygotes), and *over-dominance* if $|d| > a$.

In real data, there is of course variation around these mean phenotype values. This variation can be due to random differences in environment, individual development, or genetic variation at other loci. It is important to be aware that gene effects are always described in terms of hypothetical *allele substitutions* relative to some reference genotype. For example, using A_1A_1 as the reference genotype, replacing one A_1 allele with A_2 results in the A_1A_2 genotype. However, the effect of this substitution may vary depending on the circumstances. If the effect of an allele substitution at locus A varies depending on the individual's genotype at locus B (or other loci), this interlocus non-additivity is called *epistasis*. On the other hand, *genotype × environment interaction* means that the effect of an allele substitution depends on the environment. A famous example is sickle-cell anaemia, where heterozygotes have partial resistance to malaria because the parasite cannot reproduce inside the abnormal blood cells. However, this is only a net advantage in areas where malaria is common, and only for heterozygotes. For homozygotes, the drawbacks of sickle cells always outweigh any benefits.

A key tool in quantitative genetics is *variance decomposition*, which uses statistical regression models to estimate the proportion of phenotypic variance that can be attributed to additive gene action, dominance, and epistatic interactions between genes (collectively called *genetic variance*), and *environmental* variation. The latter typically includes all variation not otherwise represented in the regression model.

You should take a moment to convince yourself that the proportion of genetic variance depends on the frequencies of the different genotypes in the population. For example, if almost all individuals have the same genotype, most of the phenotypic variance will be due to environmental differences.

Although a full understanding of variance decomposition is outside the scope of this book, genetic variance components form the basis for many central concepts in genetics. For instance, the broad-sense heritability of a trait is defined as the ratio of total genetic variance to the total phenotypic variance, whereas the narrow-sense heritability is the ratio of *additive* genetic variance to the total phenotypic variance. In well-fed human populations, body height is one of the most heritable traits, with narrow-sense heritability often in excess of 80%. This means that 80% of the variance in height can be attributed to differences in genotype, while the remaining variation is due to environmental factors such as nutrition. Thus, heritability refers to *variation between individuals*; the preceding example does *not* mean that you as an individual owe, say, 34 cm of your height to the environment!

3.2.3 Linkage Mapping and Genome-Wide Association Studies

There are two main classes of method for inferring quantitative trait loci from genotype data and phenotype records for a particular trait in a study population. *Linkage mapping* relies on information about the inheritance patterns or *pedigree* of the study population and capitalizes on recombinations that occur within the genotyped population. There are many software packages for linkage mapping available, including for instance the well-documented open-source package R/qtl[3]. Linkage mapping is most powerful in line crosses (crosses of inbred lines), commonly used in plant and animal breeding, and in model organisms such as fruit flies and mice.

Linkage mapping of human data, however, is often not very informative, because humans have complex pedigrees, long life cycles, and small family

[3] http://www.rqtl.org

sizes. Therefore, quantitative trait locus mapping in human populations typically uses genome-wide association studies (GWAS). These association methods do not try to track recombinations in the study population; instead, they search for correlations between alleles and phenotypes. GWAS gives greater statistical power and better resolution than linkage mapping, but at the price of many false positives due to the high LD between many loci.

3.3 Systems Genetics

In the previous section we set out the bare bones of the theoretical background of classical quantitative genetics. You may have noted that DNA was scarcely mentioned; the whole framework can be understood using the concepts of traits, loci, and alleles without specifying their molecular nature. This is because this theoretical foundation was laid out many decades before the dawn of molecular biology. Indeed, it is possible to become an excellent quantitative geneticist without much molecular knowledge. This framework, however, does have its limitations, and classical quantitative genetics has sometimes been denigrated as 'beanbag genetics', treating alleles simply as beans of different colours drawn at random from a bag, and naively ignoring the mechanistic interplay of genes and the proteins for which they code.

A new branch of genetics that has been termed *systems genetics* is emerging to challenge this century-old, 'black box' model of gene action [3]. The stated goal of systems genetics is to integrate quantitative genetics with the causal models of biological systems that are being developed. In empirical studies, this means that GWAS and other methods of searching for genotype–phenotype correlations are being complemented by causal inference methodology such as Bayesian networks. Ideally, systems genetics should go on to develop a theory that explains how the phenomena of heredity and variation arise from genes that are continually interacting in multi-scale, non-linear dynamical systems within the human body.

In the rest of this section we review some pioneering attempts at integrating non-linear dynamical systems theory with quantitative genetics and introduce the concept of the causally cohesive genotype–phenotype modelling approach.

3.3.1 **The Development of Systems Genetics**

The history of systems genetics can be traced back more than 40 years to the first explicit use of the term 'genotype–phenotype mapping' by Jim Burns in 1970 [4]: 'It is the quantitative phenotype, arising from the genotypic prescriptions and the environment, which is of critical importance for the cell's survival and which therefore features in population genetic theory. A study of this synthetic problem would thus, by providing genotype–phenotype mappings for simple synthetic systems, help to connect two major areas of biological theory: the biochemical and the population genetic.'

As you will see in subsequent chapters, the research programme of computational biomedicine is fleshing out this vision by integrating the physiological consequences of genetic and biochemical variation across scales of cells, tissues, and organs.

Henrik Kacser and Jim Burns [5] later studied the genotype–phenotype maps of metabolic systems and pointed to molecular properties of these systems that could provide a molecular basis for dominance, paving the way for studies of more complex genetic concepts such as pleiotropy, in which a single gene affects several different phenotypic traits [6]. Later studies have shed light on the quantitative genetic and evolutionary properties of generic models for systems of gene regulation; see Section 3.5 for an example of this.

We next describe two examples of systems genetics studies of specific biological systems.

Flesh Colour in Chinook Salmon

Flesh colour in salmon is an economically important trait that is under strong genetic influence. The Pacific chinook salmon (*Oncorhynchus tshawytscha*) occurs in both white- and red-fleshed varieties, but the mutations that cause this difference are not yet fully understood. Furthermore, the published data on crosses between red- and white-fleshed salmon cannot be reconciled with a simple genetic model. It has therefore been assumed that flesh colour in salmon must be affected by a complex inheritance pattern involving interactions between genes.

Hannah Rajasingh and her collegues built a dynamic mathematical model to describe how the red pigment astaxanthin is taken up from the intestine, transported in blood and deposited in muscle, skin,

and other organs [7]. The rates of these processes depend on the current distribution of pigment in the body, and also on age and sexual maturity. The model was capable of reproducing pigment uptake and retention over timescales of hours, days, and years and to shed light on individual variation in pigment utilization and meat colour. By introducing *in silico* genetic variation in model parameters and simulating crosses between white and red salmon, the study authors showed that variation at a single gene involved in the muscle uptake of carotenoid pigments could explain the experimental data.

Sporulation Efficiency in Yeast

Our second example concerns baker's yeast (*Saccharomyces cerevisiae*), which can grow asexually in both a haploid form (with a single copy of each chromosome) and a diploid form (with homologous chromosome pairs). The diploid form is the most common in nature, but under stressful conditions this single-celled organism often undergoes *meiosis* (that is, reductional cell division with recombination) to form haploid spores that can survive until growth conditions improve. Jason Gertz and his coworkers developed a model of the yeast regulatory pathway that controls its sporulation efficiency; that is, the fraction of cells in a culture that undergo meiosis and form spores when starved [8] (Figure 3.3). This thermodynamic model contains parameters that describe the concentrations and binding affinities of RNA polymerase and regulator proteins to the IME1 promoter region (see Chapter 2 for a brief description of gene expression). The model predicts the level of active Ime1 protein, which is assumed to be proportional to sporulation efficiency. The researchers studied a cross between two yeast strains and estimated the strain-specific parameters values for four causative genetic variants in the sporulation pathway. They found that their mechanistic model captured the functional dependencies between the causal variants quite well and could therefore explain more of the observed variation in sporulation efficiency than a classical linear model having twice as many parameters.

These studies exemplify the insights and patterns that can be discovered at the interface between nonlinear dynamic models of biological systems and

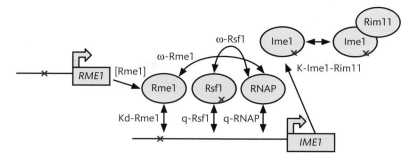

Figure 3.3 Illustration of a gene-regulatory model of the pathway that controls sporulation in yeast. Red crosses represent causative genetic variants that differ between yeast strains. Sporulation efficiency was assumed to be proportional to the relative amount of the protein Ime1 that is bound to Rim11 (on the far right of the figure). Reprinted from *Theoretical Population Biology*, Vol. 77, Iss. 1. J. Gertz, J.P. Gerke, B.A. Cohen, Epistasis in a quantitative trait captured by a molecular model of transcription factor interactions, pp. 1–5, © 2010 with permission from Elsevier.

classical genetic theory. We describe the theory that underlies these studies in more detail in the following sections.

3.3.2 Causally Cohesive Genotype-to-Phenotype Modelling

It can be a useful mathematical abstraction to view life as a genotype–phenotype map, which assigns a phenotypic outcome to each possible genotype. This enables us to characterize and compare the genotype–phenotype relationship for different biological systems under different environmental conditions. However, standard population genetic models simply assign phenotypic values directly to genotypes, without any intervening explanation of the causal relationship between them.

In this section, we describe a framework to incorporate physiological models in the genotype–phenotype map concept. Well-built mathematical models condense physiological knowledge into a succinct set of parameters. The key is to map genotypes to physiological model parameters, and to do so at the most fundamental physiological level practicable, rather than map genotypes directly to the high-level phenotypes that are the eventual target of understanding. Then, the physiological models account for the variation in phenotypes at intermediate and high levels. The simulated data can be collected and subjected to classical population and quantitative genetic analyses. Because these *in silico* data are derived from mechanistic models, they act to frame

genetic analyses in terms of observable quantities and testable hypotheses. This approach points the way towards a real mechanistic understanding of how the interplay between genetics and the environment leads to variation in organisms, eventually with important biomedical applications. By increasing understanding of how complex diseases develop based on individual patients' genetic background, lifestyle, and age, it holds great promise for the development of new, individualized treatment regimens and personalized medicine.

This framework is called *causally cohesive genotype–phenotype modelling* (or cGP modelling; Figure 3.4). This term denotes a mathematical conceptualization of how genetic variation is manifested in phenotypes at various levels of the biological system, from the molecular through the cell and organ and up to the whole organism. The basic premise of cGP modelling is that in a well-validated model that is capable of accounting for the phenotypic variation in a population, the genetic variation that is responsible for this will be reflected in the parameters of the model.

The distinguishing feature of cGP models is the explicit representation of the individual's genotype and its direct and indirect consequences. cGP models seek to account for the effects of genetic variation by linking the genotypic and phenotypic domains in terms of regulatory principles and mechanisms. This enforces logical consistency and coherence between the components of a proposed explanation for a genotype–phenotype map (which is why such models

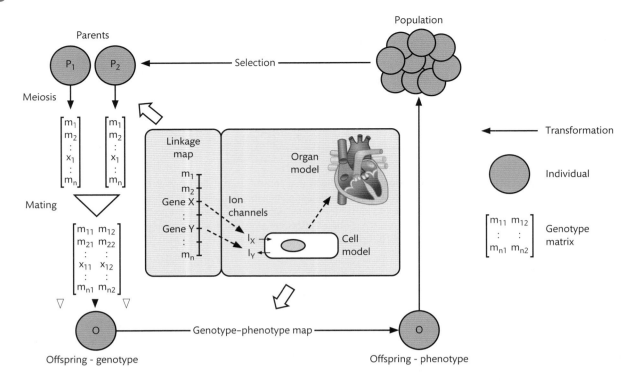

Figure 3.4 Integrating genetics, genomics, and multi-scale models in a population context. This illustration shows genes coding for the ion channels that carry the transmembrane currents that make up the action potential of a heart cell. Genetically determined variation in low-level parameters propagates through multiple levels of electrophysiological, mechanical, and fluid dynamic processes. Phenotypic variation emerges at each level of organization. A multi-scale model of this biological system is connected to a linkage map via the genes encoding ion channels. The resulting cGP model thus describes the creation of new genotypes as well the phenotypes arising from them. By simulating populations of cGP models, whose parameters arise by recombination of virtual genomes, we can obtain a deeper understanding of the high-dimensional genotype–phenotype. Figure made by Arne Gjuvsland.

are termed 'cohesive'). Of course, the explanation will still be simplified (that is a necessary consequence of its being a model), and it may still be inadequate or downright wrong, but at least it is internally coherent and expressed in terms of observable quantities as far as practicable. A concrete example of a cGP model for the electrical activity of a heart muscle cell is described in Box 3.1.

In a mathematical model of physiology, a phenotype can be any observable characteristic of the entities included in the model. Physiological models usually deal with dynamical systems, so that their *change over time* (or lack of change over time, if forces or processes are balanced in dynamic equilibrium) is the essential output. Whereas the full multivariate dynamics of the system over space and time can be considered as its 'raw' phenotype,

researchers usually focus on a specific set of features that are relevant to the purpose of the model. For instance, the model of a heart muscle cell described in Box 3.1 aims to explain the time course of voltage across a membrane (called the 'action potential') from its constituent ion currents. Each of those currents involves several state variables. The action potential is often summarized in descriptive statistics such as duration and amplitude, whereas voltage-dependent ion currents can be characterized by how quickly they respond to changes in the transmembrane voltage (which can be either spontaneous or experimentally induced). Whole-body and whole-organ heart phenotypes might include the heart rate, stroke volume, deformation of the heart muscle, or the propagation speed of the electrical activation wave.

Box 3.1 A cGP Model of the Action Potential of a Heart Muscle Cell

This box gives concrete examples of the components of a cGP study [9]. It is based on a detailed model of a mouse heart muscle cell, and the cGP analysis is described in Section 3.5.2.

Biological System

The model describes the flow of ions across the cell membrane and between different compartments of the cell. This flow of ions achieves the two main functions of the heart muscle cell: to contract the heart muscle and to propagate an electrical signal. The cell 'charges its battery' by transporting many positive ions out of the cytoplasm. Once it is charged, an electrical impulse will cause the cell to 'fire', as ion channels open to allow rapid depolarization, allowing ions into the cell and producing a signal that propagates to neighbouring cells. Muscle contraction, on the other hand, is triggered by the release of calcium into the cytosol. This calcium is initially sequestered into special compartments until depolarization triggers its release.

Multilevel Phenotypes

The main cell-level phenotypes of this model are the *action potential* and *calcium transient*; that is, the time courses of the transmembrane potential and the cytosolic calcium concentration, respectively. However, this model includes many subcellular phenotypes that can be observed and studied experimentally. These include the *ion currents* that make up the action potential and calcium transient and that are carried by specialized protein complexes called *ion channels* (see Box 3.2). There are many types of ion channel, and each cell includes many molecules of each type. Ion channels help or hinder the passage of ions (particularly calcium, sodium, and potassium) across the cell membrane, opening or closing in response to conditions such as ion concentrations or transmembrane potential. Ion channels differ in aspects such as voltage thresholds and in how fast they switch between states. Their structure and function is well understood and can be experimentally observed and manipulated. In particular, stepwise changes in transmembrane voltage can be induced to study the voltage-dependent conformation switching behaviour and 'memory' of ion channels, offering a common basis for comparing the ion-channel behaviour of different cell types, models, or parameter scenarios.

Disease Phenotypes

The heart needs to be able both to relax (to fill the heart chambers with blood) and to contract (to pump blood out to the body). If the calcium transient is used as a proxy for this contraction and relaxation, cell dynamics can be categorized as 'failed' if, for example, the peak is below 50% of the baseline (illustrating failure to contract); if the base is more than 200% of the baseline (failure to relax); if the amplitude is less than 50% of the baseline; or if dynamics fails to converge within 10 min of simulated time.

Virtual Experiments

Muscle cells require external stimuli to exhibit their characteristic dynamics, and different experiments can be designed to explore different aspects of the behavioural repertoire of the system. These simulated heart cells were subjected to four different protocols: pacing at regular intervals, to mimic the normal function of the heart; two-step voltage clamping to observe the voltage-dependent activation of ion currents; variable-gap voltage clamping to study the recovery from inactivation; and quiescence (no stimulus) as a 'null experiment'.

Mapping Genotypes to Parameters

The model incorporates a simple genetic model, in which it is assumed that each model parameter could be determined by one locus with two alleles. The relationship between each genotype and its related parameter (a low-level phenotype) is the same in each case, with the heterozygote A_1A_2 being assigned 100% of the baseline parameter estimate and the homozygotes A_1A_1 and A_2A_2 being assigned 50 and 150% of that estimate respectively.

Genotype Generation

The effects of the hypothetical genes were estimated by an initial sensitivity analysis describing the percentage change in each scalar phenotype per percentage change in each model parameter. For a subset of the parameters, all possible combinations (i.e. a full factorial design) were generated to study higher-order parameter interactions.

The cGP approach views the genotype–phenotype map as a composite of mappings between phenotypes from low to high levels (grounded in a mapping from genotypes to parameter values, see below).

Each layer of the multilevel genotype–phenotype map can be described and analysed, shedding light on phenomena in genetics such as robustness, 'sloppiness' (referring to the phenomenon that many parameter combinations may give rise to similar phenotypic outcomes), and incomplete penetrance; that is, where mutations that have been related to complex traits and diseases do not always manifest clinically.

We noted in Chapter 1 that physiological models are often built 'from the middle out'. This means that they emphasize certain aspects of system dynamics at a particular spatial and temporal scale, or level, while incorporating connections to levels above and below. The essence of these connections is that one model's phenotype is another model's parameter, i.e. a summary measure of system behaviour at one scale can enter as a parameter in a model that focuses on a different scale. This goes both from lower to higher levels (for instance, a heart cell's ion channels combine to form the action potential that propagates the signal to contract the heart muscle), and from higher to lower levels (for instance, some ion channels are stretch-sensitive, so that the combined deformation of muscle tissue affects the behaviour of ion channels in the cells). You will learn much more about this in later chapters: Chapters 5 and 6 describe models at the cell and tissue levels, whereas Chapter 7 describes how to combine models across scales.

As noted earlier, one of the benefits of mathematical modelling is that knowledge gets distilled down to a succinct set of model parameters. Therefore, such parameters may even be especially *relevant* phenotypes in a search for the genetic causes of high-level phenotypes. For example, the heart cell model in Box 3.1 includes submodels for conformation switching in ion channels that can be derived mechanistically from protein structure and dynamics, as illustrated in Box 3.2.

A fundamental feature of cGP studies is that genotypic variation must be mapped to variation in parameter values, assigning each simulated individual with a given genotype to a point in the parameter

---

Box 3.2 Refining the Genotype-to-Parameter Map: From Nucleotide Mutation to Protein Conformation to State Switching in Ion Channels

Cardiac ion channels are prime candidates for realistic genotype-to-parameter mapping, as both their physiology and their genetic variability have been well studied. One example comes from Silva and colleagues[10], who used protein-folding models to predict the effects of amino acid substitutions on conformation stability and transition rates in ion channels. They developed Markov-type state-transition models as a higher-level approximation and verified the results experimentally for some mutations. For small perturbations in overall structure, and particularly for the voltage-sensing region of voltage-sensitive ion channels, this marks a major advance in techniques for mapping genetic variation onto physiological parameters.

Whole-cell models of cellular electrophysiology usually lump together the combined activity of many ion channels. They generally describe changes in transmembrane voltage due to the ion currents arising from different types of channel using an equation like this one:

Rate of change in voltage =
(current 1 + current 2 + ...) /
(capacitance of cell membrane)

The current is driven by the difference in ion concentration between the inside and outside of the cell, as well as differences in electrical charge concentration. At a certain transmembrane voltage, those forces cancel and there can be no net current. Thus, a typical ion current is modelled as:

Current = (amount of ion channel) ×
(proportion of channels that are open) ×
(voltage − equilibrium voltage)

As indicated, not all ion channels are open at any given time; instead, they flicker between open and closed states. In *voltage-sensing* ion channels, rates of opening and closing depend on the transmembrane voltage in a way that is dependent on the amino acids in the voltage-sensing domain of the ion channel. Thus, the sequence of DNA that codes for this protein domain has a very direct effect on the model parameters.

This multi-scale model of a cardiac potassium channel has produced accurate results on the molecular and macroscopic levels; the stages of the modelling exercise are described more fully in the panels of Figure 3.5. The potassium channel KCNQ1 consists of four subunits, each of which has six segments (colour-coded in panels (a) and (b) of Figure 3.5). Here, panel (a) shows the three-dimensional structure of this channel (viewed from outside the cell), and panel (b) shows alternative meta-stable conformations of

... continued

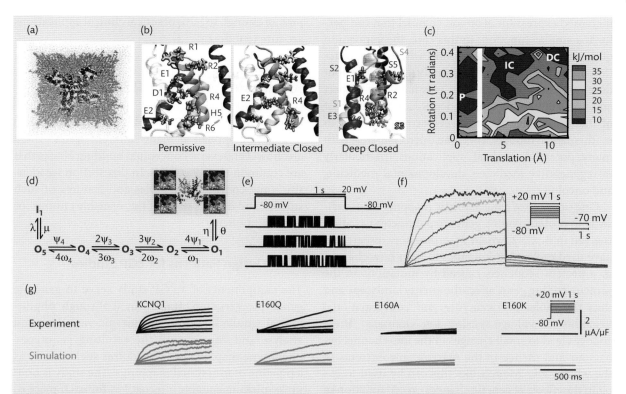

Figure 3.5 Bridging the genotype–phenotype gap with data and models at multiple phenotypic levels. Details are given in Box 3.2. Figure from Gjuvsland et al. (2013) Bridging the genotype–phenotype gap: what does it take?, *Journal of Physiology*, 591, 2055–2066, based on [10].

the channel, only one of which—the 'permissive' conformation shown here—allows ions to pass through the channel.

Panel (c) shows the 'energy landscape' of this conformational change, which is characterized by the translation and rotation of the segment labelled S4 in panel B, for transmembrane voltage V_m = 0 mV. Current through the channel occurs only when all subunits are in the permissive conformational state. Panel (d) shows a schematic diagram of a Markov model, simulating the conformational changes in the ion channel into changes between discrete states that

correspond to energy minima; the transition rates between these are voltage dependent. Panel (e) shows experimentally recorded traces of voltage across single channels showing stochastic switching between the open (permissive) and closed conformations, and panel (f) shows the macroscopic current obtained by summing the current through 1000 separate single channels. The model accounts well for the macroscopic current of both wild-type and mutated ion channels across a range of imposed voltage steps, as is shown in panel (g).

space of the physiological model. This mapping is a model in itself, a simplification that imitates the main effects of genes and their interactions in determining low-level physiological parameters. Even more fundamentally, the genotype–phenotype map relies on the existing infrastructure within the cell that is inherited from the mother and often implicitly taken for granted. The eventual value of a physiological parameter (or phenotype) in a cell, tissue, or organ depends on a complex interplay between the cellular

machinery, gene expression, and cues and signals from both inside and outside the body. Cellular and bodily development involves an impressive mixture of *phenotypic plasticity* (the ability to adjust a phenotype depending on signals and cues from within and outside the body) and *robustness* (the ability to produce a consistent phenotype in spite of perturbations in environment, development, or genetic background). For instance, a loss-of-function mutation in one ion-channel gene might be compensated by increased expression of an ion channel that can per-form a similar function. In this case, the expression of the latter ion channel is plastic, whereas the func-tion of the cell is robust. We already know a lot about developmental processes and plasticity (see Gilbert and Epel 2009 in Recommended Reading), but most of this knowledge has not yet been formulated in mathematical models. This is an open challenge with many potential medical applications, particularly in regenerative medicine, developmental medicine, and oncology.

One example of a mechanistic mapping from vari-ation in protein-coding genes to parameter variation is given in Box 3.2, which describes how researchers have used molecular dynamics to infer how changes in a protein sequence, caused by mutations in the protein-coding gene, affect the energy landscape and thus the molecular structure and conformational flexibility of voltage-sensing ion channels. These changes lead to small changes in the ion-channel be-haviour and thence in the macroscopic currents of the whole cells. Models like this can be used to ex-plain the effects of genetic changes on cell function in terms of the cells' fundamental chemistry.

Figure 3.5 also illustrates how cGP models can enable the integration of genetics, genomics, and multi-scale models in a population context; this forms a meeting point for many of the themes that are covered this book. We have seen already, in Chapter 2, how the high-throughput acquisition of data on genomic structure and genotypic variation has important consequences for our understanding of protein expression and metabolism, and that this can feed into computational models. We describe how both molecular and genetic data feed into models of cells, tissues, and organs in later chapters.

The next two sections give a practical introduc-tion to implementing cGP studies, followed by ex-amples of its application. Although applications of this type of modelling that are of use in the clinic still seem a long way off, the approach has proved useful in more limited applications. The level of de-tail and realism that is obtainable in each study will depend both on the purpose of that study and on the amount and range of relevant data that are available. We have seen already how the volume and quality of data are growing, and improved prospects for im-plementing genotype–phenotype models that are

precise enough to have practical clinical applications are sure to follow.

3.4 Implementing cGP Models

As you have seen, cGP modelling has a wide scope and can be applied at many different scales. There are, however, several conceptual building blocks that are common to cGP studies. In this section we iden-tify some common principles and features of such studies, and discuss some of the available options for each component.

3.4.1 Design Patterns for cGP Studies

The components of a cGP study serve to generate genotypic variation, transform it through physio-logical simulations into phenotypic variation, and analyse and characterize the resulting genotype–phenotype map. Figure 3.6 pictures such a study as a pipeline of functional mappings. The study is built up by picking suitable alternatives for each piece of the pipeline and ensuring that the output format of each piece matches the input format of the next. Analysis focuses on characterizing these mappings, for instance by comparing the sensitivity of pheno-types to different parameters or to genetic varia-tion at different loci. On a more general level, a cGP model enables biological systems to be compared in terms of characteristics of their mapping by, for instance, applying the same analysis to different cell types.

cGP modelling is a multidiscipline approach, and the various pieces of a cGP pipeline are rarely built from scratch. Rather, they use existing physiological models that can be obtained from research col-laborators or open repositories; genetic and other molecular data from a range of genomic and other omics databases; and third-party genetics or statis-tical software for analysis. This interchange of sub-models, analysis methods, and other components of a cGP model is greatly facilitated by open stand-ards for data and model formats, and by interfaces to workflows. The concepts of open data and com-mon data formats are discussed briefly in Chapter 2 and the Appendix, and workflows are discussed in Chapter 8.

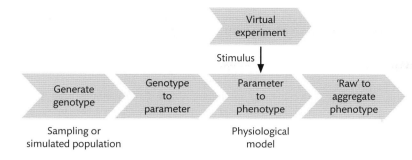

Figure 3.6 Simulation pipeline for cGP studies. Peach arrows denote functions that generate genotypes or transform them through successive mappings, genotype to parameter to 'raw' phenotype to aggregated phenotypes. Figure modified from [9].

Each of the steps in a cGP pipeline such as that shown in Figure 3.6 will now be described in more detail.

Generation of Genotypic Variation

There are many different reasons why you, as a computational modeller, might need to simulate genetic or genotypic variation. You might, for example, be evaluating statistical genetics methods and therefore need to use models to mimic population-level patterns such as allele frequency distributions and the degree of LD. Alternatively you may simply need to characterize the main gene effects and interactions involved in the genotype–phenotype map itself. Simulating phenotypes for all possible genotypes may be feasible when the model is computationally inexpensive and involves relatively few loci and alleles. If this is not the case, however, it is often necessary to generate and sample a subset of genotypes. This can either be a random sample or one generated according to some experimental design, perhaps based on statistical efficiency.

In population simulations, genotypes arise by *in silico* 'mating' of the parent genotypes. Genetic recombination can be modelled as statistically independent between loci (which assumes that the loci involved reside on separate chromosomes or at least are widely spaced on a single chromosome). Alternatively, it is possible to use linkage maps and information about chromosome structure to model genetic recombination more realistically. Observed (or hypothetical) pedigrees and genotypic information such as HapMap data may be used as input for Monte Carlo methods (see Box 3.3) to generate new genotypes with realistic allele frequencies and levels of linkage disequilibrium. Such simulation methods can also make use of linkage maps, which contain information about the ordering of loci on chromosomes and the probability of genetic recombination during the formation of gametes.

The Genotype-to-Parameter Map

In an ideal situation, the parameters involved in a model will be reliably measurable and their effects will generalize to other individuals or circumstances with little confounding. However, many parameters can be difficult to estimate, so it is likely that hypothetical assumptions must be made for at least some of them. Furthermore, model parameter scenarios cannot usually be estimated directly for all possible genotypes. Any genotype-to-parameter mapping is therefore a model in itself, a simplification to imitate the main effects and any major interactions between genes in determining low-level physiological parameters.

A realistic genotype-to-parameter mapping will be tightly linked to the particular system that is being modelled. However, some simplifications or generalizations can often be reused between models. Examples of these include the assumption that each parameter is governed by a single gene, and the specification of a fixed percentage of variation between genotypes at a locus.

The Parameter-to-Phenotype Map

Physiological models for cGP studies can be obtained from model repositories, and some of these are described briefly elsewhere in this book (particularly in the Appendix). The standardization and curation of models is most mature for the large class of single-cell models that have been formulated as differential equations; it is less so for more complex modelling paradigms such as continuum mechanics or neural

---

Box 3.3 Monte Carlo Methods

The term 'Monte Carlo' refers to the use of repeated random sampling to obtain a range of possible, generally numerical values and probabilities for an outcome. Monte Carlo methods have applications in many different disciplines, including mathematics, the physical sciences, engineering, and finance, as well as computational biology. They are suitable for any situation that is amenable to mathematical modelling but where there is significant uncertainty in both the input and the output states; that is, where the number of degrees of freedom is large.

A typical Monte Carlo method will involve the following steps:

1. defining a range of possible input values for the simulation;

2. selecting a large number of random (or pseudorandom) values from within that range;

3. running a simulation starting from each of the random input values and recording the results of each;

4. aggregating the final results to yield a probability distribution for the output.

In molecular modelling, Monte Carlo methods have historically been used as an alternative to the more common molecular dynamics to simulate the behaviour of complex molecules and model conformational changes. In genetics they are often used to sample the possible range of genotypes based on the likelihood of each occurring, simulating the process of genetic recombination. You will come across these methods again from time to time throughout this book, particularly in Chapter 11.

You may be interested to know that this technique was originally invented in the 1940s within the US nuclear weapons programme. The term Monte Carlo selected as a code name for the technique makes reference to the analogy between the many runs of a simulation and (for example) many throws of dice at a casino.

---

network simulations. Many multi-scale frameworks for simulation are available, and many of these are open source, but the models and data themselves are often not standardized or freely available.

There are usually several candidate models for any system or phenomenon and a serious modeller is likely to submit a series of comparable models to similar analyses in order to compare how the systems behave. This critical scrutiny is also facilitated by the open standards, open software, and open data that we mentioned briefly in Chapter 2 and that will be discussed at greater length later in the book.

Virtual Experiments

The excitable cells described in Box 3.1 form an example of a model system where the most important phenotypes are defined by the model system's response to some stimulus or perturbation in the form of a *virtual experiment*. Like real-world experiments, virtual experiments are designed to bring out essential aspects of the system's behaviour that may not be evident from passive observation. For example, two types of experiment are routinely applied to the heart muscle cells that are modelled in Box 3.1, both *in silico* and in real life: regular pacing to elicit voltage spikes (action potentials), and voltage clamping

to study the voltage-dependent opening and closing of ion channels, decoupled from the effect of ion current on voltage.

Through their applicability to whole classes of similar systems, these virtual experiments offer a basis for making comparisons between parameter scenarios for a given model; between candidate models for a given system; or within a class of biological entities such as different related cell types.

Aggregation of Phenotypes

Even if you have a lot of data—in fact, especially then—you need to summarize it in a meaningful way before you can use it to compare physiological systems, whether real or simulated. In many cases, interest may centre on the eventual equilibrium of the system (e.g. in a model of blood sugar regulation) or its steady dynamics (e.g. in models of the suprachiasmatic nucleus in the brain that controls daily biological rhythms, and of the cardiac pacemaker cells that spontaneously generate action potentials).

Some aggregated measures, such as heart rate, are known from experience to be clinically relevant. These may be at different physiological levels; examples from cardiology include pressure/volume diagrams at the whole-heart level, action-potential duration at the

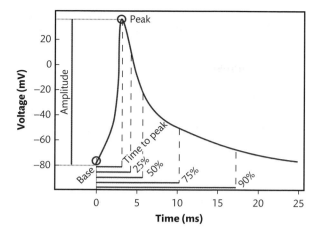

Figure 3.7 Summary statistics for an action potential. For example, action potential duration to 90% repolarization (APD$_{90}$) is an aggregated phenotype of known clinical relevance. Modified from [13].

cell level (Figure 3.7), and the current decay rate at the ion-channel level. A complementary approach that is growing in importance as phenotypic data becomes richer is multivariate analysis, a technique in which many variables can be analysed simultaneously. This can simplify complex data by identifying and describing patterns of covariation between variables.

Population-Level Simulations: Selection, Mating, and Recombination to Generate New Genotypes

In real populations, new genotypes arise from the selection and mating of parents, and occasionally by mutation. Gametes are formed by recombination of the parent genomes and fuse to produce the genome of the offspring, initially as a single fertilized egg (a *zygote*). The modelling of selection, mating, and recombination is important for many disciplines within biology, including evolutionary ecology as well as plant and animal breeding.

Furthermore, population-level simulations are useful in evaluating methods to detect genetic variants underlying human genetic diseases. For instance, many methods assume statistical independence between genetic variation at different loci, an assumption that is commonly violated by high LD (e.g. in human populations). In population simulations, one can pretend to know which markers do in fact predispose for disease and then simulate the genotypes of new 'individuals' from a population

that has realistic allele frequencies and levels of LD. The method being evaluated is then applied to the simulated data, and the findings are compared with the known answer. This makes it easier to distinguish true from false positives, to evaluate alternative methods, and to optimize experimental designs.

Tools for cGP Modelling: The cgptoolbox

Several groups worldwide have put together suites of mainly open-source software that can be used for genotype–phenotype modelling. One example of this is the *cgptoolbox*[4], a set of software and worked examples that glue together numerous open-source tools and resources that are relevant to this area. We would advise our readers that it is much easier for novice modellers to start cGP modelling using a package like this one than to work from scratch, even if they are experienced programmers.

The cgptoolbox forms the basis for many of the examples in the next section. It follows the workflow and design pattern that is illustrated in Figure 3.6 and facilitates the use of existing tools and standards where these are convenient, mixing and matching interchangeable options for each of the building blocks in the workflow. This toolbox and other similar resources use open standards and resources wherever these are relevant, including the omics databases reviewed in Chapter 2 and the model repositories described in the Appendix and elsewhere. Section 3.6 describes how data can be obtained from different sources and linked together within a model.

Many other software tools that can be useful for cGP modelling are available in the public domain. These include the HDF5[5] Hierarchical Data Format and associated tools, which are useful for efficient storage, retrieval, navigation, and subsetting of huge datasets.

The well-known statistical software suite R[6] can be used in cGP modelling for experimental design, statistical aggregation, and visualization. It can be interfaced with other scripting tools and languages, for example using rpy2 as a bridge between R and Python[7]. Modelling population processes such as selection, mating, and recombination can involve

---

[4] https://github.com/jonovik/cgptoolbox
[5] http://www.hdfgroup.org
[6] http://www.r-project.org
[7] http://rpy.sourceforge.net/rpy2_documentation.html

simuPOP[8], which provides functions for simulation of mating and recombining genomes, keeping track of chromosomal organization and genetic distances (map units) between genes and thereby simulating parameter values for the offspring. Section 3.5.3 gives an example of an *in silico* GWAS that uses PLINK[9] to analyse simulated population data obtained using the package simuPOP.

Box 3.4 Models of Gene Regulation

Recall from Figure 2.7 that several processes are involved in gene expression in eukaryotes. The first is transcription, in which DNA is copied into pre-mRNA. Next, pre-mRNA is processed into mature mRNA. This mRNA is then transported from the nucleus into the cytoplasm, where it is finally translated into protein. Regulation of gene expression is dominated by transcription factors that affect the rate of initiation of transcription by binding to a *regulatory promoter sequence* of DNA that precedes the protein-coding part of the gene.

The dose–response relationship between the concentration of a transcription factor and the transcription initiation rate is called a *gene-regulation function*. These form the connections in gene-regulatory networks, and the shape of gene-regulation functions determines key features of cell behaviour such as switching and oscillations. In many cases, these functions are smooth and switch-like (S-shaped or sigmoid; see Figure 3.8), reflecting cooperative activation around an initial concentration threshold but saturation at higher concentrations.

Gene-regulatory networks can be modelled using several different mathematical frameworks. We describe one of these, the so-called sigmoid modelling formalism, in some detail below. (Alternatives not described here include graphs, Bayesian networks, Boolean networks, differential equations, and stochastic master equations.)

Consider a simple gene-regulatory network with two genes, X and its regulator Y, and let x and y denote their protein expression levels. The rate of change of x with time, written dx/dt or $\dot{x}$, can be modelled as

$$\dot{x} = \alpha GRF(y) - \gamma x, \tag{3.1}$$

where α is the maximal production rate and γ is the relative decay rate of x.

If y acts as an activator, the gene-regulation function GRF is given by the Hill function

$$H(y, \theta, p) = \frac{y^p}{\theta^p + y^p} \tag{3.2}$$

where the threshold θ is the regulator concentration needed to obtain 50% of the maximal production rate and p determines the steepness of the response (see Figure 3.8).

If instead y acts as a repressor, then

$$GRF(y) = 1 - H(y, \theta, p). \tag{3.3}$$

This simple framework can be used to model large gene-regulatory networks, and can be extended in many directions, adding, for example, multiple regulators per gene, basal transcription rates, and diploid systems.

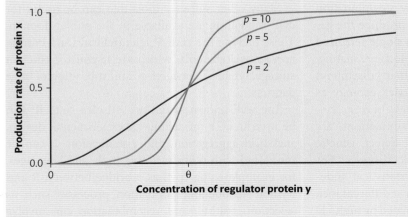

Figure 3.8 Examples of sigmoid gene-regulation functions. The production rate of protein x (relative to the maximum, called α in the text) is described as a function of the concentration of the regulatory protein y. This example shows the Hill function (eqn 3.2) which is characterized by its threshold (θ) and the steepness (p) with which it switches from no production to full production around the threshold.

... continued

---

[8] http://simupop.sourceforge.net
[9] http://pngu.mgh.harvard.edu/~purcell/plink

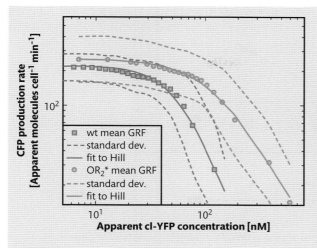

Figure 3.9 A mutation (OR$_2$, orange) in the P$_R$ promoter changes its gene-regulation function compared to the wild type (blue). CFP and YFP denote cyan and yellow fluorescent proteins, which are used to measure the rate of protein production from the promoter P$_R$ (y axis) and the concentration of the repressor protein cI (x axis), respectively. The solid lines are fitted Hill functions, decreasing as in equation 3.3 because cI represses production from the promoter. Note that the scales are logarithmic, so that zero production is infinitely far down. Experimental measurements are shown as dots. Reproduced from [11].

Variation in the regulatory sequence of a gene can tweak the gene-regulation function by modifying the molecular interactions between DNA and the relevant transcription factors. Thus, the gene-regulation function of a gene is a trait that is determined by that gene's regulatory sequence. Regulatory mutations can also affect parameters such as transcription rates, mRNA processing rates, and mRNA decay rates. For example, Nitzan Rosenfeld and his coworkers characterized the gene-regulation function of the bacteriophage lambda P$_R$ promoter [11]. They controlled the concentration of a repressor protein (cI) while reading promoter activity from fluorescence levels. These experiments showed that the wild-type gene-regulation function fit equation 3.3 very well, with a maximal production rate (α) of 200 molecules/cell per min, a regulation threshold (θ) of 55 nM, and a steepness parameter (p) of 2.4. Furthermore, they showed that a single point mutation in one of the binding sites in the PR promoter changed the shape of the gene-regulation function (Figure 3.9).

3.5 Some cGP Applications

This section provides further examples of how the cGP approach ties together physiological knowledge and genetic concepts. One example shows how gene-regulatory networks can give rise to different patterns of dominance and genetic interaction; another explores the parameter-to-phenotype map in cell electrophysiology; and the last studies cGP models in the population context of GWAS.

3.5.1 Diploid Models of Gene-Regulatory Networks

Gene-regulatory networks are an important class of systems in which cGP studies have yielded conceptual links between regulatory biology and quantitative genetics. This section describes a formalism for modelling such networks (Box 3.4) and some example applications of this framework.

The way of modelling switch-like gene regulation networks that is outlined in Box 3.4 was used by Omholt and coworkers [12] to propose a unified explanation for various observations concerning additivity, dominance, and epistasis (see Section 3.2). This study modelled small gene-regulatory networks using one differential equation for each of the two alleles of a gene. The genotype-to-parameter map consisted of allele-specific parameters for the gene-regulation functions (illustrated in Figure 3.8), and the phenotypes of interest were the equilibrium concentrations of the gene products. The analysis showed that several patterns of gene action can arise from the same network, depending on the details of gene regulatory parameters. In particular, a single locus with negative auto-regulation (i.e. one that encodes a transcription factor that represses its own transcription) can show either additive gene action or partial or complete dominance, whereas positive auto-regulation (where the transcription factor

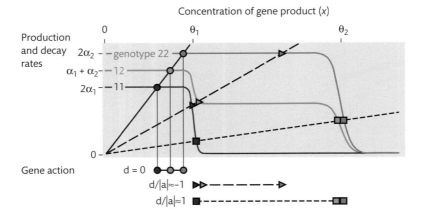

Figure 3.10 Sigmoid gene regulation can implement many types of gene action. This example shows how negative auto-regulation can give rise to additivity, recessivity, or dominance depending on the regulatory parameter values. The red, green, and blue curves show the production rates for genotypes 11, 12, and 22, given that the two alleles differ in their regulation thresholds θ and maximal production rates α. The black straight lines show decay rates for different values of γ. The equilibrium expression level (solution of eqn 3.5) is found when the production rate equals the decay rate. In the case of high decay rates (solid black line) the equilibrium values (circles) for the three genotypes show completely additive gene action (see Figure 3.2). With medium decay rate (long-dashed black line), allele 2 is recessive, in that genotypes 11 and 12 give very similar equilibrium concentrations (triangles) of the gene product. For low decay rate (short-dashed black line, squares), allele 2 is dominant. Simplified from [12].

enhances its own transcription) can enable over-dominance. Analysis of a network including three loci showed that, contrary to the assumptions of classical genetics, both dominance and over-dominance may arise from interactions between loci.

We now work through one of these examples of what you can do with the concepts in Box 3.4 as building blocks. Consider a diploid, negatively auto-regulated gene X with two alleles, indexed 1 and 2. The equations describing the rates of change in the allele-specific expression levels x_1 and x_2 are:

$$\dot{x}_1 = \alpha_1(1 - H(x, \theta_1, p)) - \gamma x_1,$$
$$\dot{x}_2 = \alpha_2(1 - H(x, \theta_2, p)) - \gamma x_2, \quad (3.4)$$

where $x = x_1 + x_2$ is the total expression level of X. The parameters α_i and θ_i are allele-specific maximal production rate and regulation threshold for allele i ($i = 1,2$), as in Figure 3.8. For simplicity, we assume that the regulation steepness p and decay rate γ are the same for both alleles. The equilibrium expression levels for the three genotypes 11, 12, and 22 are given by the condition that decay rates equal production rates, as follows:

$$\gamma x^{11} = 2\alpha_1(1 - H(x^{11}, \theta_1, p_1)),$$
$$\gamma x^{12} = \alpha_1(1 - H(x^{12}, \theta_1, p_1))$$
$$\quad + \alpha_2(1 - H(x^{12}, \theta_2, p_2)), \quad (3.5)$$
$$\gamma x^{22} = 2\alpha_2(1 - H(x^{22}, \theta_2, p_2)).$$

Figure 3.10 shows graphical solutions of these equations for three distinct scenarios. Depending on the specific decay rate parameter γ, gene action at locus X can be additive, allele 1 can be dominant, or allele 2 can be dominant.

3.5.2 Characterizing the High-Dimensional Genotype–Phenotype Map of an *In Silico* Heart Cell

The cGP model of a heart cell described in Box 3.1 was able to exhibit many well-known genetic phenomena, including intralocus dominance, interlocus epistasis, and varying degrees of phenotypic correlation. Furthermore, it showed how variation at a locus could manifest strongly or hardly at all (i.e. having high or low penetrance, respectively), depending on the genotypic variation at other loci. Thus, even within the same physiological model, a trait can appear monogenic, oligogenic, or polygenic (i.e. determined by one, a few, or many genes), depending on the genetic background.

Sensitivity analysis showed that there was a modular and fairly sparse structure to the high-dimensional mapping from inputs (experimental and model parameters) to outputs (measures of model dynamics and behaviour). This reflected a combination of the model's modular structure and whether the simulated genetic effects on parameters were able to penetrate to higher-level phenotypes. Furthermore, the study showed that virtual experiments can improve sensitivity screening by providing *relevant* measures of model behaviour, in that parameter variation that might go undetected in some situations may manifest clearly in others. This implies that cGP models may be used to suggest experimental perturbations and measurements that may reveal crucial genetic variation, and that this may have implications for the development of personalized medicine.

Although much of the phenotypic variation simulated in this model was found to be additive, the simulated disease phenotype (i.e. failure to contract, relax, or converge to steady dynamics) was found to depend on multiple parameters in complex epistatic patterns. Highly complex interactions between genetic factors and environmental challenges are likely to be a common feature of most complex diseases. Therefore, cGP models can have the power to shed light on the interplay between genetic, age-related, and lifestyle factors based on the way a disease manifests at multiple phenotypic levels.

3.5.3 GWAS: Model Parameters as Containers of Missing Heritability

In recent years, the worldwide biomedical research community has been surprised and disappointed to discover that GWAS of human traits generally fail to explain more than a small proportion of the heritable variation in traits such as susceptibility to disease. The variance that is not explained by DNA polymorphisms identified through GWAS has been dubbed 'missing heritability' or 'the dark matter of genetics' and is considered a major problem in current biomedical genetics. Its very existence shows

that the architecture of the genotype–phenotype map is quite different than had been expected, and that new methodology is needed. The idea of viewing the parameters of physiological models as intermediate-level phenotypes and aggregating the effects of genetic variation that are related to a physiological function offers a possible way forward, and computational simulations of GWAS studies provide a way of doing this.

Figure 3.11 illustrates virtual GWAS of populations of parameter scenarios for the heart cell model described in Box 3.1 [13]. For each simulated population, hypothetical effect sizes of each SNP on a parameter were randomly generated. Then, genotypes were simulated, and the resulting parameter scenarios computed from the SNP effects. Genotypes were generated by the simuPOP package, using data from the HapMap genome database to ensure realistic LD between SNPs.

A GWAS analysis was performed on this simulated population, picking out candidate SNPs to explain the targeted phenotypes. Although the ultimate study objective was to explain the variation in cell-level phenotypes such as action potential duration and calcium transient amplitude, the GWAS was much more successful when it was targeted at the variation in model parameters than at cell-level phenotypes directly. Furthermore, the GWAS results for the cellular phenotype groups were predominantly a consequence of the sensitivity structure of the dynamic model. SNPs associated with traits that are sensitive to only a few parameters will have a higher penetrance than SNPs associated with traits that are sensitive to many parameters for a given model resolution. Thus, using model parameters as target phenotypes for GWAS, rather than targeting the trait directly, can be especially helpful for traits that are sensitive to many parameters.

3.6 Linking cGP Models to Data

We start this section with a word of caution. Most omics data on phenotypes are not directly usable for the physiological models that are a main theme of this book. That is because the fundamental notion of dynamical systems theory is that *rate of change* is a function of *state*. As we show in Box 3.2, the voltage

(a)

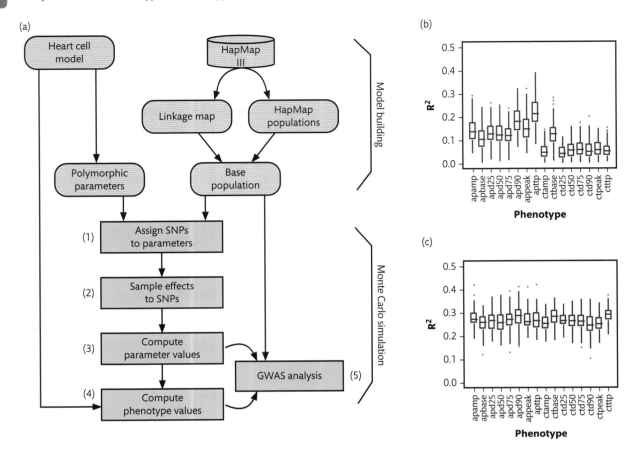

Figure 3.11 Model parameters can be useful phenotypes for GWAS. In this way, cGP models can improve the sensitivity of GWAS. Here, the heart cell model from Box 3.1, a linkage map, and a virtual population are tied together by randomly selecting heart model parameters hypothetically assumed to be under influence of genetic variation and associating the parameter variation to DNA variation on virtual genomes. GWAS analysis was performed on the virtual population, picking out candidate SNPs to explain the targeted phenotypes. Targeting model parameters as intermediate phenotypes proved more efficient than targeting top-level phenotypes alone. Panel (a) shows how the heart cell model, a linkage map, and a virtual population are tied together by selecting heart model parameters assumed to be under influence of genetic variation and associating the parameter variation to DNA variation (SNPs) on virtual genomes. Individual genotypes are mapped into heart model parameters (steps 1–3) and by running the heart cell model parameters are mapped into cell-level phenotypes (step 4). Finally, GWAS analysis is then performed on the virtual population (step 5). The boxplots in panels (b) and (c) show the Monte Carlo variability in the total amount of variance that can be explained by GWAS for 100 simulations of the model (see Box 3.3). Each box shows the median (dividing line) and interquartile range over the 100 runs, whereas the whiskers show most of the rest of the data and the dots represent outliers. Panel (b) shows the variance in cellular phenotypes that can be explained using causal SNPs detected in GWAS targeting these phenotypes directly; panel (c) shows the more consistent results that are obtained by using causal SNPs obtained from GWAS targeting all genetically controlled parameters. Figure modified from [13].

of a cell at a given moment depends on what it was a short while ago, and how fast it is changing. And it changes because of ion currents, whose rates depend on what the ion concentrations and transmembrane voltage are at the given moment. You saw the same thing in the differential equations for gene expression

in Box 3.4, and you will see the same in Chapter 5 on cell modelling.

However, this means that the key data you need concerns process rates and the state of the system, at a time resolution appropriate to the phenomenon that you study. Most omics data are more aggregated than that, smeared out over space or time so that you cannot easily infer dynamic models from them. cGP modelling must therefore always involve compromises to make the best possible use of the data that is available.

The vast knowledge embodied in genomic and phenomic databases (presented in Chapter 2) and in the model repositories that are described in the Appendix can only be used effectively if all the different types of data and models can be used together. They must therefore be properly and consistently labelled, allowing data and model resources to speak the same language. This requires formal representations of knowledge, called *ontologies*, which are domain-specific lists of concepts and of the relations between them. Thus, labels such as 'abnormal enlargement of liver cells' can be given a precise technical meaning. The ability to consistently navigate and query a set of data and model resources using terms taken from one or more ontologies is called *semantic interoperability*, where semantic means 'relating to meaning'.

In knowledge representation terms, genotypic and phenotypic data are *qualities* (measurements or observations) of biological *entities*, which are instances of *entity classes*. Ontologies define and arrange entity classes, describing relationships between them that form hierarchies or networks of concepts. As an example, consider the phenotype 'mass of a liver cell'. This is the *quality* 'mass', pertaining to a liver cell, which *is a* cell and *is located in* the liver, which *is an* organ. Actual measurements of this phenotype for specific individuals can then be stored in a database and retrieved in response to biologically meaningful queries. In the next section, Figure 3.12 shows how ontologies can be combined to encode biological knowledge, and how databases labelled in this fashion can be queried in a biologically meaningful way. The Recommended Reading at the end of this chapter includes a formal introduction to the concept of ontologies, but also an overview of ontologies as they are used in bioinformatics and in biology more widely. There are also websites that provide facilities for the visual browsing of biological ontologies[10,11].

Structuring biological knowledge in an ontological format helps researchers to add additional information about a biological system into a study and integrate it into a comprehensive modelling framework. Ontologies can also be helpful in visualization, as they lend themselves naturally to entity-relationship diagrams, and can also be used as anatomical (see Figure 3.13 below) or conceptual maps onto which to drape biological data.

In the following section, we describe some of the standards and tools that allow data and model resources relevant to genotype–phenotype studies to be used together in an integrated fashion, with some examples.

3.6.1 Tools for Semantic Interoperability and Knowledge Management in Biology

We have already discussed how and why the standardization of data and model formats using ontologies is very useful. This is essential if these resources are to be used in an interoperable way. Even when standardization is in place, however, efficient research using these resources requires user-friendly tools that take the drudgery out of knowledge management and semantic reasoning over data. Finally, it is usually necessary for the results of the investigations to be visualized. We now describe some ongoing efforts to develop such for assisting the modeller in bridging and managing diverse datasets and models.

Scientists and programmers working on the RICORDO project[12] are among those who are developing methods for the annotation, sharing, and reasoning over 'semantic metadata' of data and model resources (or DMRs) that are relevant to the computational modelling of human physiology and biomedicine. They use the term *metadata* to refer to machine-readable documentation that is linked to a DMR element, indicating how the content of that element should be interpreted. Thus, *semantic* metadata ascribes meaning to elements of data and to model resources. This enables the automatic identification of related DMR elements on the basis of their metadata alone, regardless of differences in format or accessibility. The RICORDO toolkit therefore includes features for:

1. the creation of composite terms to describe phenotypes;

---

[10] http://bioportal.bioontology.org
[11] http://www.ebi.ac.uk/ontology-lookup

[12] http://www.ricordo.eu/

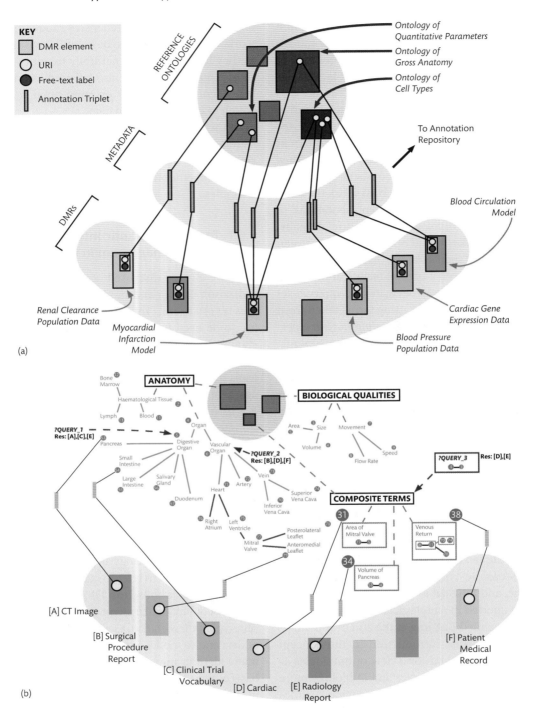

Figure 3.12 Semantic interoperability means the transparent linking of data and model resources, using standardized ontologies to annotate data with biological meaning. (a) The role of semantic metadata. For example, abnormal enlargement of the cells lining blood vessels in the liver could be represented as as an entity (yellow dot) in a 'reference list of pathology terms,' annotated using ontologies for biological quality (*size*), cell type (*endothelial*), and gross anatomy (*liver*). This links the pathology entity URI via distinct relations to URIs respectively from PATO (biological qualities), CellType and FMA (gross anatomy). (b) The repository of annotations can be queried using simple or composite ontology terms. Three examples (labelled Query_1, etc. in panel (b) are detailed in the main text. Figure reproduced from De Bono et al. (2011) Open Access.

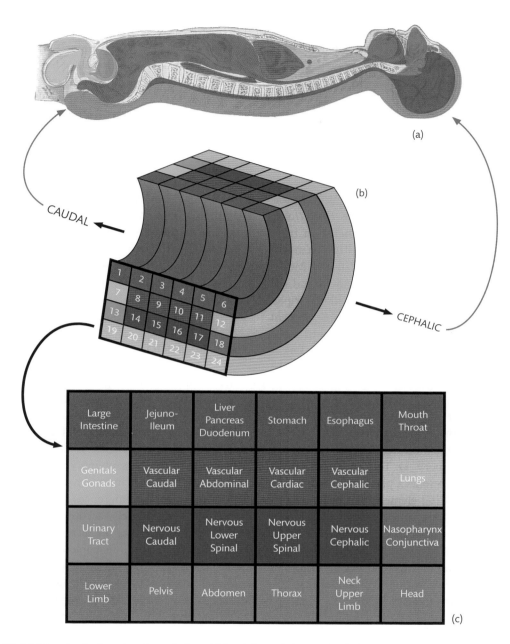

Figure 3.13 (a) A longitudinal section through the middle of the male human body showing the external and internal surfaces and organs. The colour coding overlaid onto the picture follows the same colour scheme as (b) and (c) below. (b) An idealized radially symmetric body plan in 24 cylindrical regions. This homunculus has a central longitudinal axis of rotation located in the idealized gut lumen which runs in the cephalocaudal (i.e. 'head to tail') direction. (c) The 24-tile FMA 'Body' template describes tissues located in similar regions. This template is applied by the ApiNATOMY tool to constrain the layout of anatomy term tiles. Figure modified from De Bono et al. (2012) with permission.

2. the annotation of DMR metadata using either such composite terms or individual terms from selected reference biomedical ontologies;

3. the semantic integration of DMRs, and
4. the retrieval of these DMRs from databases based on complex queries over biomedical ontologies.

To illustrate how RICORDO tools work together, we describe below the reasoning tools based on the Web Ontology Language (OWL) that have been developed to carry out logical operations over the graph structure of ontologies to automate the classification of DMR annotations.

Figure 3.12 illustrates how the components of RICORDO can fit together to annotate an element that represents the phenotype 'Volume of Pancreas' (label 34 in Figure 3.12B) in a radiology resource. A composite ontological term can be defined by combining the term 'Volume' from the PATO ontology of qualities relating to phenotypes (http://obo.sourceforge.net/); the relationship 'inheres_in' from the OBO Relationship Ontology; and the anatomical term 'Pancreas' from the Foundational Model of Anatomy (FMA). The resource element and the corresponding composite term are linked in a triplet that consists of an identifier for the resource element (e.g. from a radiology report), a relation and a reference to the composite term (label 34). This link is deposited in a metadata store. This annotation can then be retrieved using complex queries over both the composite terms and the reference ontologies. For example, Query_1 (Figure 3.12B, upper left) finds annotations involving *digestive organs* (ontology Uniform Resource Indicator (URI) '5', pointed to by the arrow from Query_1). This returns (1) DMR [A] element that represents a pancreatic region on a CT image; (2) DMR [C] element that bears a clinical trial vocabulary term; and (3) DMR [E] element on the volume of the pancreas. In another example, Query_2 (Figure 3.12B, upper middle) searches for annotations to all known vascular organs and their parts. This query takes as input the anatomy ontology URI for Vascular Organ (URI: 6) and returns: (1) DMR [B] element pertaining to the a surgical procedure report about the mitral valve; (2) DMR [D] element for an cardiac echo Doppler; and (3) DMR [F] for a medical record about the flow of blood from the patient's central veins to the right atrium. In our final example, Query_3 (Figure 3.12B, upper right) poses the composite query: 'Are there annotations that explicitly describe the size of organs?' Results: DMR [D] as well as DMR [E]. Both 'Area' and 'Volume' are subclasses of 'Size' in the quality ontology. 'Pancreas' is a subclass of 'Organ', while 'Mitral Valve' is a part of the 'Organ' subclass 'Heart'.

Anatomy is a particularly important conceptual framework for communicating physiological knowledge, because biophysical processes communicate according to the physical layout of body structure.

The ApiNATOMY toolkit (Figure 3.13) provides an effective projection of the three-dimensional human anatomy onto two-dimensional diagrams while preserving the spatial and functional meaning of physiological knowledge. The principle behind ApiNATOMY is to reduce the bilaterally symmetrical body plan of a human into an idealized radially symmetric one, in which the gut provides the central longitudinal axis around which co-centric layers of body parts are laid out. The resulting layout is intuitively understandable and preserves key aspects of anatomy, such as which organs are located next to each other. On a smaller scale, the anatomy of individual cells can be similarly laid out.

3.7 Conclusions

Computational physiology is an essential supplement to traditional genetics in accounting for the genotype–phenotype relationship in terms of physiological knowledge. Instead of focusing solely on end-result phenotypes such as disease or size, it is valuable to include as phenotypes the parameters that are used in mathematical models of physiology. This allows genetic differences to be incorporated into multi-scale models, accounting for the genotype-to-phenotype map in terms of a more direct genotype-to-parameter map in conjunction with our understanding of physiology. This link is becoming essential in an era of personalized medicine, where the acquisition of high-dimensional phenotypic data from patients will—hopefully—eventually become routine, with models and data routinely integrated across multiple spatial and temporal scales.

Successful construction of such models requires the linking of data and models from disparate sources. This, in turn, requires standardized vocabularies that give precise technical meaning to data and model resources and make them easier to integrate. Such ontological annotation simplifies tasks such as searching genomics databases for polymorphisms related to a given model parameter, extracting candidate models for a given cell type in a repository, or finding relevant experimental data for testing a model. Many omics databases are already annotated with ontological information, and the similar annotation of model repositories is now beginning. Once model and data resources speak the same language, it will be possible to process biologically meaningful database and repository queries automatically, thus making bioinformatics and computational modelling ever more useful in biomedicine.

This chapter has discussed the central role of computational physiology in relating genotypic to phenotypic variation at multiple scales. However, models at this level of detail require massive amounts of physiological data. The next chapter presents one important source of such data: biomedical imaging.

 Recommended Reading

Baclawski, K. and Niu, T. (2005) *Ontologies for Bioinformatics (Computational Molecular Biology).* Cambridge, MA: MIT Press.

Benfey, P.N. and Mitchell-Olds, T. (2008) From genotype to phenotype: systems biology meets natural variation. *Science* 320(5875), 495–497.

de Bono, B., Grenon P., and Sammut, S.J. (2012) ApiNATOMY: a novel toolkit for visualizing multiscale anatomy schematics with phenotype-related information. *Human Mutation* 33, 837–848.

de Bono, B., Hoehndorf, R. et al. (2011) The RICORDO approach to semantic interoperability for biomedical data and models: strategy, standards and solutions. *BMC Research Notes* 4, 313.

Gilbert, S.F. and Epel, D. (2009) *Ecological Developmental Biology.* Sunderland, *MA:* Sinauer Associates.

Gjuvsland, A.B., Vik, J.O., et al. (2013) Bridging the genotype-phenotype gap: what does it take? *Journal of Physiology* 591, 2055–2066.

Gómez-Pérez, A. and Benjamins, V. (eds) (2002) *Knowledge Engineering and Knowledge Management: Ontologies and the Semantic Web,* Lecture Notes in Computer Science, vol. 2473. Berlin: Springer.

Latchman, D.S. (2005) *Gene Regulation: A Eukaryotic Perspective,* 5th edn. New York: Taylor & Francis.

Lynch, M. and Walsh, B. (1998) *Genetics and Analysis of Quantitative Traits.* Sunderland, MA: Sinauer Associates.

Nadeau, J.H. and Dudley, A.M. (2011) Systems genetics. *Science* 331(6020), 1015–1016.

Neale, B.M., Ferreira, M.A.R. et al. (eds) (2008) *Statistical Genetics: Gene Mapping Through Linkage and Association.* New York: Taylor & Francis.

Omholt, S.W. (2006) From bean-bag genetics to feedback genetics: bridging the gap between regulatory biology and classical genetics. In *Biology of Dominance*, Veitia, R.A. (ed.), Chapter 10. Georgetown, TX: Landes Bioscience.

Phillips, P.C. (2008) Epistasis: the essential role of gene interactions in the structure and evolution of genetic systems. *Nature Reviews Genetics* 9(11), 855–867.

Wang, Y., Gjuvsland, A.B. et al. (2012) Parameters in dynamic models of complex traits are containers of missing heritability. *PLoS Computational Biology* 8, e1002459.

 References

1. Bateson, W. (1909) *Mendel's Principles of Heredity.* Cambridge: Cambridge University Press.

2. Punnett, R.S. (ed.) (1928) *Scientific Papers of William Bateson.* Cambridge: Cambridge University Press.

3. Mackay, T.F.C., Stone, E.A. and Ayroles, J.F. (2009) The genetics of quantitative traits: challenges and prospects. *Nature Reviews Genetics* 10, 565–577.

4. Burns, J. (1970) The synthetic problem and the genotype-phenotype relation in cellular metabolism. In *Towards a Theoretical Biology. 3. Drafts. An I.U.B.S. Symposium,* Waddington, C.H., pp. 47–51. Chicago: Aldine Publishing Company.

5. Kacser, H. and Burns, J.A. (1981) The molecular basis of dominance. *Genetics* 97, 639–666.

6. Gibson, G. (1996) Epistasis and pleiotropy as natural properties of transcriptional regulation. *Theoretical Population Biology* 49, 58–89.

7. Rajasingh, H., Gjuvsland, A.B. et al. (2008) When parameters in dynamic models become phenotypes: a case study on flesh pigmentation in the chinook salmon (*Oncorhynchus tshawytscha*). *Genetics* 179, 1113–1118.

8. Gertz, J., Gerke, J.P., and Cohen, B.A. (2010) Epistasis in a quantitative trait captured by a molecular model of transcription factor interactions. *Theoretical Population Biology* 77, 1–5.

9. Vik, J.O., Gjuvsland, A.B. et al. (2011) Genotype-phenotype map characteristics of an in silico heart cell. *Frontiers in Physiology* 2, 106.

10. Silva, J.R., Pan, H. et al. (2009) A multiscale model linking ion-channel molecular dynamics and electrostatics to the cardiac action potential. *Proceedings of the National Academy of Sciences USA* 106, 11102 –L 11106.

11. Rosenfeld, N., Young, J.W. et al. (2005) Gene regulation at the single-cell level. *Science* 307, 1962–1965.

12. Omholt, S.W., Plahte, E. et al. (2000) Gene regulatory networks generating the phenomena of additivity, dominance and epistasis. *Genetics* 155, 969–980.

13. Wang, Y., Gjuvsland, A.B. et al. (2012) Parameters in dynamic models of complex traits are containers of missing heritability. *PLoS Computational Biology* 8, e1002459.

Chapter written by
Jon Olav Vik, Centre for Integrative Genetics, Department of Animal and Aquacultural Sciences, Norwegian University of Life Sciences, Ås, Norway;
Arne B. Gjuvsland, Centre for Integrative Genetics, Department of Animal and Aquacultural Sciences, Norwegian University of Life Sciences, Ås, Norway;
Bernard De Bono, Auckland Bioengineering Institute (ABI), University of Auckland, New Zealand; Centre for Health Informatics & Multiprofessional Education (CHIME), University College London, 3rd Floor, Wolfson House, Stephenson Way, London NW1 2HE, UK;
Stig W. Omholt, Centre for Integrative Genetics, Department of Circulation and Medical Imaging, Norwegian University of Science and Technology, Trondheim, Norway.

4 Image-Based Modelling

Learning Objectives

After reading this chapter, you should:

- know something of the different methods of biomedical imaging and the parameters that these methods provide;
- understand that image- and signal-based modelling of the human body relies on the quantitative parameters that these image analysis techniques can provide;
- understand how image-based models can be used to simulate the structure and function of physiological systems, including organs;
- describe some medical applications that illustrate different aspects of image-based modelling;
- know of some open-source software tools for image-based modelling.

4.1 Introduction

We have already seen how the core of computational biomedicine is concerned with the development of integrative computer models of human biology at all levels of biological organization. These levels range from the genes to the whole organism via regulatory networks, protein pathways, integrative cell functions, and models of entire tissues and organs. Modelling aims to transform current medical practice and should support the development of innovative P4 (personalized, predictive, participatory, and preventive) medicine [1].

In this context, computational analysis is already playing an important role in providing methods to quantify and fuse together data produced using different imaging techniques. The broad research areas discussed here include the transformation of generic computational models of anatomical structures to represent those from specific individuals, which is helping to pave the way to personalized computer models [2].

There are three complementary methods in which imaging data are used to construct a personalized model of an individual's physiology:

1. modelling the anatomy of a specific domain (organ) and specific subdomains (tissue types) using data obtained from static imaging;
2. defining the boundary and initial conditions of a model using dynamic imaging;
3. using imaging to characterize individual structural and functional properties of a tissue.

Imaging also plays a pivotal role in the evaluation and validation of computational models using clinical and animal data, and in the translation of research models into the clinic. These issues will be discussed at much more length in Chapter 11 of this book. In this specific context, however, preclinical imaging data can often yield an understanding of human physiology that is essential for both the development and validation of clinical models.

Models based on imaging data, therefore, have enormous potential in both basic medical research and clinical applications. It can even be said that image-based modelling promises to become an entirely new 'virtual imaging technique'. It will become possible to integrate sparse and sometimes inconsistent observable data obtained from one or more different imaging modalities into more comprehensive images *in silico*. Computational models will help researchers and clinicians to interpret imaging measurements in a way that is consistent with the underlying biology of the physiological or pathological processes that are under investigation. These new investigative tools and systems should help with data interpretation, and also in our detailed understanding of disease processes and the effect of therapeutic interventions on the course of such diseases.

In future years, it is likely that models based on specific imaging data obtained from particular patients will be combined with models of the effects of medical devices and pharmacological therapies, allowing the development of a new discipline of predictive imaging. In this, the clinician will use this combination of models to understand a patient's condition and plan and optimize the necessary interventions *in silico*. As such, every patient will be offered the 'best available' combination of treatments for his or her condition.

This chapter consists largely of a number of case studies of image-based modelling techniques. These will start by setting the medical application to be discussed in context, and then describe the techniques used to acquire the appropriate data from the images, to analyse those data, and to construct a relevant computational model.

We start, however, with a short introduction to each of the imaging techniques or modalities that this type of modelling is based on: magnetic resonance imaging (MRI); computed tomography (CT); positron emission tomography (PET); and ultrasound. These techniques work in different ways but all are capable of producing three-dimensional images of parts of the human body, and they are all now in routine clinical use. In each case we describe the physics behind the production of the images, their main applications in the clinic, and their respective advantages and disadvantages. Should you require any further information about the techniques we recommend you to consult the textbooks listed under Recommended Reading at the end of this chapter.

4.1.1 Magnetic Resonance Imaging

In magnetic resonance imaging (MRI) the signal originates from the intrinsic magnetic moment of hydrogen nuclei (protons) in the body. When manipulated within a strong static magnetic field, the protons 'spin' and can create a time-varying magnetic field. By means of Faraday induction, these drive currents through receiver coils placed around the body and can be detected electronically.

In clinical systems, the strong magnetic field is created by a superconducting magnet and typically has a value of 1.5 or 3 tesla (more than 10^4 times stronger than the Earth's magnetic field). Protons in such a field can be considered to have a spinning magnetic moment with a spin frequency that is proportional to the strength of the magnetic field experienced by the proton. This is called the Larmor frequency. While these processes have their origins in quantum mechanics, what we observe is the aggregate of many protons and a classical model is able to explain the magnetic resonance signal.

When a magnetic field oscillating at the Larmor frequency (an 'RF pulse') is applied using dedicated coils, a resonant effect occurs and the direction of the proton spins can be tipped from being aligned with the external static field (the longitudinal direction) to being perpendicular to it (in the transverse plane). These 'excited' protons rotating in the transverse plane generate an oscillating magnetic field that induces a current in receiver coils. For a signal to be detectable there must be components of the spins in the transverse plane and these must have phase coherence (see Figure 4.1).

The rotation frequency is governed by the local magnetic field and this can be intentionally manipulated by the use of imaging gradients. Here, a current passed through gradient coils within the bore of the scanner causes the field strength, and hence Larmor frequency, to vary in a known way as a function of position. The receiver coils detect a time domain signal from all excited positions and when this *k-space* is Fourier transformed we obtain the power at each frequency (spatial position) and hence can form an image.

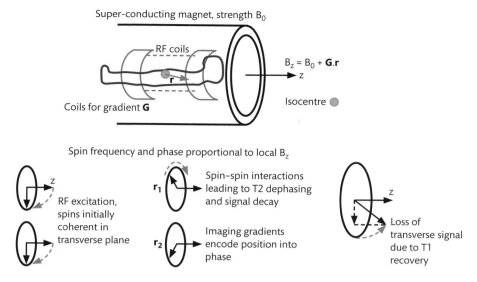

Figure 4.1 MRI. The MRI scanner provides a strong magnetic field B_0, magnetic field gradients G, and RF pulses. An excitation RF pulse tips proton spins into the transverse plane where they can be detected by receiver coils on the patient (not shown). Signal changes due to T1 and T2 effects provide contrast to different soft tissues. Applied gradients encode position information that is used in image reconstruction.

The RF excitation is a resonant effect, which implies that the excitation frequency has to closely match the Larmor frequency to tip the spins into the transverse plane. Recovery of spins back to the longitudinal direction also requires resonance and this comes from the interaction with other molecules at the Larmor frequency. The effectiveness of this interaction governs the rate at which the signal relaxes back to the longitudinal direction and has a characteristic time constant called T1 (Figure 4.1). This rate depends upon factors such as viscosity and varies between soft tissues, providing one of the mechanisms for soft tissue contrast. Another important mechanism is T2 decay, which describes the rate at which signal is lost due to spins in the transverse plane losing phase coherence. This spin dephasing occurs due to local variations in the magnetic field that give differing Larmor frequencies and hence cause different spin phases. The random component of these local fields originates from interactions with other molecules and is tissue dependent, which provides another source of contrast within soft tissue types.

The order, strength, and timings in which the RF pulses and gradients are played out form the magnetic resonance 'sequence'. Multiple sequences will be used during a patient visit to provide images that are sensitive or insensitive to T1, T2, and proton density. This gives images with an exquisite soft-tissue contrast that is unrivalled by any other imaging modality.

MRI can also provide a wide range of functional information in addition to this structural information. If spins move through a magnetic field gradient, they experience a changing field strength and thus gain a phase that is proportional to the distance they travelled. For coherent motion such as blood flow, the change in phase can be used to measure velocity. Flow directions in vessels and heart chambers can be observed by using flow-encoding gradients in different directions. For incoherent motion due to molecular diffusion, the phase changes are random and the net signal drops due to a loss of phase coherence. The larger the distance molecules can diffuse during a sensitization period, the greater the drop in signal. Thus, in diffusion-weighted MRI, the image contrast is related to diffusion distances and these can change with cellularity in tumours, and are different for diffusion parallel or perpendicular to fibres in the brain.

Many other types of MRI scans can be obtained. Two of the most interesting are functional MRI (fMRI) and contrast-enhanced studies. fMRI uses the local T2 changes caused by deoxygenated haemoglobin to map activity in the brain when a subject performs a task. In contrast-enhanced studies, the patient receives an injection of a gadolinium-based contrast agent into the blood. The gadolinium locally lowers the T1 and T2, which can be used to image the position and width of vessels, to look for enhanced uptake in regions of inflammation, or to study the uptake and washout in tumours which tend to be more 'leaky' than normal tissue due to their disordered growth.

MRI has the advantages of being free of ionizing radiation (the frequencies are of the order of 100 MHz); the gradients and hence imaging planes can be in any three-dimensional (3D) orientation; it is extremely safe when operated correctly, with side effects to contrast being very rare; and it provides excellent structural and functional information. The main drawbacks are that it is noisy, slow, susceptible to motion artefacts, and unsuitable for patients with some implants, and the images obtained are often qualitative rather than quantitative. It is also expensive due to the specialized equipment that is required, and well-trained staff are needed to operate the scanner and interpret the images.

4.1.2 Computed Tomography

In computed tomography (CT), X-rays are transmitted through a patient to provide projections of tissue X-ray attenuation. Volumetric images are reconstructed from projections acquired at multiple angles.

The covers of a CT scanner conceal a gantry that rotates around the patient at approximately 2 rpm (although this may not seem fast, the stored energy is very high). Mounted on the gantry is an X-ray source, with a detector panel diagonally opposite (Figure 4.2). The source creates X-rays by using a high voltage to accelerate electrons from a hot cathode filament to an anode. When these electrons are decelerated in the anode they give out Bremsstrahlung X-ray radiation that has a broad energy spectrum. The upper energy of this spectrum corresponds to the energy of the incident electrons and is quantified in electron-volts (eV). Atoms are composed of electrons in orbital shells around a nucleus. An additional source of X-rays occurs when an outer orbital electron fills the space of an inner K-shell electron that has been ejected by an incident electron from the cathode. This produces characteristic radiation with a line spectrum dependent upon the target material. The X-rays generated may be filtered and directionally collimated into a beam before leaving the source.

The X-rays that are not attenuated in the patient form a projection that is imaged by an array of solid-state detectors composed of a scintillation layer, where the X-ray energy is converted into light, and a photodiode that converts this light into an electrical signal. To generate a 3D volume (rather than a single slice) the patient bed can be moved during

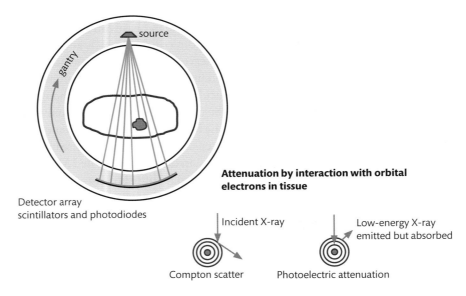

Figure 4.2 CT. X-rays from a source are attenuated by interactions with tissue. The detector array measures projections of attenuation within the patient. Images are reconstructed from projections at multiple angles obtained by rotating the gantry during scanning.

scanning to achieve a spiral or helical sampling pattern. Additionally, if the detector array contains many rows in the head–foot direction, projections at multiple slice positions are recorded simultaneously. Image reconstruction is based on the inverse radon transform but modified to account for the beam shape (fan or cone), detector geometry, and bed movement.

Within the patient, there are two dominant sources of attenuation that lead to contrast in the image. In both cases, the X-ray photons interact with the electrons of the tissue material. Electrons with a binding energy much less than the X-ray photon can cause Compton scattering, in which part of the energy of the photon is transferred to the electron. Electrons with a binding energy similar to the photon energy can result in absorption of the X-ray photon due to the photoelectric effect. Attenuation is more effective for materials that are thicker, denser, and have a higher atomic number (because they have more deeply bound electrons). The calcium in bones and arterial calcifications has a higher density and effective atomic number than most tissue and is thus more rapidly attenuated and easily visualized, for example in the coronary arteries.

Absorption as a function of photon energy differs between materials and thus useful information can be obtained by imaging with two energies. Dual-energy systems may have two source/detector pairs mounted on the same gantry acquiring data simultaneously but with different energy sources (e.g. 80 and 140 kV). Processing of the two datasets can provide improved contrast with applications including removal of bone signal to reveal vessel structures.

A contrast agent can be given to add information. In the gastrointestinal tract, barium-based compounds may be used orally or via the rectum, and for other studies an iodine-based compound may be injected into a vein. Both barium and iodine have a K-edge at about 35 keV which means that X-rays with energies a little over this value can more easily cause the photoelectric effect, eject an electron from the K shell, and attenuate the beam. Iodine-based contrast can reveal a narrowing of vessels or leakiness at the sites of tumours.

The advantages of CT include its ease of use (particularly when compared to MRI), the lack of safety issues relating to ferrous objects in or near the patient, its speed, measurement of tissue attenuation in standardized, quantitative units, and good spatial and temporal resolution. Typically CT provides first-line imaging in emergency situations such as trauma and stroke. The disadvantages of CT are that the soft tissues have similar contrast, the iodine-based contrast agents can cause side effects in some patients, and the radiation dose received by the patient can be quite large. The dose and its possible effect depend upon the anatomy exposed, patient size and age, and scan parameters. Dose can be a serious concern for conditions that require repeated imaging or if children are involved.

4.1.3 Positron Emission Tomography

In positron emission tomography (PET) a pharmaceutical labelled with a radionuclide is injected in trace amounts into the patient and is designed to accumulate at specific sites of interest. The radionuclide decays by emitting a positron (the antimatter equivalent of an electron) which travels a short distance before annihilating with an electron and then emitting two gamma rays travelling in opposite directions, each with an energy of 511 keV. The gamma rays are detected by rings of detectors around the patient, enabling an image of tracer uptake to be obtained (Figure 4.3). When two gamma rays are detected in different detectors, but close in time, there is a coincidence count and we know the emitter must lie along a path joining the detectors: the line of response. In some modern machines, the difference in arrival times at each detector can also be measured, providing information about the distance of the source along the line of response (so-called time of flight).

The most commonly used radionuclide is an isotope of fluorine (fluorine-18) which is produced in a cyclotron and has a half life of 110 minutes. A commonly used radiopharmaceutical is ^{18}F-fluorodeoxyglucose (FDG). This is a glucose analogue that accumulates at sites of glucose utilization. Many cancers exhibit increased glucose uptake and thus become visible in FDG PET. Although FDG is the most common tracer, other tracers incorporating fluorine-18 are possible; for example, a relatively new group of tracers can bind to the amyloid plaques found in the brains of people with Alzheimer's disease.

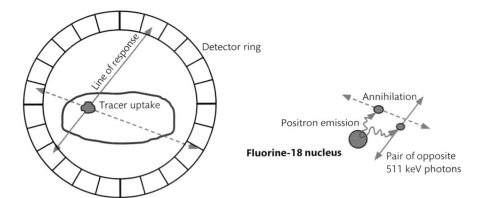

Figure 4.3 PET. Positrons emitted from the radioactive isotope (here shown as fluorine-18) travel a short distance before annihilating with an electron and emitting a pair of 511 keV photons. Image reconstruction uses detection of photons coincident in time to compute an image of tracer uptake.

Other radionuclides can also be used, such as carbon-11 or gallium-68. Carbon-11 has a half life of only 20 minutes and so requires an imaging site close to a cyclotron. Gallium-68 has a half life of 67 minutes but can be obtained on site from a gallium generator. This consists of a shielded vessel containing the parent radionuclide germanium-68 that decays with a half life of 271 days to gallium-68. The gallium-68 is removed by elution and can be used to label a range of diagnostically useful radiopharmaceuticals.

The image-reconstruction process uses knowledge of the lines of response, and any available time-of-flight information, to compute the reconstructed 3D image from the measurements. The gamma rays are partially attenuated by tissue and bone and it is necessary to account for this in the reconstruction by including an attenuation map, which is usually obtained from a CT image. The reconstruction can also account for random coincidences and scatter of photons in the patient: both physical effects that can reduce the observed contrast. The final image is often expressed in standardized uptake values (SUV), which provide a semi-quantitative measure that can be used to compare uptake in serial studies.

The advantages of PET are its very high sensitivity to small concentrations of tracers that could be toxic at larger amounts, the wide variety of tracers that are becoming available, and the quantitative metabolic or molecular information provided. Applications of PET include cancer imaging (including whole-body scanning for metastases or more detailed characterization of tumours), studies of cardiac perfusion,

or investigation of various neurodegenerative disorders. The disadvantages are a poor contrast to different tissues, poor spatial resolution (of the order of 4–6 mm), and the radiation dose received by the patient.

4.1.4 Ultrasound

The reflections of high-frequency sound waves (or ultrasound) from structures in the body can be detected using a hand-held transducer. The delay between transmission and detection determines the depth of the reflecting structure, and this is the basis of ultrasound image formation (Figure 4.4).

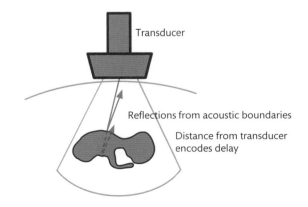

Figure 4.4 Ultrasound. A transducer produces ultrasound waves that are reflected at the boundaries of tissues with different acoustic properties. The delay between transmission and receiving the reflections enables the depth of the boundary to be calculated. The image shows the reflection strengths at different depths.

Ultrasound waves are pressure waves within a material at frequencies of the order of megahertz, which is well above the audio range. The signal is generated and detected by transducers that are typically made from a piezo-electric material whereby an oscillating electric signal is converted to mechanical oscillations and vice versa. Efficiency is gained through the use of transducer arrays whereby phased arrays can focus the beam. A linear array can be used to form a two-dimensional (2D) image and 2D arrays are used to give 3D images. To reduce losses at the transducer–skin interface, a gel is used to couple the signal into and out of the body.

The sensitivity to structures in the body originates from changes in acoustic impedance. At one of these acoustic boundaries an incident sound wave will be partially reflected and partially transmitted. The depth of the structure from the transducer is determined by the time it takes for the signal to make the round trip from transducer and back. Biological contrast comes from the changes in acoustic impedances of different materials. Acoustic impedance is a measure of how the tissue responds to the mechanical vibrations caused by the sound wave. Impedance increases with the speed of sound and density of the material. Large changes in impedance—for example, air to soft-tissue boundaries—lead to high reflections and limit the visibility of any tissues behind air.

When the acoustic boundary is irregular compared to the wavelength of the ultrasound (typically of the order 1 mm), non-specular reflection or scattering can occur. The interference between reflections from many small structures leads to a characteristic 'speckle effect' that provides a textured noise within the images. The structure seen in the speckle is not the underlying tissue structure, although its intensity does relate to the tissue.

Reflection of sound from moving structures, especially blood, causes a frequency shift of the reflected signal which is related to the velocity and direction of the flow. In pulse wave Doppler, the round-trip time of a pulse determines the depth from the transducer and the frequency shift provides velocity information. Typically Doppler is used in cardiovascular studies to look at blood flow in major vessels or the performance of the heart.

The advantages of ultrasound are its ease of use, low cost, the portability of the equipment, safe operation,

and the ability to image in real time, which is particularly useful in cardiac imaging and obstetrics, and for guiding biopsy needle placement. The drawbacks are limited access to acoustic windows due to the absorption of bone and high reflectance from tissue/air interfaces. Images are very operator- and patient-dependent and the angle of the probe is not usually recorded so positional information is lost. Fat in the skin also limits image quality. There is much less contrast between different soft tissues than there is with, for example, MRI.

4.1.5 Production and Storage of Imaging Data

In order to promote interoperability between imaging modalities, equipment vendors, standardize storage, and image presentation, for most clinical imaging equipment it is necessary to adhere to a particular set of standards. The Digital Imaging and Communications in Medicine (DICOM[1]) standard was set up to provide this. DICOM specifies how the image and its associated data are stored in files, the network protocols for data transmission, how the data are stored, and how the images should be displayed. The extra information that is stored with the image includes data such as patient name and date of birth, pixel resolution, scan time, unique equipment identifier, dose information, and scanner settings. This standard specifies a minimum amount of this header information for each imaging modality, along with other optional publicly defined fields. Manufacturers can add their own information in private fields. Unfortunately the minimum requirements do not keep pace with technological developments, and this has led to some information being hidden in private fields and to ambiguities in interpreting the data. A multi-frame version of the DICOM standard is now in use in which slices and time frames can be grouped into a single file. This relieves the problem of a previous version which had one file per image, resulting in many thousands of files for some studies. However, the header information for multiframe DICOM can itself be difficult to interpret.

In hospitals, images and some other patient data are stored in Picture Archiving and Communications

---

[1] http://medical.nema.org/

Systems (PACS) that enable images to be retrieved electronically by clinicians at multiple locations.

Frustrations with DICOM and the inertia of some image-processing code mean that other formats for images are sometimes used, especially in research settings. For example, in neurological studies a format known as NIfTI is often used. In clinical settings, where it is essential to ensure that the image data are associated unambiguously with the correct patient, that scan-related information is not lost, and that geometrical information is accurate, there is no practical alternative to using DICOM. Researchers will find that there are many free online systems that are able to read DICOM images; for example, OsiriX, K-PACS, and RadiAnt.

4.2 Image-Based Modelling

Most of the approaches being developed for image-based modelling are specific for a single technique or modality in terms of both the image processing and analysis methods, and the medical problems that they are designed to address. In each of the four case studies discussed in this section we will guide you through a process that starts with the acquisition of the image and proceeds through image analysis, parameter extraction, and model design to the use of the model and the report of the results obtained.

4.2.1 Modelling Bone Micro-Structure

Osteoporosis is a disease of bone fragility that is associated with an increased risk of fracture, particularly of the hip and spine. It primarily affects the elderly, more often women than men, and it presents a major economic burden to all developed countries. However, the parameters that determine the risk of fracture are not yet completely understood. It is clear that bone mass is a major determinant (the more bone lost, the greater the fracture risk) but it is not the only parameter involved. It is also important to consider the micro-structure and tissue properties of the bone, which can be lumped together under the term 'bone quality' [3]. It is now possible to image bone micro-structure using a laboratory CT known as X-ray micro-CT and to extract parameters from the images that can be used for developing patient-specific models of bone micro-structure.

Bone structure has a hierarchical organization, so any physiologically correct model for bone will be based on data obtained at different levels of spatial resolution. This case study shows how 3D X-ray CT imaging can be used to image bone at different scales and how parameters can be derived from these images. You should note, however, that not all the CT systems that we discuss are yet in routine clinical use.

You should remember throughout this section that bone is divided into two types based on its structure at the micro-scale: trabecular bone and cortical bone. Trabecular bone is composed of a complex network of small structures called trabeculae, each of which has a thickness of about 100–500 μm. Cortical bone is denser and more compact, and as its name implies it makes up the cortex, or outer structure, of most bones. Its matrix is packed with solid organic and inorganic salts, leaving only small spaces (lacunae) for the bone cells.

X-Ray Micro-CT

Structures on the scale of trabecular bone are ideal for imaging using a technique called X-ray micro-CT, which is a variation of CT that has a higher spatial resolution of between 1 and 30 μm. This technique enables the non-destructive and accurate imaging of the three-dimensional network of trabeculae either *ex vivo* in human bone or *in vivo* in animals. Commercial systems for micro-CT are now becoming more widely available, and these machines are now progressively replacing the previous clinical gold standard of 2D histomorphometry.

X-ray micro-CT can be implemented by using the high-intensity, single-wavelength (monochromatic) beams of X-rays that emanate from synchrotron sources. The images that are obtained using this technique, which is known as synchrotron radiation (SR) micro-CT, have a high signal-to-noise ratio and so images of small and complex structures can be detected clearly. These images can also provide simultaneous information about the micro-structure of the bone and about bone mineralization, which is an important determinant of the strength and quality of bone [4]. The technique has been applied to imaging both trabecular and cortical bone in both humans and animals, and, recently, to determining 3D images of micro-cracks in bone and of the lacunae, the spaces within the bone structure in which bone cells are stored. Previously such ultra-structure was only studied in 2D.

Characterization of Bone Micro-Structure

Bone micro-structure can be characterized by quantitative parameters that are extracted from the volume of an image after the image of the bone itself has been separated from its background. Trabecular bone is typically described using a set of standard morphometric parameters that include: trabecular bone volume per tissue volume (BV/TV), bone surface per bone volume (BS/BV), trabecular thickness (Tb.Th), trabecular spacing (Tb.Sp), and trabecular number (Tb.N). As micro-CT imaging is a 3D technique there is no need to make assumptions on the 3D geometric model of the structure, to calculate these parameters, as was the case when 2D histomorphometry was used.

Let us take trabecular thickness as an example. The trabecular thickness at each point of a 3D image is defined as the diameter of the largest sphere that can be included in the object. It can be computed from digital images by using fast discrete distance algorithms [5].

Three-dimensional images of bone can provide unbiased information about the bone topology in addition to morphometric parameters. It is now possible to characterize the degree of connectivity or even the complete geometry of the topological network of trabecular bone.

The micro-structures that make up trabecular bone can be classified into 'plate-like' and 'rod-like' structures, and both osteoporosis and normal ageing are associated with the conversion of plates into rods. It is therefore useful to compute the Structure Model Index or SMI of a sample of bone, which describes whether it is more plate-like or more rod-like. Local topological analysis, which indicates the percentage of volume elements (voxels) in the image that show plate and rod structures, can also be useful. In Figure 4.5, image (a) illustrates an SR micro-CT structure of trabecular bone at 10 µm resolution and part (b) shows the same image with the micro-structures divided into plates and rods and colour-coded using this type of analysis.

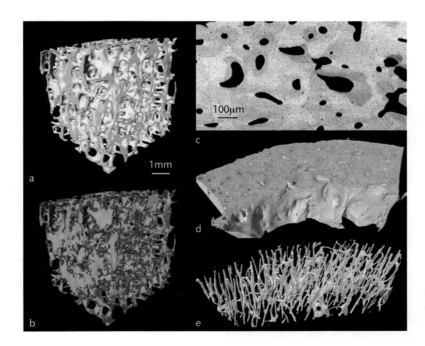

Figure 4.5 Images of bone samples acquired with SR micro-CT (voxel size: 10 µm) showing the different types of bone micro-structure: (a) 3D display of a trabecular bone, (b) plate/rod analysis of (a) with plate voxels in green and rod voxels in purple, (c) region of interest in a slice of cortical bone sample: levels of grey reveal differences in mineralization, (d) 3D display of a cortical bone sample showing its dense micro-structure, and (e) cortical pore network extracted from (d) showing the Haversian canals.

X-ray micro-CT is also used to analyse the structure of the cortical or compact bone tissue that commonly forms the outer shell of bones [6]. The same morphological parameters that are calculated for trabecular bone may be applied to cortical bone. In addition, however, it is now possible to use SR micro-CT to identify the cylindrical structures called osteons that are the primary functional units of compact bone. Each osteon surrounds a pore that is characterized by a lower degree of mineralization in the surrounding tissue. Figure 4.5c illustrates a slice of cortical bone in which the pores are shown in black, newly formed osteons are in dark grey, and interstitial tissue between the osteons in lighter grey. Part (d) of this figure shows a basic 3D image rendering of this type of bone sample and part (e) shows the pore network of cortical bone. The Haversian canals described in the caption to part (e) are narrow channels containing blood vessels that are found within the pores.

Figure 4.6 illustrates a 3D rendering of bone at an even higher spatial resolution. It shows the 3D morphology of a micro-crack in trabecular bone that is surrounded by osteocyte lacunae (voxel size: 1.4 μm). More recently still, it has become possible to use this technique to display complete 3D images of the osteocyte lacuno-canalicular network that allows signals, nutrients, and waste to be transported through bone at nanoscale resolution (voxel size: 280 nm) [6].

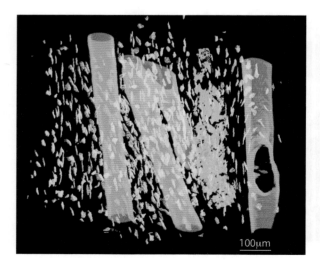

Figure 4.6 3D rendering of bone osteocyte (yellow) and a micro-crack (blue) surrounding Haversian canals (grey) in 3D human cortical bone from SR micro-CT images (voxel size 1.4 μm).

Modelling with 3D Images as Input Data

High-resolution images of bone tissue generated using micro-CT are now being used in models of the biomechanical parameters of bone. The morphometric parameters that were described in the previous section have been found to be highly correlated to experimentally measurable mechanical properties of bone, such as its elasticity and the maximum strain that it can bear. Elasticity is generally characterized using the Young modulus, which is defined as the ratio of stress (a pressure) to strain (a measure of deformation). These parameters, however, may also be obtained from 3D micro-CT images using simulations that involve finite element (FE) methods with the images used as direct inputs. These are numerical models that can be used to simulate changes in bone micro-architecture and model their effect on the physical properties of the bone [7].

The simplest biomechanical models of bone assume that the properties of the bone are uniform in all orientations (i.e. that they are isotropic) and use a binary representation of the image of the bone as an input to a FE model. There are other approaches, however, in which greyscale levels obtained from the micro-CT images are converted into an elastic modulus without any need for the binarization of data.

More information still can be obtained if these greyscale levels are calibrated with the local density or degree of mineralization of the bone matrix (DMB). This is in general easier and more accurate when images that have been obtained using SR micro-CT are used. A general law that relates the Young modulus of bone to its DMB has been used to study how the various biomechanical properties of trabecular bone are affected by its degree of mineralization.

It is possible to customize this conversion to each sample of bone (and thus to each individual patient) by coupling imaging modalities. An orthotropic biomechanical model of bone can be obtained by coupling images from SR micro-CT to results obtained using scanning acoustic microscopy [8]. This technique uses focused sound waves to generate maps of acoustic impedance. In bone, this quantity reflects elastic anisotropy and has been found to correlate with the Young modulus. It is possible to estimate local elasticity coefficients of bone from these values and the local tissue density, which in turn can be derived from the DMB.

Conclusion

We have shown in this case study that X-ray CT techniques can generate numerical data on bone at all scales from the organ to the cellular level. The images generated using this technique can be used both to derive relevant, model-independent morphometric parameters and to generate realistic geometrical models of bone micro-structure. Bone mineralization can also be mapped if synchrotron radiation data are also available, and this can be taken into account in the derivation of mechanical models. These techniques are now taking both imaging and modelling down to the cellular scale (nanoscale) in modelling the lacuno-canalicular network in bone, which plays a key part in sensing and responding to mechanical stimuli.

4.2.2 Subject-Specific 3D Modelling of Bony Structures from Projective Images

One major advantage of detailed, accurate 3D imaging modalities such as the micro-CT described in the last section, or even regular CT, is the ability of these techniques to reconstruct and model the 3D geometry of bony structures in a subject-specific or patient-specific way. There is, however, a significant drawback to the routine clinical use of this technique. CT involves a high radiation dose, and it is also expensive to use and therefore not always readily available. It is therefore rarely used when a regular follow-up is required. For example, straightforward 2D radiography (i.e. X-rays) is the modality that is still most often used in routine clinical practice for the diagnosis of conditions such as osteoporosis or scoliosis. The diagnostic

information available from 2D analyses, even from the best available technique (dual-energy X-ray absorptiometry, or DXA) is necessarily limited. A 3D representation of the bony structure can be expected to give a more accurate diagnosis of osteoporosis and assessment of fracture risk.

Methods have been proposed for recovering the 3D shapes of bony structures from one or more projective 2D X-rays. A common approach here is the use of a statistical model incorporating *a priori* information about variations in bone shape. This model involves deforming a model of bone so that the projection of its silhouette matches contours that can be observed in the X-rays. This transformation from 2D images to a 3D model is still a semi-automatic process that combines manual adjustments performed by an operator with automatic matching of the silhouette with bone contours extracted using image processing. It is now possible to use these methods to reconstruct both the shape and the bone mineral density (BMD) distribution of bone from one or several DXA projections, as shown in Figure 4.7.

This type of modelling requires a statistical model of bone structure derived from measurements obtained from a large number of individuals, which can be built from a database of CT scans of the bone or bones of interest. The principal types of variation in shape and bone density distribution are extracted from this database and used to derive a model that is then used to reconstruct the 3D structure of an individual bone from a set of 2D DXA images. This process involves optimizing the parameters obtained from the statistical model—which can include pose,

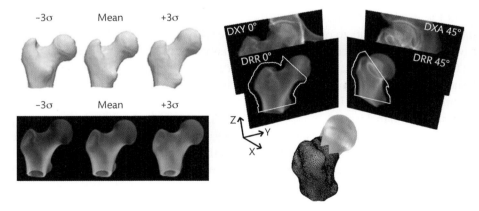

Figure 4.7 3D reconstruction achieved by maximizing the similarity between two DRRs generated from a statistical model and DXA images.

scale, shape, and BMD parameters—to maximize the similarity between the 2D images obtained from the patient and one or more digitally reconstructed radiographs (DRRs) generated from projections taken from the model (see Figure 4.7).

To conclude, 3D reconstruction methods can now provide accurate subject-specific 3D models, including representations of bone shape and BMD distribution, from a limited number of projective X-rays obtained from a patient. This development has improved the diagnosis of diseases such as osteoporosis through incorporating statistical information, while keeping radiography—which is safe, non-invasive, cheap, and widely available—as the major clinical modality.

4.2.3 Patient-Specific Evaluation of Cardiac Function

As you will see later in this book, the heart is almost certainly the organ where computational biomedicine has had the most success to date in developing precise and accurate models, and some of these are beginning to find acceptance in drug discovery and clinical medicine. One key objective of biophysical heart modelling, however, is simply to improve our understanding of how the organ works through numerical simulation. And these models have reached such a degree of realism that it is now possible to compare simulations of the whole organ quantitatively to cardiac images and signals that have been obtained from patients. This case study involves the use of cardiac images and electrophysiological signals obtained from patients to personalize an existing electromechanical model of the heart. We then estimate both the electrophysiological conductivity and the mechanical contractility of the heart using these personalized data to obtain a model that is predictive for the patient concerned. You should appreciate that these developments have already had—and will continue to have—a significant impact in clinical practice in improving both the diagnosis of heart disease and clinical decision making [9].

The personalization of a cardiac model consists of optimizing some parameters of the model so that it reproduces behaviour that is consistent with patient-specific image and signal data. Such models are produced by, firstly, generating a computational mesh from the available medical images and then optimizing the model parameters, including initial and boundary conditions, so that properties obtained

from the simulation match those observed. This can be thought of as a type of 'inverse problem', and solving it is a complex task for the following reasons:

1. it is very time consuming and convergence is not guaranteed;
2. the information extracted from the clinical data is usually very sparse in space and time;
3. usually only a subset of possible parameters may be observed, estimated, and incorporated in the model.

Furthermore, estimation of the parameters of dynamic systems is often highly mathematically involved and no one method for incorporating personal data into cardiac models is yet fully accepted. In this case study, we show how a 3D cardiac model can be personalized using both electrophysiological data and moving images obtained using magnetic resonance. The method is broken down into two successive steps: firstly electrophysiological and then mechanical personalization.

Electrophysiological Personalization

Modelling the electrophysiology of cardiac cells is one of the oldest research areas in the whole of computational biomedicine, and it is still one of the most active [10]. Incorporating these cellular models into a model of the whole organ, however, is considerably harder. This involves representing the cardiac tissue that is described mathematically by a set of partial differential equations (PDEs). Solving these dynamic PDEs is computationally very demanding; this is due in particular to the spatial scale of the electrical propagation front being much smaller than the size of the ventricles, and to stability issues arising from the dynamic nature of these equations. In this context, it can be productive to use an equation called the Eikonal equation that is more often used to compute the trajectory of light through a medium.

The Eikonal equation is a non-linear partial differential equation that is used to compute the shortest path between two points, given a local speed function. Its variable is the time at which a given point in the trajectory is reached. Therefore, in the context of cardiac modelling, the static Eikonal equation computes only the activation point of each point of the myocardium. It allows the front of an electrical impulse and its progress across the heart to be observed

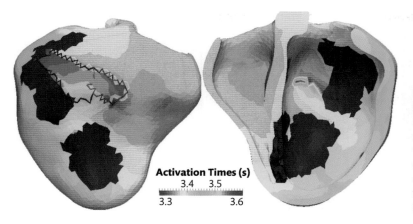

Activation Times (s)

3.4   3.5

3.3   3.6

Figure 4.8 Personalization of an electrophysiological model of the heart in order to predict points where tachycardia (fast heartbeat) may be induced in the ventricles and where ablation therapy could be applied. We see here the activation time isochrones of a ventricular tachycardia, from early- (red) to late- (blue) activated points.

at a larger scale and calculated very quickly and efficiently, enabling different models to be coupled together in a hierarchical manner. The results of such an approach, applied to testing for and predicting the results of ventricular tachycardia (or excessively fast heart rate), are illustrated in Figure 4.8.

It is necessary to make two important adjustments to this general electrophysiological model in order to personalize it. First, to set parameter values for the onset of the electrical propagation and the local conduction velocity of that electrical wave. And second, imaging plays an essential part in obtaining the data from which these parameters can be derived. It is possible to map the electrophysiology of the left-ventricular endocardium and observe the exact point at which a wave in the right ventricle traverses the septum into the left ventricle. These data can be combined with electrocardiographic (ECG) readings to estimate a mean conduction velocity (or CV) for an individual heart. The conductivity of the heart can then be estimated by matching simulated propagation times [11].

This methodology can generally guarantee that conductivity on the endocardium can be estimated with less than a 10% error and extrapolated realistically to the whole myocardium. Results with low errors can be obtained from even very partial data for the electrical propagation from the baseline. However, it is also possible to adjust this basic model to give a more detailed personalized model such as the Mitchell–Schaeffer (MS) model [12]. This biophysical model is still simplified; it involves only one input and one output current, and these are described using a system of partial differential equations.

The next stage in developing the integrated model is to set up this mathematical model of heart

electrophysiology onto a 3D anatomical model of the patient's heart, using a FE method and representing the heart anatomy with a biventricular tetrahedral mesh (Figures 4.9 and 4.10). This leads to a system of algebraic differential equations that model the passage of electrical impulses across that individual heart. Organs close to the heart in the body can impose constraints on cardiac motion or electrical interactions, and boundary conditions are used to describe how the simulated heart interacts with these surrounding structures. The model can then be run, imposing initial pacing conditions using a voltage stimulus for a set time.

Mechanical Personalization

To develop a fully personalized heart model it is not sufficient just to choose correct parameters for the electrophysiological characteristics of the patient's heart: it is also necessary to model its mechanics,

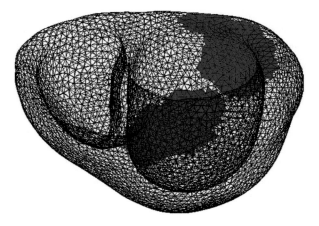

Figure 4.9 The representation of a heart structure using a biventricular tetrahedral mesh.

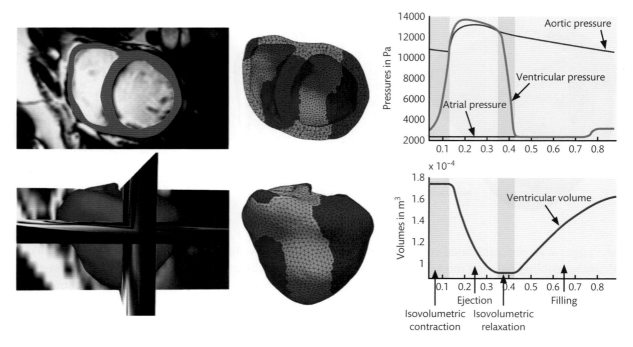

Figure 4.10 Personalized electromechanical models of the heart. Left: comparison with cine MRI. Middle: electrophysiological wave propagation, illustrated using a tetrahedral mesh. Right: the resulting pressure and volume curves.

anatomy, and physiology. In particular, it is necessary to model the varying properties of the cardiac tissue. This tissue is described as active, non-linear, anisotropic, incompressible, and visco-elastic; many of these properties can be characterized numerically, and the values will of course vary from individual to individual. Several formulations have been proposed for this, the details of which are far beyond the scope of this book. You should appreciate, however, that models should be designed as far as possible to include all clinical data that can contribute to this personalization. The most important components of observations of heart function are the heart's apparent motion and left-ventricular pressure, and models should take particular account of these if they are to be both accurate and precise. Using this approach it is possible to build a complete multi-scale electromechanical model for a patient's heart that both respects the laws of thermodynamics and allows the model fibres to maintain isometry and isotony along their length thanks to the elastic and viscous components of the model.

The next section describes one such electromechanical heart model in a little more detail.

Passive Stress: Non-Linear Elasticity

The heart is a soft tissue and as such is modelled best by assuming that there is no linear relationship between stress and strain (non-linear elasticity). The architecture of the heart fibres is known to be orthotropic; that is, to have different properties in different orthogonal directions. There are several recognized approaches to modelling this. Some models include the direction of the fibres, while others (such as the one that is mainly discussed here) model the myocardial tissue as a hyperelastic, rubber-like material, using for example the equations established for a Mooney–Rivlin solid[2]. This model takes the anisotropy of the heart fibres into account only in modelling the active myocardial tissue. This approach, in which a biological tissue (the heart muscle) is modelled with rubber-like properties, is quite a common one.

The hyperelasticity of a material can be fully characterized using a strain energy function that describes the amount of energy that is necessary to deform the material. Modelling hyperelastic materials in this

[2] http://en.wikipedia.org/wiki/Mooney%E2%80%93Rivlin _solid

way, however, is computationally expensive, and approximations need to be used. The model discussed here uses the Multiplicative Jacobian Energy Decomposition (MJED) application, which was developed for modelling liver tissue [13] and which uses tetrahedral FEs.

Active Stress: Non-Linear Electromechanical Coupling

The purpose of the heart is to contract, and any heart model must therefore be able to simulate contraction. A usual way of doing this is to add a strain energy density function to the model of passive stress discussed in the previous section. This is applied only along the cardiac muscle fibres and is controlled using a normalized transmembrane potential simulated using the electrochemical model described above. The contraction component is applied in parallel with a modelled viscosity, and a linear elastic component is applied in series with the contractile element to simulate the local motion of the ventricles.

Several strategies have been adopted to measure and estimate biophysical parameters for use in cardiac models. It is still a scientific challenge, however, to estimate patient-specific parameters and initial and boundary conditions for these models. Figure 4.10 illustrates how the model that we have been discussing in this section has been used to obtain predictions for the pressure in the left ventricle under different pacing conditions; these were found to be in good agreement with measurements obtained from the patient concerned using invasive techniques [14].

Conclusion

In this section we have discussed examples of methods for modelling the electromechanical function of the heart that can be personalized for individual patients. The relevance to this chapter is that imaging techniques such as cardiac MRI are used with electrophysiological mapping and pressure recordings to determine patient-specific estimates of, for example, the conductivity and contractility of the myocardium. A model such as this one can be used to predict, for example, whether or when tachycardia will be induced or how different pacing configurations affect the blood flow. Several of these models have been found to compare well with measured data, demonstrating how models like these can be adapted to make them specific for individual patients

and how this approach may be useful for planning therapeutic interventions.

These models allow researchers to integrate together information about the anatomy, electrophysiology, kinematics, and mechanics of the heart and to explore how these aspects correlate in the case of a particular patient. This in turn can provide an integrated view of the patient's cardiac function, enabling clinicians to evaluate the possible effects of different therapies before selecting one.

This is still a relatively new method; it is complex to apply in practice and has not yet been fully validated in the clinic. If or once this validation process is complete, however, this method should be able to provide a means of optimizing the treatment offered to each patient. This model-based approach is a promising adjunct to complex cardiac therapies that require optimization for each patient. It can also help in developing clinical databases and biomarkers that can feed into the planning and evaluation of therapeutic options for future patients. One potential problem with this approach, however, is that it is essential that clinicians have confidence in the models and that they find them easy to use. The second problem may be solved by developing automatic pipelines for running models known as workflows, and these will be discussed at length in Chapter 8. The concepts of validation and verification of computational models will be discussed in Chapter 11.

4.2.4 Model-Based Evaluation of Cerebral Aneurysms

The term cerebral aneurysm is used to refer to a bulging in the wall of one of the vessels supplying blood to the brain, and that can eventually burst and bleed. Some individuals can live with these all their lives without noticing any problems, but if they rupture, it can cause permanent disability or even death. Imaging techniques have helped us gain an understanding of the structure of these aneurysms, but our knowledge of their development, growth, and rupture is still quite limited. Nevertheless, the clear clinical advantage that would be gained by understanding this mechanism enough to be able to predict rupture is motivating attempts to model this important pathology.

In this section, we describe a data-processing pipeline that focuses primarily on extracting models of cerebral aneurysms from medical images and enriching them to produce full biomechanical models [15]. The ultimate goal is to provide a personalized assessment of aneurysm growth and rupture that is associated with an image-based characterization of the blood flow within and through the aneurysm as well as its mechanics and morphodynamic behaviour. Ideally it should prove possible to simulate how a particular patient's aneurysm would respond to treatment using a stent or coil.

Figure 4.11 shows an image-based data-processing pipeline for streamlining the creation of anatomic, structural, and haemodynamic models of aneurysms and for deriving robust and reliable patient-specific descriptors from these models. The models are built up using both 2D and 3D images obtained from the patient.

The first step in this pipeline involves the extraction of a geometrical representation of the patient's aneurysm from the images available. This requires the use of image-segmentation and surface-correction techniques, and its accuracy is crucial in building up

a representative model. This geometry will be used to obtain a complete set of 3D descriptors of the morphology of the aneurysm. Then, full computational fluid dynamics simulations will be run to simulate the evolution of blood flow inside the aneurysm during the patient's cardiac cycle. This process can be further extended to estimate the structural properties of the vascular wall so that potential areas of weakness can be spotted.

Computational models of implanted vascular devices have also been developed that should allow clinicians to plan and predict the optimal treatment for each patient [16].

In addition, the pipeline discussed here can provide a large number of patient-specific descriptors for any aneurysm, and it has many potential uses. It can, for example, be used to support the verification of a clinical hypothesis, assess the evolution of a given aneurysm over time or assist in clinical decision making, eventually improving the outcomes for patients.

Readers who would like to learn more about how a pipeline of this kind can be used in practice can

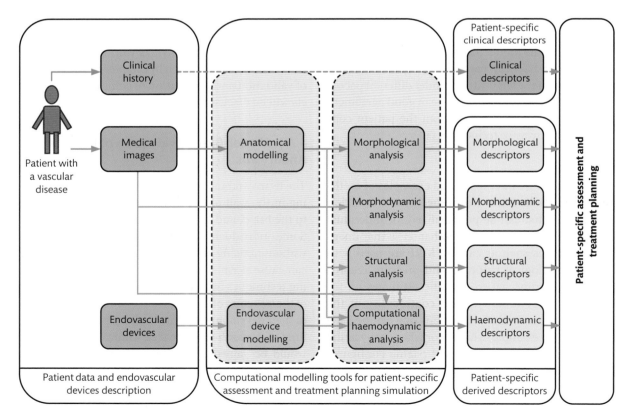

Figure 4.11 An image-based modelling pipeline for modelling intracranial aneurysms. Figure generated by Maria-Cruz Villa-Uriol.

turn to [17] which presents results of the evaluation of one that was constructed using the image-based workflow environment, GIMIAS[3].

4.3 Simulating the Physics of Image Formation

In order to design programs, pipelines, and workflows for the acquisition and analysis of medical images it is first necessary to understand how to simulate them, and that involves understanding the physics of how the techniques work. This process of simulating the physics of image formation is described in this section with an emphasis on one particular open-source simulation platform that is widely used for this purpose: the Virtual Imaging Platform (VIP) [18], available at **http://www.creatis.insa-lyon.fr/vip/**.

This imaging platform is an open web platform for accessing medical image simulators. It incorporates many of the tools and concepts that are widely used throughout computational biomedicine, including workflows and ontologies (both of which are discussed elsewhere in this book). After a brief overview, this section highlights the concepts incorporated in the VIP using the example of simulated cardiac imaging.

Medical image simulation involves simulating the physical process of image acquisition to produce digital 'synthetic images' of objects: in our discipline, these 'objects' will most often be patient. The code that is written to perform the simulation will obviously be specific for the image modality that is to be modelled. These codes can, however, be classified roughly into two types: analytical and Monte Carlo simulations. Analytical simulations are solutions of the equations that represent the physics of the imaging process (for example, Bloch equations for magnetic resonance) whereas Monte Carlo simulations model phenomena as random processes, estimating their effects using repeated random sampling. An example of this is the estimation of the energy measured by a CT detector from the trajectories of photons emitted by an X-ray source and travelling through a patient.

It should be self-evident that this type of image simulation requires knowledge of the physical parameters of objects involved in the simulation. The parameters required will again depend on the imaging modality to be simulated (see Section 4.1). For example, knowledge of magnetic properties will be required for simulating magnetic resonance; echogenicity, or the ability to 'bounce' an echo, for ultrasound imaging; X-ray signal attenuation for CT; and radioactivity for PET. In some cases it is possible to estimate these parameters directly from a real image of the object. One example of this is with relaxometry maps in magnetic resonance, which can produce simulated images that are hardly distinguishable from real ones.

These simulations, however, are of limited flexibility since a real-world counterpart must exist for each simulated image. An alternative method involves generating the required physical properties from a geometrical model of the object such as a set of labelled meshes or maps of individual volume elements known as voxels (i.e. voxel maps). Parameter values are then sampled from distributions that are associated with different regions of the model. As these are detailed distributions that are hard to estimate, the images that result are generally less 'realistic' (i.e. they compare less well with a real image) than they would be if the parameters are estimated from experimentally determined images (Figure 4.12). The advantage of this method, however, is that it can be used to simulate a wider variety of images.

The VIP allows the use of both these types of simulated image. All models and their physical parameters are annotated using an ontology that was specifically developed for this purpose, which makes it easier for all users of the system to share models. Components of models can be described using two different time frames: the longer one is designed for modelling the evolution of a patient's condition between appointments, and the shorter one to simulate movement.

Each instance or time point of a model consists of one or several layers, which may be concerned with the patient's anatomy, his or her pathology (such as a tumour), or an external agent or foreign body (such as a medical device). Models can also be annotated with descriptions of physical parameters such as parameter maps or distributions. VIP models of objects

---

[3] http://www.gimias.org

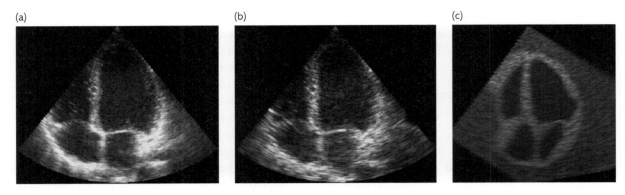

(a)   (b)   (c)

Figure 4.12 Example of simulated cardiac ultrasound images. (a) A real image acquired *in vivo*; (b) the result of a simulation using parameters that are extracted from the real image in (a); (c) a simulated image where the parameters were obtained from a geometric model of the anatomical structures shown.

such as anatomical structures can be placed in a repository that allows them to be shared in a structured way, to define simulation scenes in a graphical user interface (GUI), or to launch simulations of CT, PET, or ultrasonic imaging (see Figure 4.13).

Simulating medical images using the techniques described here is a computationally intense procedure, and it is necessary to use high-performance computational resources if realistic simulations are to be obtained. The different types of computing resource available for researchers to use are discussed at length in Chapter 9. Briefly, for now, the term high-performance computing (HPC) refers to a network of supercomputers with a large total number of central processing unit (CPU) cores that are located in a single room or building, minimizing communication costs and allowing a very high peak level of performance. Researchers will generally need to apply for access to these resources via a procedure that is not unlike a grant application, and this process, again, is discussed in more detail in Chapter 9.

In contrast to HPC, high-throughput computing (HTC) infrastructures search a wide range of resource providers for relatively small amounts of computer power. The result is a worldwide distributed system such as an academic grid (e.g. European Grid Infrastructure and Open-Science Grid) or a volunteer grid of CPUs donated by individuals to research projects such as those run through the BOINC system [19]. Specific software is deployed to make these resources usable as a single computer. HTC infrastructures are more easily accessed by

researchers for day-to-day use and for trying out experiments. This ease of use can, however, be rather compromised by communication costs and software errors. The VIP is able to exploit resources provided by the European Grid Infrastructure, which provides substantial computer power and storage to both beginners and advanced users.

The VIP includes simulators for four different image modalities:

- magnetic resonance (SIMRI),
- PET (PET-Sorteo),
- ultrasound (FIELD-II),
- CT (Sindbad).

These simulators were integrated as workflows that were chosen for two reasons. First, they are all parallel languages, so they provide a formal way of describing concurrent executions of codes without modification. In practice, this parallelism is obtained by splitting the simulation scene (spatially and/or temporarily) into chunks on which the code is iterated. It therefore becomes possible to simulate several projections of a CT image or positrons involved in a PET acquisition concurrently. This method of parallelization is called data parallelism, because the simulation is split between different pieces of data with only a minor impact on the simulation code. Data parallelism and the alternate approach of task parallelism are discussed in detail in Chapter 9.

The second reason for choosing workflows is that they provide a way of tracking dependencies between the datasets that are generated during a simulation.

(a)

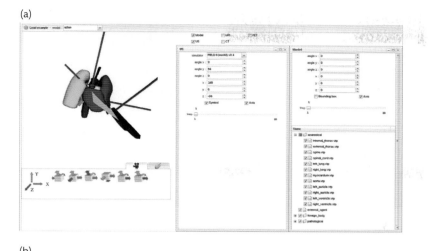

(b)

FIELD-II (model) v0.4

i Documentation and Terms of Use

Simulation Name

test for VPH

ProbeParameters_mat

| List | ⌄ | /vip/VIP (group)/TestData/FIELD-II/ProbesParams/settings.mat | + 🔍 |

SimulationDirectory

| List | ⌄ | /vip/Home | + 🔍 |

Model

| List | ⌄ | /vip/VIP/Models/1334935282352-adam.zip | + 🔍 |

Transformation

| List | ⌄ | 149 0 -59 0 -56 0 | + 🔍 |

simulation size

| List | ⌄ | medium | + 🔍 |

⚙ Launch   🖫 Save Inputs   ⚒ Save as Example

Figure 4.13 Launching an ultrasound simulation in VIP. (a) A cardiac model is selected from the VIP repository, and positioned with respect to the transducer in a 3D interface. The different model parts can be hidden or shown. (b) Simulation parameters are added to the launch panel. The model name and geometrical transformation linking the model and transducer frames are automatically added by the 3D interface. Probe parameters are collected in a single file.

The graph structure of these makes it possible to record the provenance of data automatically. It then becomes possible to know the object model and simulation parameters from which any particular dataset was derived. This allows structured databases of simulated data to be built.

All these simulations generate data in substantial quantities, whatever metric is used to measure this data: data volume, the number of files produced, or the number of different data types involved. Therefore any simulation platform or set of tools that is used for such simulations must include a range of efficient tools for managing this data. The data annotation resources in the VIP include tools for file storage, file transfer, and data annotation. Storage

is rarely an important issue on distributed computational infrastructures. In order for simulation jobs to run efficiently, however, file transfer between local and remote resources must be fast, efficient, and reliable. Two mechanisms are commonly required for this: file replication involves storing several copies of the same file for use at different locations, and file caching allows processes to keep local copies of frequently used files.

Figure 4.14 illustrates an example of the use of the VIP. It shows ultrasound and MRI slices of a cardiac model simulated using this platform. This model consists of eight types of material that were segmented using *in vivo* magnetic resonance images: fat, muscle, blood, skin, spinal cord, lung, myocardium,

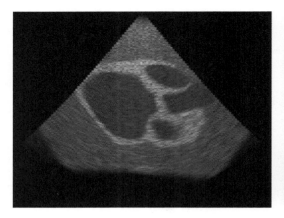 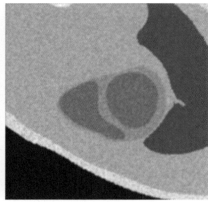

Figure 4.14 Cardiac images simulated in VIP. Left: echocardiography simulated with FIELD-II showing a parasternal long-axis view of the heart; right: cardiac MRI simulated with SIMRI showing a short-axis view of the heart.

and air. For MRI, a spin-echo sequence was simulated with SIMRI. For ultrasound, a 64-element cardiac phased array including 128 lines for a 90° sector angle was used.

In this example, the ultrasound image was simulated using a total of 750 CPU hours, which were computed on the VIP in about 22 hours of wall-clock time. The simulation of the magnetic resonance slice was much faster, taking 24 CPU minutes corresponding to about 4 minutes of wall-clock time on the VIP.

4.4 Statistical Atlases, Population Imaging, and Modelling

In this chapter we have already discussed how image-based models can be personalized for individual patients using information that has been derived from images that have been obtained from those patients using *in vivo* and, where possible, non-invasive images. It is therefore possible to build up statistical models of the structure and function of the organs and tissues modelled using sets of images obtained from large populations. These can be built up into databases or atlases of normal and pathological structures, helping in the analysis of the morphology and function of organs and systems across population groups. Another use of these large datasets is to facilitate the fusion of data between different imaging protocols and modalities and between imaging data and data from other sources.

These atlases can play important roles in helping researchers to understand and simulate the physiology of any human organs or systems, and can

provide insight into the location of structures and substructures at all levels within the body. This type of 'map' is extremely important in many areas of computational biomedicine, as different structures within an organ or system will have different electrical and mechanical properties that will therefore need to be modelled using different parameters. A statistical atlas of human anatomy and physiology such as those discussed here both provides an average layout of the structures concerned and encodes the observed deviations from this average. This provides a set of statistically justified bounds within which the encapsulated structure can move and be deformed. This alone makes two very important applications computationally possible. Firstly, it becomes possible to personalize the geometry of the organs or systems that are to be simulated without necessarily using subject-specific measurements.

In geometrical personalization, the atlas can be matched to medical imaging data. In a typical case only a subset of structures catalogued within the atlas can be matched explicitly to the image, and the remaining structures will be placed using statistical correlation. Secondly, the statistics that are generated from a population of real patients and/or volunteers and stored in such an atlas can be used to generate virtual populations for further studies. This population can be controlled by the user, for example to match an existing population statistically, to increase the size of a sample, or to generate extreme cases. In both cases the atlas will also be useful in the post-processing and analysis of any simulation results that are obtained. The fact that it is possible to encode the anatomical and physiological

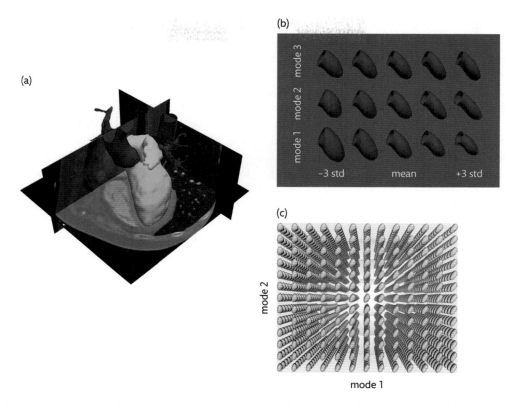

Figure 4.15 An example of a statistical atlas of the heart constructed from high-resolution CT images. (a) The average representation of various structures in the left heart. Blue: aorta; teal: left-ventricular blood pool; green: left-atrial blood pool; red: pulmonary vein trunks. (b) The statistical variability of the left ventricle as determined using principal component analysis (PCA), showing the three most significant modes of variation. (c) A virtual population of left ventricles obtained by uniform sampling of the first three modes of variation in the shape space obtained using PCA.

variation within a population in an atlas implies that the spatial relationships between individual instances are known. This ensures that it is possible to easily place the output data from a study into a common reference frame within which it can be analysed further.

The construction of a statistical image atlas of this type involves three distinct steps: spatial normalization, 'atlasing', and statistical analysis. The first spatial normalization step involves the synthesis of an 'average' image from the population of images that are to be included in the atlas. This registration step consists of aligning the images into one common coordinate system by applying geometrical transformations. In this way, registration can also provide spatial relationships between the population images and the mean.

The second step involves the transfer of information already known about the anatomy of the structure that is being mapped into the synthesized mean image. A representation of this model is then generated using a mesh, and this mesh is then warped onto the whole population using the transformations obtained in the first step of spatial normalization. Finally, the information in the atlas is analysed statistically using point distribution analysis of the mesh vertex locations.

An example of a statistical atlas of the heart that has been constructed from high-resolution CT images is shown in Figure 4.15. This figure shows an average representation of various structures in the heart in (a), the statistical variability of the left ventricle in (b), and a virtual population of left ventricles uniformly derived from the observed variation in (c).

4.5 Open-Source Tools for Image-Based Modelling

You will already be aware that computational biomedicine is a complex and multidisciplinary endeavour and that most modelling and simulation projects involve large numbers of researchers from different institutions and with different areas of expertise. If such large-scale projects are to succeed, the researchers will need to use a common infrastructure. This may include common standards, open-source software tools, and freely accessible databases and model repositories. This infrastructure has been organized in Europe by the Virtual Physiological Human Network of Excellence (VPH NoE), which is discussed in more detail in Chapter 11. The Network of Excellence has also organized a range of training and dissemination activities for researchers and other stakeholders, and has maintained links between the many scientists and clinicians involved in its projects and other researchers in computational biomedicine throughout the world.

The Network of Excellence has also maintained a collection of open-source computational biomedicine tools which are free for the worldwide community to use and modify. The tools that are stored and distributed in this VPH Toolkit portal[4] include programs for the collection, management, and integration of data; the processing and curation of this data; modelling and simulation of normal human physiology and pathophysiology; and the user-friendly deployment of these tools for end users.

However, this is only one framework that has been developed for providing the worldwide computational biomedicine community with access to the tools that they need. This short overview is restricted to collections of tools that are completely open source, that are specifically relevant to image-based modelling applications, and that provide a base from which further tools can be developed.

One useful approach is to provide a simple, ready-to-use application that may be extended using a plug-in mechanism through public Application Programming Interfaces (APIs). This is the approach taken in the workflow environment GIMIAS (see Section 4.2.4) and in MedInria[5]. Both these include basic input/output (or IO) operations and a set of specialized algorithms. Their functionality can then be extended using plug-ins, which are implemented in the same programming language as the basic infrastructure (which is most often C++). GIMIAS includes a graphical workflow manager that

allows users to chain algorithms. There has recently been an international effort to unify platforms for image-based modelling. This has been named the Common Toolkit[6] and it provides an implementation of an industry-approved plug-in design. All developers will find it helpful to use standards like these when building tools that they expect to be used in a wide variety of modelling applications.

Another useful approach involves using graphics to help users build up and create new applications. This is the strategy that is used in the CreaTools suite[7] for medical image processing and visualization. One important element of CreaTools is the black-box toolkit (BBTK), which provides a flexible framework for the design, programming, testing, and prototyping of applications. The BBTK provides users with libraries of high-level components that can be bolted together to construct (for example) GUIs, input/output, file management, and display and interaction features. These components ('black boxes') can be assembled into pipelines using a scripting language, and these pipelines will either form stand-alone applications or high-level 'meta-widgets' that can be incorporated into more complex applications. Editing and testing scripts can be carried out using either the command line or a graphical environment.

A final approach for us to mention here involves a framework of basic building blocks that can be assembled programmatically. This is used in the Multimod Application Framework[8] (MAF), which is a software framework that provides multiview, GUI-based interactive visualization and supports a vast range of data types. These include import/export formats and complex data types such as time-varying volumes to represent tensor fields.

All the examples that we have cited above use two C++-class libraries, the Insight Segmentation and Registration Toolkit[9] (ITK) and the Visualization Toolkit[10] (VTK). These are very popular packages that were funded from US government grants; each provides a range of basic and advanced algorithms for data segmentation, registration, and visualization. They were both developed in the US research

---

[4] http://toolkit.vph-noe.eu
[5] http://med.inria.fr

[6] http://www.commontk.org
[7] http://www.creatis.insa-lyon.fr/site/en/CreaTools_home
[8] http://www.openmaf.org
[9] http://www.itk.org
[10] http://www.vtk.org

community and are now maintained by a company named Kitware Inc. It is possible to use these building blocks to construct the complex workflows that are often required for image-based modelling. Workflows are discussed in much more detail in Chapter 8.

With very many people building and using modelling applications that involve the same data, it is clearly important to have a common representation for that data. There is, in fact, a common data type for medical images: it is called DICOM (see Section 4.1.5). All the toolkits and libraries discussed in this section, and many other open-source packages, support this data type. However, some types of analysis are so specialist that their requirements cannot be met only using this standard data type. Consequently, researchers developing image-based models may find that they need to use markup languages such as CellML (for cellular models) and FieldML (for hierarchical organ and tissue models) alongside DICOM. These markup languages and some others are discussed in more detail in the Appendix.

The development of standards and markup languages like these is wholly beneficial for the worldwide modelling community, as it helps avoid misunderstandings and incompatible models and encourages collaboration between groups. One important characteristic of the ways in which all the toolkits and standards mentioned in this section have been set up is that they are left open for any user to adapt and modify in new software as long as this software is in turn made available to the community. For example, models written in CellML and FieldML may be visualized using the cmgui tool[11] and/or used as input to a mathematical modelling environment such as OpenCMISS[12].

This rich range of resources is only possible because of the emphasis that the whole computational modelling community places on encouraging the reuse, adaptation, and interoperability of existing tools by open-source licensing. One good example of the range of tools that can be used within a single endeavour is given by the euHeart project[13]. Researchers involved in euHeart are aiming to develop a platform for cardiac modelling based on the GIMIAS tool and to display results from simulations run using OpenCMISS and representing data using FieldML. They both benefit from the accessibility of existing software and build functionality back into that software that can be adapted and reused by researchers in other projects. This type of cross-fertilization provides a true win/win situation for the whole computational biomedicine community.

4.6 Conclusions

We have aimed in this chapter to give a brief overview of computational physiology techniques that are based on medical imaging and of their applications. You will have seen, even from the small number of examples that it has been possible to discuss here, that information obtained from multimodal images can be crucial in generating data from individual patients to be used for personalizing generic models of human physiology. Many features of such a physiological model—its anatomy, tissue distributions, and material properties—vary between individuals, and it is important to include as much specific information as possible if a model is to be predictive for an individual patient. Information and data obtained using several different imaging modalities and on several different observational scales have been incorporated into these models, all providing useful information. We have also shown that it is possible to use imaging information to assess the predictive value and accuracy of computational models, therefore helping with their validation.

You will also have noted, however, that there are some intrinsic limitations on the use of imaging information, at least at the current state of technological development. Imaging is less able to record physiological variability accurately over a long than over a short time period, and other techniques, such as pervasive physiological sensing, will possibly provide a more useful way of introducing information about the environment and lifestyle of patients into simulations. This would clearly have enormous benefit for modelling the evolution of a chronic or degenerative disease. There is not enough space to cover sensing technologies here, but it is very likely that these and

---

[11] http://www.cmiss.org/cmgui
[12] http://www.opencmiss.org/
[13] http://www.euheart.eu/

imaging technologies will come to converge and that this will have enormous benefits for the whole discipline of computational biomedicine.

 Recommended Reading

Allisy-Roberts, P. and Williams, J. (2008) *Farr's Physics for Medical Imaging*. London: Saunders Elsevier.

Bushberg, J., Seibert, J. et al. (2011) *The Essential Physics of Medical Imaging*. Philadelpia, PA: Lippincott, Williams & Wilkins.

Cierniak, R. (2011) *X-Ray Computed Tomography in Biomedical Engineering*. Berlin: Springer.

Gibbs, V., Cole, D., and Sassano, A. (2009) *Ultrasound Physics and Technology*. Oxford: Churchill Livingstone.

Granov. A., Tiutin, L., and Schwarz, T. (eds) (2013) *Positron Emission Tomography*. Berlin: Springer.

McRobbie, D.W., Moore, E.A., Graves, M.J., and Graves, M.R. (2007) *MRI from Picture to Proton*. Cambridge: Cambridge University Press.

Smith, N. and Webb, A. (2011) *Introduction to Medical Imaging. Physics, Engineering and Clinical Applications*. Cambridge Texts in Biomedical Engineering. Cambridge: Cambridge University Press.

Suetens, P. (2009) *Fundamentals of Medical Imaging*. Cambridge: Cambridge University Press.

 References

1. Winslow, R.L., Trayanova, N. et al. (2012) Computational medicine: translating models to clinical care. *Science Translational Medicine* 4(158), rv11.
2. Ayache, N., Boissel, J.-P. et al. (2005) *Towards Virtual Physiological Human: Multilevel Modelling and Simulation of the Human Anatomy and Physiology*. http://ec.europa.eu/information_society/activities/health/docs/events/barcelona2005/ec-vph-white-paper-2005nov.pdf.
3. Seeman, E. and Delmas, P.D. (2006) Bone quality–the material and structural basis of bone strength and fragility. *New England Journal of Medicine*, 354(21), 2250–2261.
4. Nuzzo, S., Lafage-Proust, M.H. et al. (2002) Synchrotron radiation microtomography allows the analysis of three-dimensional micro-architecture and degree of mineralization of human iliac crest biopsies: effects of etidronate treatment. *Journal of Bone and Mineral Research* 17(8), 1372–1382.
5. Martin-Badosa, E., Elmoutaouakkil, A. et al. (2003) A method for the automatic characterization of bone architecture in 3D mice microtomographic images. *Computerized Medical Imaging and Graphics* 27(6), 447–458.
6. Pacureanu, A., Langer, M. et al. (2012) Nanoscale imaging of the bone cell network with synchrotron X-ray tomography: optimization of acquisition setup. *Medical Physics* 9(4), 2229–2238.
7. Haiat, G., Padilla, F. et al. (2007) Variation of ultrasonic parameters with microstructure and material properties of trabecular bone: a three-dimensional model simulation. *Journal of Bone and Mineral Research* 22(5), 665–674.
8. Raum, K., Cleveland, R.O. et al. (2006) Derivation of elastic stiffness from site-matched mineral density and acoustic impedance maps. *Physics in Medicine and Biology* 51, 747–758.
9. Smith, N., de Vecchi, A. et al. (2011) euHeart: personalized and integrated cardiac care using patient-specific cardiovascular modelling. *Journal of the Royal Society Interface Focus* 1(3), 349–364.
10. Noble, D. (2004) Modeling the heart. *Physiology* 19, 191–197.
11. Chinchapatnam, P., Rhode, K.S. et al. (2008) Model-based imaging of cardiac apparent conductivity and local conduction velocity for diagnosis and planning of therapy. *IEEE Transactions on Medical Imaging* 27(11), 1631–1642.
12. Mitchell, C.C. and Schaeffer, D.G. (2003) A two current model for the dynamics of cardiac membrane. *Bulletin of Mathematical Biology* 65, 767–793.
13. Marchesseau, S., Heimann, T. et al. (2010) Fast porous visco-hyperelastic soft tissue model for surgery simulation: application to liver surgery. *Progress in Biophysics and Molecular Biology* 103(2–3), 185–196.
14. Sermesant, M., Chabiniok, R. et al. (2012) Patient-specific electro-mechanical models of the heart for the prediction of pacing acute effects in CRT: a preliminary clinical validation. *Medical Image Analysis* 16(1), 201–215.
15. Villa-Uriol, M.C., Larrabide, I. et al. (2010) Toward an integrated management of cerebral aneurysms. *Proceedings of the Royal Society A* 368, 2961–2982.
16. Larrabide I., Kim M., et al. (2012) Fast virtual deployment of self-expandable stents: method and in-vitro validation for intracranial aneurysmal stenting. *Medical Image Analysis* 16(3), 721–730.
17. Larrabide, I., Villa-Uriol, M.-C. et al. (2012) Angio-Lab—a software tool for morphological analysis and endovascular treatment planning of intracranial aneurysms. *Computer Methods and Programs in Biomedicine* 108(2), 806–819.
18. Glatard, T., Lartizien, C. et al. (2013) A Virtual Imaging Platform for multi-modality medical image simulation. *IEEE Transactions on Medical Imaging* 32(1), 110–118.
19. Sansom, C. (2011) The power of many. *Nature Biotechnology* 29, 201–220.

Chapter written by
Alejandro Frangi, Center for Computational Imaging & Simulation Technologies in Biomedicine, Information & Communication Technologies Department, Universitat Pompeu Fabra, Barcelona, Spain and Department of Mechanical Engineering, University of Sheffield, Sheffield, UK;

Denis Friboulet, Université de Lyon, CREATIS, CNRS UMR5220, Inserm U1044, INSA-Lyon, Université Lyon 1, France;

Nicholas Ayache, Inria Asclepios Team, Sophia Antipolis, France;

Hervé Delingette, Inria Asclepios Team, Sophia Antipolis, France;

Tristan Glatard, Université de Lyon, CREATIS, CNRS UMR5220, Inserm U1044, INSA-Lyon, Université Lyon 1, France;

Corné Hoogendoorn, Center for Computational Imaging & Simulation Technologies in Biomedicine, Information & Communication Technologies Department, Universitat Pompeu Fabra, Barcelona, Spain;

Ludovic Humbert, Center for Computational Imaging & Simulation Technologies in Biomedicine, Information & Communication Technologies Department, Universitat Pompeu Fabra, Barcelona, Spain;

Karim Lekadir, Center for Computational Imaging & Simulation Technologies in Biomedicine, Information & Communication Technologies Department, Universitat Pompeu Fabra, Barcelona, Spain;

Ignacio Larrabide, Center for Computational Imaging & Simulation Technologies in Biomedicine, Information & Communication Technologies Department, Universitat Pompeu Fabra, Barcelona, Spain;

Yves Martelli, Center for Computational Imaging & Simulation Technologies in Biomedicine, Information & Communication Technologies Department, Universitat Pompeu Fabra, Barcelona, Spain;

Françoise Peyrin, Université de Lyon, CREATIS, CNRS UMR5220, Inserm U1044, INSA-Lyon, Université Lyon 1, France;

Xavier Planes, Center for Computational Imaging & Simulation Technologies in Biomedicine, Information & Communication Technologies Department, Universitat Pompeu Fabra, Barcelona, Spain;

Maxime Sermesant, Inria Asclepios Team, Sophia Antipolis, France;

Maria-Cruz Villa-Uriol, Department of Mechanical Engineering, University of Sheffield, Sheffield, UK;

Tristan Whitmarsh, Center for Computational Imaging & Simulation Technologies in Biomedicine, Information & Communication Technologies Department, Universitat Pompeu Fabra, Barcelona, Spain;

David Atkinson, UCL Centre for Medical Imaging, 250 Euston Road, London NW1 2PG, UK.

Modelling Cell Function

Learning Objectives

After reading this chapter, you should:

- understand why the cells of all organisms are central to their regulation and how they respond to perturbations in their environment;
- have a broad appreciation of some general and specific aspects of cell anatomy;
- be able to summarize the more important activities common to all cells;
- understand the principles of cell division, growth, and development;
- understand the principles of some of the different techniques used to model cells;
- be able to look for information and locate cell models in repositories and databases.

5.1 Introduction

In this chapter, we focus on a key part of biology: the cell. All organisms apart from viruses consist of either one cell or many cells. A human, or indeed any other vertebrate, consists of many trillions of cells of a wide range of different types. This chapter will concentrate almost entirely on cells in the human body.

Human cells are therefore extremely important as parts of the *physiome* and are, possibly, key to how computational physiologists look at and understand the human body. The cell is in the middle of the hierarchy of levels that make up the 'system' of the human body. As such, modelling cells is modelling at a higher level than molecules and networks, and at a lower level than tissues and organs. Furthermore, cell systems have been studied in depth for many levels, so researchers wishing to model cells are able to draw on a great depth and breadth of knowledge.

Modelling the human body from the 'middle out', as proposed from the very beginning of the Physiome Project [1], will almost self-evidently draw on and involve modelling the cell.

All living organisms need to be able to respond to and to survive in a range of environments. Living cells need to be robust in their responses to their environment in order for organisms to remain in a viable state in conditions that may be stressful and rapidly changing. It is therefore essential for signals to pass into cells from their environment and for those cells to be able to respond. Biochemical and electrical signals passing into cells from the environment cause them to adjust the rate at which different proteins are expressed. This will alter the metabolic and physiological functions of that cell in response to its differing needs. The functions of all organs and of organisms themselves are derived from the functions of their constituent cells.

If you are thinking of using modelling to explore the function of an organ or organism, therefore, the cell will very often be a good place to start. It is possible to start by treating a cell of a particular type as an isolated entity, surrounded by a membrane that separates its 'inside' from its 'outside'. This may be unrealistic, but it is perhaps no more unrealistic than thinking that cells in an isolated culture *in vitro* (in a Petri dish, for example) will mimic the behaviour of the same cells *in vivo* in a living organism. And modelling an isolated cell can itself be a very complex procedure involving a large number of calculations to simulate the many biochemical and biophysical processes that go on in that cell. Modelling communication between cells is computationally much more demanding.

Most, although not all, human cells are between 10 and 30 μm (0.01–0.03 mm) in diameter. These cells are small enough, and diffusion through them rapid enough, for it to be possible to model biochemical processes within cells fairly realistically using the approximation of a 'stirred tank' or 'mixing chamber'. This implies that the composition of the liquid contents of the chamber (the cytoplasm in the cell) stays constant throughout the volume of the cell.

Treating the cell as a stirred tank is a relatively accurate assumption in calculations involving electrophysiology; that is, in calculating changes in the potential energy across the membrane of excitable cells. (This potential energy is known as the membrane potential or E_m.) Electrical currents across cell membranes arise because charged ions can pass into and out of cells through a set of transmembrane proteins called ion channels; each ion channel is essentially independent of all others. In modelling ion currents into cells, it is possible to represent the function of each transmembrane ion channel, and therefore each transport process, by an independent module and to derive the behaviour of the whole cell by integrating these modules together. This modular way of constructing cell models has many advantages, not least the fact that it is relatively easy to substitute one module with another, to improve it, to simplify it for running on a slower machine, or to adapt it for a slightly different cell type.

Modelling the cytoplasm within a cell as a stirred tank implies that its contents are mixed in a uniform manner. This, however, is only an assumption, and there are plenty of scenarios in which it does not apply. Modelling enzyme-catalysed reactions within cells, for example, will need to include the concentrations of the reactants and the products as parameters, and these will not be constant throughout a cell or over time. Furthermore, cells are never in true equilibrium. Rather, they settle into a steady state in which nutrients such as glucose are continually entering and being transformed by metabolic processes into structural and functional proteins; into substances that will be exported from the cell for use elsewhere in the body; into the biochemical 'energy store', adenosine triphosphate (ATP); and into waste products such as carbon dioxide and water.

The human body is remarkably resilient, in that it can continue to function under a wide range of adverse circumstances, whether external events or internal problems. For example, a bruise to the heart muscle does not prevent the heart from pumping blood, just as a bruise to a leg muscle does not prevent the affected individual from walking. The body of a human or, indeed, any other organism can therefore be described as robust.

When a part of the body is injured, compensatory mechanisms come into play. Some of these protect the damaged organ, for example by reducing the contraction strength in a bruised muscle so that it can rest. Other mechanisms involve enhancing the function of cells, tissues, and organs in response to stress, for example by switching to anaerobic metabolism (in which glucose is consumed and lactic acid produced) in an environment in which oxygen has been depleted. All these changes, and changes during long-term processes such as athletic training, ageing, heart failure, and pregnancy, begin with changes in the metabolism of cells.

Let us take another example from the function of the heart. When a coronary artery is blocked suddenly, a part of the heart will have its blood supply cut off. This section of heart tissue therefore lacks oxygen and nutrients and is said to be *ischaemic*. The remaining, healthy part of the heart will immediately begin to pick up the load on the heart and therefore allow cardiac output to be maintained. This sets up a complex series of events involving the cells and tissues of the heart.

In the first few minutes after the blockage, the ischaemic region—for example, in the left ventricle—fails to contract, is stretched, and balloons outward in a kind of dilation. In this state, the heart fails to

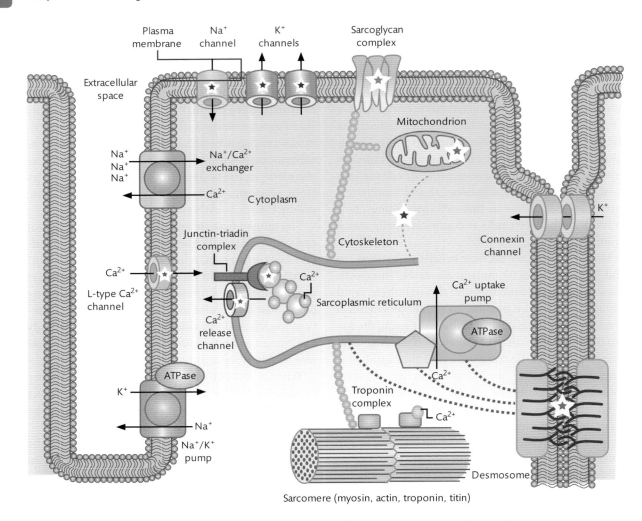

Figure 5.1 Diagram of a cell, showing the organelles and processes involved in contractile cell energetics and excitation–contraction coupling. Image provided by Dr Vanessa Díaz-Zuccarini. Redrawn from Knollman and Roden (2008) *Nature* 451, 929–936.

empty completely with each beat as it does normally. It therefore slightly over-fills each time the atrium contracts, which stretches the normal part of the ventricle. In turn, this stretch (or 'prestretch') causes a shift in the balance of ion currents and metabolic processes in the heart cells a few milliseconds before the next beat, which causes that next beat to increase in strength. This effect has been known for many decades; it is called the Starling effect, after Ernest Henry Starling who first studied it a century ago in 1914 [2].

This effect causes heart cells to respond to the electrical and chemical signals that indicate an extra load by increasing in volume, in a process that is known as hypertrophy. This increase in volume arises from an increase in the transcription rates for myofilament proteins within the cells. These cells lay down more of the basic muscle cell units known as sarcomeres at both the ends and sides of the cells. A good analogy is that of enlarging the size of an army by both widening the columns and lengthening the files of soldiers. This process goes on until the heart tissue mass has grown so that the energy requirement per gram of tissue is back to normal.

This is an example of a process in which the rates of gene and protein expression—in this case, the expression of genes that code for muscle proteins—are set by the cell, rather than being simply coded in the genome. The synthesis of the cell machinery is controlled by the needs of the organ and, therefore, of the organism at a particular time through a combination of intracellular feedback mechanisms and external stimuli.

This relatively simple process is mainly controlled by ion currents into and out of the heart cells. It is important to realize, however, that no cell is just a mixed bag of dissolved chemicals inside a membrane, however complex. The interior of each cell is crowded, containing a *nucleus* and many other organelles. Each of these organelles is bounded by a membrane, which is a highly organized, fluctuating, interconnected network of components (Figure 5.1).

All human cells contain a different set of components and organelles depending on their type, but some organelles are found in almost all of the 200-odd cell types. These include:

- transport networks composed of *microtubules*, which deliver proteins to the locations where they are needed;
- *lysosomes*, which degrade proteins that are no longer needed;
- lipid-enclosed 'bubbles' or *vesicles*, such as the Golgi apparatus and endoplasmic reticulum, which sequester proteins away from the general cytoplasm;
- *mitochondria*, the cell's 'energy factories', which are positioned to produce ATP as a source of energy in the places in the cell where it is most needed, such as beside bundles of micro-filaments in muscle cells.

All intracellular reactions are slowed by the 'crowding' that results from this high concentration of both organelles and macromolecules in the cell. Such crowding slows down the diffusion of all molecules, even small ones, through a cell, causing the apparent equilibrium concentrations for reactions to be higher than they would be in the steady state.

We hope that this introduction has shown you that the cell is central to human physiology and that cell modelling must therefore be a central part of computational physiology. Cells, however, are built up into tissues and organs in complex ways. No organ—no tissue even—is composed of just a single cell type; each comprises a family of different, but often related cell types that are in balance with each other and that communicate with each other through signalling to enable the tissue or organ to function. The processes through which cells communicate with each other and the functions they control will also be discussed in Chapters 6 and 7, which cover tissue and organ modelling and multilevel (multi-scale) modelling respectively.

5.2 General Functions of Cells

In an organ such as the heart, each cell will be located very close to a capillary: generally within 1 μm (see Figure 5.2). Capillaries are the smallest blood vessels and their walls are only one cell thick. These walls consist of a single layer of endothelial cells, which forms a barrier to the exchange of blood and solutes between the circulation and the cells of the organ.

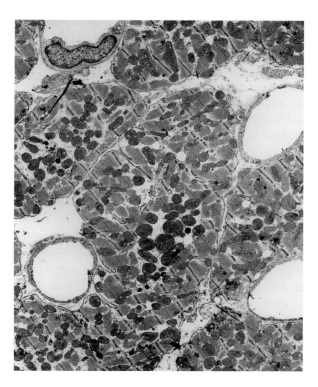

Figure 5.2 Transmission electron micrograph showing four capillaries and seven cardiac cells. Gap-junctional connections between endothelial cells are visible on all but the left upper capillary; none are visible between cardiomyocytes in this particular section. The capillary endothelial cell thickness is less than 0.4 μm except at the nucleus (upper left capillary) where it is more than 1 μm. For scale, the left lower capillary is 5 μm in diameter, and the sarcomere lengths are about 2 μm.

Solutes that cross the capillary endothelium can reach all the cells in an organ by diffusion within a few milliseconds. Each capillary is accompanied by sympathetic nerves and microscopic lymphatic vessels.

Toxic compounds, or compounds that are simply not needed in one part of the body, are bound tightly to a protein called albumin in the blood. This prevents them from leaving the bloodstream until they reach an organ that is equipped with special receptors and transporters that can remove them from their binding proteins. Copper ions are similarly bound to a specialist copper-binding protein called ceruloplasmin. One of the functions of the endothelial cells in the capillary walls is to protect the cells of the organ from these toxic or inappropriate substances. These cells also send signals into the organ and to the smooth muscle cells of larger blood vessels called arterioles to regulate the local flow of blood. Other processes, signals, and fluxes of substrates and ions are driven by cells within the organ with the endothelial cells acting in a way as 'slaves' in response to the signals.

These regulatory processes are all fantastically complex. In a normal, healthy person, each organ acts as a harmonious community, functioning normally without suffering any breakdowns or 'traffic jams'. Consequently, we can discuss in this chapter individual cells as entities in their own right while remembering that they always act within a community of cells forming a tissue or organ. Similarly, however, the cell can be viewed as a tiny 'command centre' which organizes the expression of genes and deployment of proteins, often in response to signals coming in from outside. In computational biomedicine, where all levels of biological organization can be viewed as equally important and equally worthy of modelling, the cell is often a sensible place to start from, working 'up' the hierarchy towards the organs and the whole organism and 'down' towards organelles, networks, and molecules [3].

All cells require energy for all their functions, which include, but are not necessarily limited to, metabolism, growth, cell division, processes of uptake and secretion, movement, communication with other cells, and programmed cell death (apoptosis). In molecular terms, energy is needed in the process of protein synthesis for opening up the paired DNA strands, transcribing DNA into mRNA, processing the RNA molecules and translating them into protein at the ribosome, and moving the newly synthesized proteins into their places in the cell (see Chapter 2 for more details of these processes). This can be thought of as a manufacturing system in which raw materials are brought in to the factory and made into products, and the products are exported. In a factory, the plant must be maintained, worn parts replaced, and the plant must be kept clean by removing waste products. The 'currency' that is used in maintaining the cell factory is the molecular energy store, ATP.

Cells, however, are much more dynamic than the average factory. All the different 'molecular machines' that are needed for cell processes, enzymes, membranes, and structural proteins are continually being synthesized, degraded, and replaced. When proteins are degraded, their component amino acids are reused in new proteins. This remarkable process, which occurs mostly in lysosomes, is known as autophagy [4]; Aaron Ciechanover was awarded a part share of the 2004 Nobel Prize for Chemistry for its discovery.

We show in Box 5.1 that even relatively simple cellular processes can produce complex dynamics.

5.3 Fundamentals of Reactions in Cells

5.3.1 Introduction to Reactions in Cells

All chemical reactions are fundamentally reversible, which means that, in principle, they can proceed in either direction in given (often unusual) circumstances. This reversibility applies also to reactions that are catalysed by enzymes. Consequently, the net forward rate of an enzyme-catalysed reaction will be reduced by any build-up of the product of the reaction. Furthermore, all enzymes, as proteins, are subject to breakdown by proteases (proteolysis) and must therefore be replaced by new enzyme molecules synthesized through gene expression. Enzymes are often more resistant to proteolysis when a substrate molecule is bound than when it is not. This has the effect of reducing the rate of proteolysis when substrate is abundant, and allowing the concentration of the enzyme to decay when there is little substrate available.

Most membranes within and around cells are impermeable to most proteins. However, we have already heard how some newly translated proteins can be carried along the Golgi bodies to their required sites of action, carry signal sequences that allow them to be exported through membranes, or cross membranes via transporters. Other proteins are directed by dynein to penetrate the nuclear membrane and enter the nucleus. The complex processes through which protein expression is regulated and proteins

reach their different destinations within or outside the cell is not yet fully understood in detail.

These complex processes can be summed up, however, by saying that it is theoretically possible for proteins to go anywhere they are needed within or

Box 5.1 The Cell as a Complex System

The metabolic reactions that take place in cells are complex systems in both their dynamics and their kinetics. In parts of metabolic networks, such as those that control gene expression, the relationship between the supply of a substrate and the rate of a reaction is steep and there is a 'switch-like' effect with a function being effectively turned on as a concentration passes a particular value. This contributes to some instability in the metabolic systems. Although these systems seldom become chaotic, it is possible for chaotic behaviour to occur over a small scale, with small parameter ranges.

One example of this type of system is the glucose/insulin/ATP system. This is a simple system that consists of one substrate (glucose) and one enzyme. The enzyme is produced at a constant rate by gene transcription, and is degraded only when it is not occupied by substrate. The substrate can be assumed to be always present in the cell, so it can be supplied continuously. This system (shown schematically in Figure 5.3) can become chaotic within a particular parameter range.

In this enzymatic reaction, substrate S is supplied at a rate v_S that varies during the day and is also affected by the concentration of the product P. The enzyme E is synthesized at a constant rate v_E and degraded in a process that shows first-order kinetics with rate constant k_E when it is not bound to its substrate. This system allows adaptive behaviour: when the concentration of substrate [S] rises, more enzyme is present in the stable, substrate-bound [ES] form and the total amount of enzyme rises, allowing more substrate to be used. This is a form of adaptation to supply, rather than demand; it therefore has some characteristics in common with an electrical accumulator (e.g. a rechargeable battery).

These changes in the concentrations of enzyme, substrate, and product can be modelled using the following equations.

$$\frac{d[S]}{dt} = v_S + k_{-1} \cdot [ES] - k_1[E]$$

$$\frac{d[E]}{dt} = v_E + (k_{-1} + k_2) \cdot [ES] - k_1[S] \cdot [E] - k_E[E]$$

$$\frac{d[ES]}{dt} = [ES] \cdot k_1[S] - (k_{-1} + k_{21})[ES]$$

$$E_{tot} = [ES] + [E]$$

This system demonstrates a property known as resonance, in which small daily changes in v_S give rise to quite dramatic oscillations in the total concentration of enzyme (see Figure 5.4). The periodic response shown in the upper panel of this figure shows that the changes in the enzyme-substrate concentration [ES] are proportionally larger than those of the substrate, which varies over the course of the day in a sinusoidal way: S(t). This is like an electrical inductor in that it can give resonance or amplification to the system. Further increases and further changes in the substrate concentration with time result in larger changes in [ES] and in the rate of formation of the product P. This can induce a qualitative change in the behaviour of this system known as period doubling. This means that the heights of the peak concentrations alter so that the cycle repeats with a basic period that is twice the length of the period of the input function. Further increasing the amplitude doubles the period again so it is four times the original length. This is characteristic of a non-linear dynamical system in that period doubling occurs 'on the route to chaos' [6].

This example shows that chaotic behaviour can occur in a system that involves a single enzyme and substrate and no reverse reaction. It is easy to imagine the potential for chaos in a large network of biochemical reactions. But actually, chaotic behaviour is not common in complex metabolic networks, simply because the change in reaction rate with increasing substrate concentration is quite low. Even chaotic systems can still be stable, as chaotic biological systems are constrained so the extent of the variation is limited.

... continued

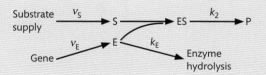

Figure 5.3 Diagram of an unstable system in which the reaction velocity k_2[ES] varies widely and unpredictably [5].

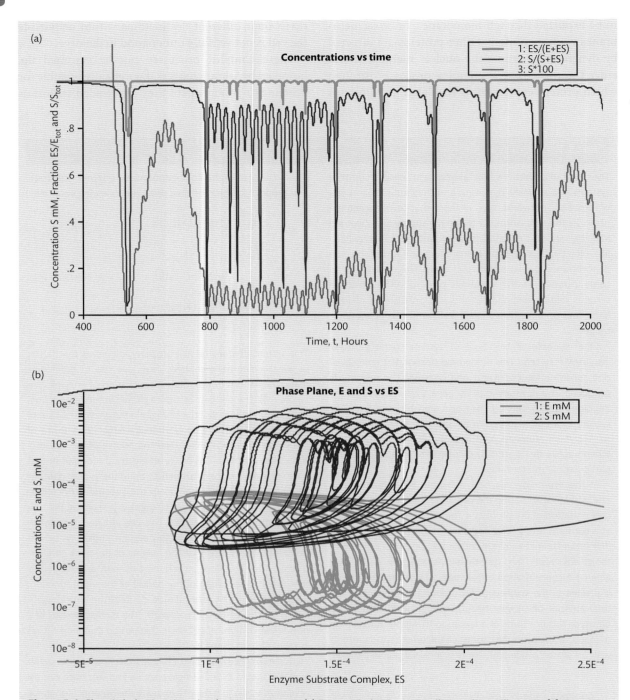

Figure 5.4 Chaotic behaviour in a simple enzyme system. (a) Enzyme induction: a small periodic oscillation in [S] gives rise to large oscillation in the E_{tot}, [ES], and the rate of production of P, which is k_2[ES]. Parameters are: $v_S = 3e^{-4} \cdot (0.435 + \sin(2\pi t/24))$ mM/h, $v_E = 5.51e^{-7}$ mM/h, $k_E = 0.6$ h^{-1}, $k_1 = 1e^6$ mM$^{-1} \cdot$h^{-1}, $k_2 = 2$ h^{-1}, and $k_{eq} = k_{-1}/k_1 = 1e^{-7}$ mM. Initial conditions were: [S] = $2.216e^{-5}$ mM, [E] = $5.542e^{-6}$ mM, [ES] = $5.8e^{-5}$ mM, and [P] = 0. The two cycles shown are different. (b) Phase plane showing chaotic attractor. The trajectory is highly sensitive to v_E.

outside cells, but that, like most molecules, they normally need to use special mechanisms to cross membranes. Enzymes are also found in many locations within the cell, wherever they are needed: within the cytoplasm, bound to, or embedded in membranes.

As we have seen, all metabolic processes in cells are dynamic, with their rates changing depending on the concentrations of enzymes, substrates, and products. It is also possible for the activity levels of the enzymes themselves to change. These changes result from modifying their active sites to alter the affinity for substrates, by modifying the rate at which the product forms from the enzyme–substrate complex, or, most drastically, by simply inactivating the enzyme. Some enzymes require metals such as iron, zinc, or copper to function, and others require small-molecule cofactors that are not changed during the reaction. Others can have their activity reduced or even be reversibly or permanently inactivated by metals such as thallium and mercury.

Metabolic processes require energy, which is, as we have mentioned, stored in the cell in the form of the nucleotide ATP, which is generated in the mitochondria within the cell. This process is self-evidently necessary for all metabolic biochemistry and—therefore—all physiology, and the biochemistry and mechanism of the mitochondrion is therefore considered to be an important system to model. Alternative energy sources do exist, in the form of nucleotides such as guanine triphosphate (GTP), nicotinamide adenine dinucleotide (NAD), and flavin adenine dinucleotide (FAD); these are functionally important but less abundant than ATP. Generally speaking, mitochondria are small enough to be satisfactorily modelled as stirred-tank models (see Section 5.1) with the cytoplasm as the external milieu. The free energy inside a cell can be characterized by its phosphorylation potential, which is defined as the ratio of the concentration of ATP to that of ADP times that of free phosphate. When cells lack oxygen temporarily (in transient hypoxic states) they adapt their purine nucleoside balances to maintain this phosphorylation potential.

5.3.2 Modelling Cellular Reactions

Modelling the biochemistry of a living cell is a very complex process, even when the models incorporate metabolic pathways that have been well known for half a century. For example, we do not yet know enough about the exact mechanisms that control the switch between burning glucose and fatty acids for energy. While we know that if insufficient nutrients are available (in starvation) proteins will degrade so that the resulting amino acids can be used as fuel, the mechanisms that regulate this process are still obscure.

Until now, most computational models of biochemical reactions have involved modelling steady states, for example in using the technique of flux balance analysis to estimate the fluxes through metabolic pathways. However, the fact that these systems are highly dynamic means that this is really too simplistic and modelling should really involve a more complex set of equations.

Many researchers have, over several decades, attempted to improve the precision and accuracy in calculating the fluxes through biochemical networks while increasing the speed of the computations and enhancing their stability. These calculations no longer assume that the concentration of each enzyme is constant throughout the simulation, but model the dynamic processes of transcription and proteolysis through which enzyme concentrations are altered and maintained.

We can illustrate this with an example: modelling the coupling between the action potential in muscle cells and the resulting contraction of the muscle. This is multi-scale modelling at the level of a single cell, and involves modelling processes that take place at dramatically different timescales from less than a millisecond (the cellular action potentials and ion currents) to days and weeks (changes in gene regulation that arise from adaptation to injury or training). This process can be modelled in a modular way, with separate biochemical and biophysical modules describing each process, and these can incorporate mechanisms that allow the model to respond to a broad range of physiological conditions. More and less complex module variants can be produced to gain either computational speed at the cost of adaptability or vice versa. The least complex modules will be specialized for fast calculations but will be accurate only within a very limited range of conditions. In the future it might be possible to substitute more or less complex modules for each other 'on the fly' during a modelling exercise as parameters change, if more computational facilities become available or (for example, in a practical application in the clinic) rapid results are required.

Modelling is also being used to study the placement of proteins in membranes and their trafficking around the cell. This involves modelling the transport of proteins along the microtubules from the Golgi apparatus using kinesin or back the other way using dynein; both processes use ATP. During this process, the actin filaments that comprise the microtubules are undergoing constant polymerization and/or depolymerization. You can think of this as being similar to laying down the tracks of a railway just before the train comes. Cell division is an even more complex mechanistic process, and, therefore, even harder to model, and we are only now beginning to understand, let alone model, the complex networks of processes that regulate RNA transcription and protein translation.

Over the past 30 years, molecular biologists have produced an immense amount of information about how proteins interact together in the cell (in what is sometimes termed the *interactome*, by analogy with terms such as genome and proteome). So far, we have identified a large number of potential protein–protein interactions with relatively few connections between them. Researchers are attempting to decipher cause-and-effect relationships from these associations between proteins. This is an example of a bottom-up approach to modelling, and it is difficult: so difficult that, by 2009 and after a decade of computation, about 50 different protein–protein interaction (or PPI) networks had been published but none had been fully validated.

Computational physiology, however, is being driven as we have seen from the 'middle out'. The cell is seen as the lynchpin of this type of modelling, as it becomes possible to integrate models both from cell to gene and from cell to organ. This is a very complex procedure, and one that has really only just started. Researchers have access to an enormous amount of molecular data obtained from (for example) mRNA arrays and proteome analysis. It should eventually prove possible to define and understand the cause-and-effect relationships of the complete set of regulatory pathways in each cell type using this data, but this has been equated to 'finding a needle in a haystack' [7]. Through the efforts of many thousands of researchers, both experimentalists and modellers, it is very likely that the needle will eventually be found.

5.4 Formalisms and Abstractions in Cell Modelling

Researchers involved in cell modelling will, necessarily, end up using formalisms and abstractions that define the processes that they are modelling. They will first need to make a choice of which type of formalism to use, and will, therefore, end up making an active choice not only about how to do the modelling but also about what to model.

When setting up a cell model, it is first of all necessary to select between two different approaches: to focus on modelling either *cellular function described by simple rules* or *cellular processes described by physical mechanisms inside the cell*. Although these may at first glance look similar, the assumptions behind them are clearly different, as are the results that are obtained from them and the conclusions that can be drawn. In the rest of this chapter we will examine these formalisms, focusing largely on models of one individual cell.

Almost all physiological processes and disease states can be understood through modelling the activity of cells. Furthermore, we have learned an enormous amount during recent decades about the structure, function, and mechanism of cells as parts of biological systems. Cells are discrete and autonomous; they behave independently although their behaviour is profoundly influenced by their local environment. Cells have a finite range of behaviours that can each be studied in relative isolation, and living cells can easily be studied in controlled environments *in vitro*.

These properties of cells mean that they can be modelled as autonomous agents with their behaviour regulated by external stimuli. This view of the cell as a functional element allows us to model each cell type using a set of algorithms or modules that capture its essential behaviours. It allows researchers to produce and validate mechanistic simulations of cells, and the results of these simulations provide knowledge and data that can feed back into the models, hopefully to make them more precise and accurate.

5.4.1 Characterizing Properties of Cells

Cells are also capable of altering their environment—part of a living organism—in ways that depend on their type. Some can contract or move; others synthesize proteins for export from the cell (for example for the production of extracellular tissue); and others can regulate the electrical potential of their environment or signal from one cell to another. Almost all cells are also capable of replicating themselves (in mitosis) and they can all self-destruct (in apoptosis). Many cells can also alter their phenotype in a process known as differentiation, taking on a new and generally more specialized set of functional characteristics. This list of behaviours is quite limited, yet the regulation and execution of these behaviours have dramatic effects on biological systems as a whole (Table 5.1). Disease often results when one or more of these cell behaviours is deregulated or becomes dysfunctional. The most notable of these is probably the way in which cell proliferation is deregulated in cancer.

Cells change their behaviour in response to stimuli from their external environment. There are many types of such stimuli. Many cells can sense changes to their mechanical environment by stretching their membranes and changing shape. They can also sense the levels of oxygen and nutrients in their environment, other chemicals present in that environment, and the number and types of neighbouring cells (Table 5.2). These stimuli are processed by the cells based on properties that are specific to their type.

However, the behaviour of cells is not just determined by their external environment, but also by internal variables that are intrinsic in the cells themselves (Table 5.3). These include the type of the cell, its age (measured in the number of cell divisions or mitoses), senescence, the numbers of molecules such as actin myofilaments, ion channels, and integrins inside the cell, and the genotype of the organism. Many, although not all, of these properties are not fixed, and can change over time.

From these descriptions, and from the properties listed in the tables, it is clear that modelling cell activity will require the development of individual models of specific cell behaviour based on both the cell type and the stimuli received by the cell. There is, however, another type of model. If we are interested in describing the processes inside a cell using continuous functions modelled by ordinary or partial differential equations, then the problem will require an additional level of abstraction. We will need to model the physical processes inside the cell as continuous functions of time, and sometimes also as continuous

Table 5.2 Cell stimuli

Cell stimuli
Biophysical stimuli: stretch, shear stress, hydrostatic pressure
Biochemical stimuli: growth factors, signalling molecules, toxins
Oxygen tension
Electrical charge
Nutrient concentration
Substrate composition and structure

Table 5.1 Cell behaviours

Cell behaviour
Proliferation
Migration
Apoptosis
Protein synthesis
Contraction
Differentiation

Table 5.3 Cell-state variables

Cell-state variable
Phenotype
Cell age: number of cell cycles, cell divisions
Senescence: ability to proliferate
Number of actin myofilaments
Concentration of ion channels and cell adhesion molecules
Cell size, shape, and orientation

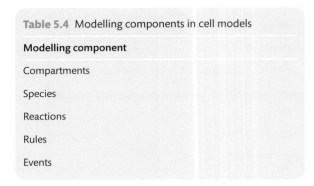

Table 5.4 Modelling components in cell models

Modelling component
Compartments
Species
Reactions
Rules
Events

functions of space. Software for this type of modelling traditionally uses five sets of components (Table 5.4).

The standard markup language, Systems Biology Markup Language (SBML), which is described in detail in the Appendix, uses the components listed in Table 5.4 with other additional information (i.e. kinetic parameters, initial values, etc.) as a structure on which models can be built. The merits and shortcomings of this approach will be described briefly in later sections of this chapter.

5.4.2 Simulating the Cell

The issue of how to simulate the cell is non-trivial, and a comprehensive description of all the ways in which cells can be modelled and cellular processes incorporated into mathematical models would take a whole book. We can only describe a few key examples here.

One key difficulty with producing realistic cell models is that models that are based on physics and physical chemistry may not be directly applicable to cell biology. This has little to do with the fact that cells are alive, but more simply because cell biology occurs at multiple scales of time, space, and complexity. Many of these issues that are associated with multi-scale modelling will be discussed in much more depth in Chapter 7.

It is very difficult to understand how molecular actions give rise to behaviours on the scale of the cell, because of the large numbers of different molecules, molecular interactions, and networks within the cell that are involved. Furthermore, cells contain organelles and biological systems formed from thousands of molecules, so they fall between the better understood microscopic and macroscopic levels.

Many parameters that are needed for developing cell models are not measurable by experiments and therefore must be estimated.

5.5 Modelling Approaches

We have seen that cell modelling is a difficult as well as a necessary component of computational biomedicine. It is particularly difficult to produce 'good' cell models that predict cell behaviour accurately and precisely, but this does not mean that it cannot be done. It is simply that, in order to produce models that are 'fit for purpose', it is necessary to understand a few things about the systems that are to be modelled. The solution to the choice of models for a given cell system is often a practical one.

In this section we describe two different approaches to modelling the cell that are used in state-of-the-art techniques. These focus on two different formalisms: firstly differential equation models and secondly agent-based or cellular-automaton-based approaches. We assume throughout this section that you have the mathematical background required to understand these formalisms.

Following these two detailed studies, we will briefly mention other approaches and then give some information about databases and repositories where these models are stored. The final part of this chapter explores some practical examples.

5.5.1 Continuous Models: Ordinary and Partial Differential Equations

Many biological processes can be represented mathematically by characterizing how one property depends on space, time, or both. This type of dependence is easily modelled using differential equations. If there is only one independent variable—that is, if the property changes in time but not in space—the system can be modelled using ordinary differential equations (ODEs). ODEs involve one independent variable and one or more ordinary derivatives of the dependent variable with respect to this independent variable. This type of equation can be described as deterministic, which means that no randomness is introduced into its development. It will therefore always produce the same output from a given initial condition.

An ODE of order n is an equation that can be described by the following general equation. In this representation, the independent variable is assumed to be time, t.

$$F(t,x,x^{(1)},x^{(2)},\ldots x^{(n)}) = 0$$

In this equation, $x^{(k)}$ designates the kth derivative of x with respect to t; the highest derivative is the nth. In this formulation, once an ODE has been defined to describe each biological process that is to be modelled, the task of solving the equations will be left to a numerical solver. Analytical solutions are used merely as checks that the model solutions are correct (the *verification* process) in some limiting cases. We will present some concrete examples of this later in this chapter.

ODEs are by far the most widespread and widely used mathematical method in cellular models for systems biology and computational biomedicine. They do have limitations, however. A clear example of this is that they are by definition able to describe changes in one variable; an ODE or set of ODEs that are set up to describe changes in time will not be able to describe changes in space as well. This is a fundamental problem in computational physiology as a whole, as properties, physical quantities, and biochemical variables are often dependent on space. It is, however, not necessarily too stringent a limitation in the specific case of modelling a single cell.

If you need to use differential equations to describe variables that change simultaneously in two dimensions at once—in space as well as time—it can be done, but it will be necessary to use partial differential equations (PDEs). Some simple one-dimensional systems, such as permeation, diffusion, and flow through capillaries, are best described using PDEs. Like ODEs, these may be stochastic or continuous.

Most software that uses differential equations (both ODEs and PDEs) to model cells represents the cellular components listed in Table 5.4 using simple mechanical principles such as mass conservation. This is a relatively intuitive approach, although it does lead to some inconsistencies [8]. A brief description of each of the processes in Table 5.4, given in mathematical terms, is outlined below.

- *Compartments.* Biological models typically define compartments as locations where reacting entities (species) may be located. If a cell (or, less logically, another physiological entity such as an organ) is considered as a well-stirred reservoir, the volume of the compartment will need to be defined by the modeller. The volumes of different cells will have an impact on the behaviour of the model.

- *Species.* The term species is used to refer to any chemical entities that are present in the model and that may interact or react. Ions, small molecules and macromolecules such as nucleic acids, polysaccharides and proteins are all classed as species.

- *Reactions.* A reaction is any cellular process that is described by the differential equations that represent the dynamics of the model. In a simple example:

$$reaction\ 1 = -k_1 \cdot A$$
$$reaction\ 2 = -k_2 \cdot A$$

implies the following ODE:

$$\frac{dA}{dt} = -k_1 \cdot A + k_2 \cdot A$$

This category is quite large and includes many different biochemical processes within the cell: bidirectional reactions, state transitions, degradation, transport, translation, catalysis, inhibition, and others.

- *Rules.* Rules are used to describe algebraic relations between variables; for example:

$$A + B = C$$

You should note that this concept is different from the concept of rules that is used in agent-based modelling, which is described in the next section.

- *Events.* This is basically a mathematical or logical way of representing conditions, specifically if/then conditions. For example, in pseudocode:

```
if [the value of variable] A is greater
than [condition]
then
[the value of variable] B is [value]
```

Box 5.2 illustrates a case study of a simple compartmental model of a cell using ODEs.

Box 5.2 Compartmental Models of the Cell Using ODEs

Let us develop a simple example of a reaction between two chemical species. Figure 5.5 shows a generic volume such as, for example, the interior of a cell with a volume of V mL. There is a constant input into this cell of a substance A at a rate of A_0 mol/litre s^{-1}. Therefore, if A did not react with any other substances in the cell, the concentration of A would rise by A_0 mol/litre per s. However, once A is inside the cell it is converted into species B at the rate k_{AB} per second. Both A and B are degraded or removed irreversibly (which can be kinetically considered as a *clearance* from the cell) at rates *ClearA* and *ClearB*. Therefore, the concentrations of A and B in the cell in mol/litre can be defined by the following equations:

$$\frac{dA}{dt} = A_0 - k_{AB} \cdot A(t) - ClearA \cdot A(t)$$

$$\frac{dB}{dt} = k_{AB} \cdot A(t) - ClearB \cdot B(t)$$

This shows that the influx of A is balanced by its consumption by two processes. One is conversion to B with rate constant,

k_{AB}, which defines the fraction of A that is converted to B in each second. The other is clearance from the system where *ClearA* is the fraction of A that is cleared from the cell per second. This part of the system comes to a balance when dA/dt becomes zero as long as A_0 remains constant, reaching a steady state of concentration A_{ss} in which

$$A_0 = k_{AB} \cdot A(t) + ClearA \cdot A(t)$$

and

$$A_{ss} = \frac{A_0}{(k_{AB} + ClearA)}$$

Likewise, when A reaches steady state, so does B(t) as dB/dt goes to zero and

$$k_{AB} \cdot A(t) = ClearB \cdot B(t)$$

so that B(t) takes the constant (steady state) value of B_{ss}:

$$B_{ss} = A_{ss} \cdot \frac{k_{AB}}{ClearB}$$

Equations can also be used to define an event such as the end of the input of A. In this case:

If 'time' < 100 s then $A_0 = 1$ mol/s

If 'time' ≥ 100 sec then $A_0 = 0$ mol/s

Thus after $t = 100$ seconds the equations above all lose the A_0 term and become:

$$\frac{dA}{dt} = -k_{AB} \cdot A(t) - ClearA \cdot A(t)$$

$$\frac{dB}{dt} = k_{AB} \cdot A(t) - ClearB \cdot B(t)$$

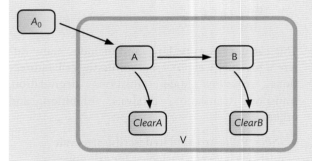

Figure 5.5 Simple model for a chemical reaction between two chemical species A and B within a cell, constructed using Jdesigner.

If in this situation the clearance coefficients *ClearA* and *ClearB* are both set to zero, then A will be converted wholly to B.

5.5.2 Discrete Models: Cellular and Agent-Based Cell Modelling

Many different methods have been used to simulate biological systems at the level of the cell, and these can be grouped together broadly by how they represent physical space. The cellular automaton (CA) method is a common one that uses a regular, orthogonal lattice and defines values of state variables at points on this lattice. These state variables can represent the presence of a single cell, or they

can represent properties of that cell, such as its age or type. The automaton updates the positions of all state variables within the model simultaneously according to rules that are derived from the environment outside the cell.

This is quite closely related to another type of lattice-based model, the cellular Potts model, which was originally developed to model annealing phenomena in metals [9]. In this case, cells each occupy several lattice points and the model involves the minimization of an energy function. This is much more fine-grained than a CA model in that off-lattice points are also considered, and this clearly carries an associated cost in terms of computational power and time.

These model types are grouped together as agent-based models, and all agent-based models have some common characteristics that distinguish them from continuum-based approaches [10]. These are summarized below.

- *Spatial representation*: in traditional engineering approaches space is described as a continuum and a variable (such as cell concentration) is assumed to be defined at every point in space. This makes it difficult, if not impossible, to represent cell morphology or the structure of cell aggregates, for example. The CA approach introduces a representation of discrete cells in which only one cell can occupy a lattice site; there is therefore a maximum concentration of cells in any region. In the similar cellular Potts method the morphology of each cell is represented by a collection of lattice points, which allows fluctuations in cell size and shape to be modelled. Off-lattice methods have used spheroidal cell shapes in which the cell centre and radius defines the morphology. This allows more interaction behaviours such as adhesion and physical growth stress to be included in the model.
- *States*: all agent-based models use state variables. These can be discrete, as with strict CA; alternatively, they can be defined using a continuous function based on mathematical rules. One example is the use of discrete cell states to represent the cell phenotype in a model of skeletal development [11]. In this process, stem cells of connective tissue commit to differentiate into diverse broad types of cell or lineages, with no dedifferentiation (return to stem cell phenotype) or transdifferentiation (change to another lineage). In other cases, for example for smooth muscle cells, the cell phenotype can be described as a continuum from contractile cells to cells that are specialized for synthesizing tissue.
- *Dynamics*: in strict CA, each state variable is updated during each increment. Another method of simulation that may be applied is the

Monte Carlo method in which only one lattice point, which is chosen at random, is updated in each increment. There are also off-lattice models in which each cell state is updated either sequentially or all at once.
- *Rules and behavioural models*: these rules and models dictate how the simulation is updated. In all agent-based models, these rules are local, that is, they are based on stimuli local to the cells. They may either be the same at each point in the simulation (i.e. homogeneous) or differ throughout the model, as in some models of early tumour growth. These rules may be implemented either using simple Boolean statements or as energy functions tending towards equilibrium values, as in Monte Carlo simulations. One advantage of this type of modelling is that it is possible to simulate the behaviour of different cells using different algorithms that represent, for example, cell migration and cell proliferation. This makes it easy to alter the model for different applications (for example, different cell types) or if a more complex mathematical model of a specific cell behaviour becomes available.

5.5.3 Case Study of an Agent Model of Skeletal Mechanics

Cell-centred modelling is an extremely versatile technique that has been used to study a wide variety of biological systems. In this example, we show how models of many individual cells can be used to simulate the behaviour of a tissue, or even of a whole organ.

The physiological process of skeletal tissue differentiation is important in many clinical applications. These include fracture healing, the behaviour of tissue around orthopaedic implants (in particular when they fail), and in the new discipline of tissue engineering. In all these cases, the tissue involved must adapt to a situation in which it will come under considerable mechanical load. The first step in skeletal repair is generally the production of granulation tissue. This consists of a type of connective tissue stem cells known as mesenchymal stem cells (MSCs) that can differentiate into bone, cartilage, or fibrous tissue depending on the nature of the mechanical stimuli that the tissue is exposed to. The currently accepted model for this process involves defining the biophysical stimulus on the tissue (which is assumed to be both porous and elastic) as a combination of octahedral shear stress and fluid flow. Lower levels of stimuli induce the formation of bone, medium levels induce the formation of cartilage, and tissue under very high stimuli levels will differentiate into fibrous tissue. This theory can explain how the amount of movement in fractured tissue or around an orthopaedic implant may prevent rapid healing.

This theory, which is known as the mechanoregulation theory, has been applied to more complex applications, and this has led to the explicit modelling of individual cells. During fracture healing or tissue growth around an implant, mesenchymal stem cells from the bone marrow begin to proliferate and infiltrate the injured area. These cells then differentiate into different types of bone and cartilage cells and fibroblasts depending on their environment. The supply of oxygen and nutrients and chemical signals affect the differentiation process, as well as mechanical stress. The formation of different tissue types alters the mechanical properties of the environment for incoming stem cells, forming a feedback loop. If we are to capture the complexity of this system in a model, we must first be able to simulate the behaviour of the cells that influence this process; that is, the MSCs. Producing an accurate model of how these cells respond to mechanical and other stimuli will allow us to introduce many interacting cells into the model and, therefore, to simulate tissue differentiation on a larger scale.

In this case study we focus on an application of this method in simulating the healing of a fractured human tibia (the larger and stronger of the two bones between the ankle and the knee in all vertebrates) [12]. This study aimed to find out the effect of mechanical stimuli, as regulated externally, on the type of tissue that is formed in the gap created by the fracture, and therefore to determine the ideal conditions of rest and/or exercise for fracture healing.

The researchers in this study generated a three-dimensional computer model of the geometry of a human tibia, including a fracture site, a callus containing the simulated differentiating cells, and an external fixator (Figure 5.6). All tissues except cortical bone (see Chapter 4) were modelled as biphasic materials (in two phases) with porous and elastic qualities. This enabled interstitial fluid flow and octahedral shear strain to be calculated; both these stimuli are

known to influence the differentiation of skeletal tissue. A regular orthogonal lattice was superimposed over the finite element geometry, representing the domain of cell activity, with each lattice point representing a possible position for a single cell. This lattice was initialized with populations of MSCs added to the membranes that line the outer and inner surfaces of the bones (which are known as the periosteum and the endosteum respectively) and to the bone marrow cavity. The migration and proliferation of the stem cells during the simulation were defined as random-walk processes using rule-based algorithms. At each time point the lattice was checked for active cells, and each cell either migrated to or produced a new cell at a vacant neighbouring lattice site (see Figure 5.7).

The process was modelled as a stochastic (i.e. random) one, with the probability of each cell either migrating or dividing at any time point based on the defined migration and proliferation rates of that cell. Differentiation into different cell types occurred at a finite, measurable rate in the presence of and influenced by biophysical processes. The differentiated cells produced different tissue types and the effect of these new tissues on the environment was modelled by changing its mechanical properties. In particular, the properties of the tissue in each part of the model were defined from the ratio of cell types there. It was assumed that cells that were consistently outside their favoured environment (e.g. bone cells trapped in an environment that favours cartilage development) would die through the process of apoptosis (that is, programmed cell death).

This simulation was able to predict some key aspects of the process of tissue evolution and fracture healing *in vivo*, and, in particular, how the different tissue types developed over time (Figure 5.8). Initially, the granulation tissue (composed of the stem cells) produced cartilage and fibrous tissue in the fracture gap. If the mechanical stimulus in that fracture gap is sufficiently low, an external callus made of bone can form, and this is able to bear more and more load. This shielded the tissue between the bone segments, which reduced the stimulus further, promoting further bone growth. The final step in the simulation showed the return of the bone marrow cavity in the healed tibia as the callus is resorbed from interior bone segments that are under-loaded.

When this model was compared to experimental data obtained from macroscopic measurements

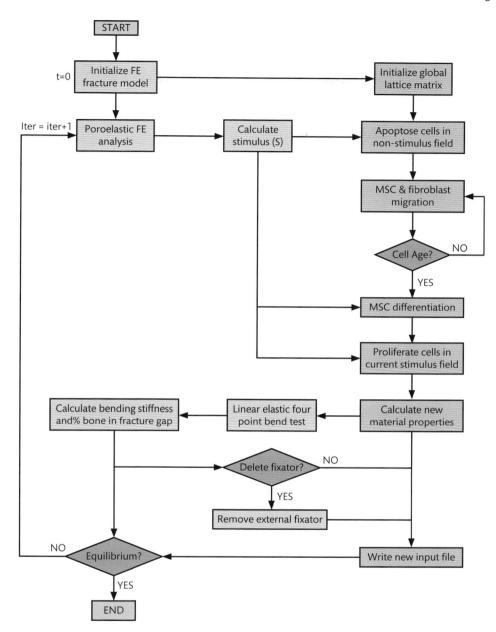

Figure 5.6 Flow chart of the simulation technique to predict tissue differentiation within a fracture callus. Biophysical stimuli from the tissue level regulate the differentiation of MSCs. The granulation tissue evolves into fibrous connective tissue, bone, or cartilage, depending on the biophysical stimuli of fluid flow and shear strain. The phenotype of cells within the callus dictate the tissue produced, and alter the mechanical properties. As the tissues evolve in the callus, the mechanical stimuli are changed. FE, finite element. Taken from [12].

of bone stiffness, it showed a qualitative agreement. It was able to predict how bone stiffness changed over time during the healing process, and how this was affected by fixation and the level of mechanical stimulus. The model can also be useful in exploring the role of the external fixator that holds the bone in

place while it heals, allowing the duration and degree of fixation to be optimized.

Very similar simulation techniques have also been applied to other situations in bone mechanics and remodelling, including simulations of the failure of orthopaedic implants. These examples show

```
FOR (each lattice point) DO
  IF (Lattice point is occupied by a migrating cell) THEN
    WHILE (Jumps realized by the cell < Number of desired jumps) AND (All cell positions surrounding the cell are not already occupied)
      Create list of possible directions: [left, right, up, down....]
      WHILE (Free position has not been found) AND (List of possible directions is not empty)
        Choose randomly a direction of migration from the list
        IF (new position = free) THEN
          Move cell to new position
          Increase by 1 the number of jumps realized by the cell
          Free position = Found
        ELSE
          Remove from list of possible directions the direction checked
          Increase by 1 the number of positions surrounding the cell already occupied
        END IF
      END WHILE
    END WHILE
  END IF
END FOR
```

Figure 5.7 An algorithm for the random-walk migration of cells in a regular lattice. Adapted from Checa and Prendergast (2009). [13]

Medial side, tibia included (in grey)   Lateral side, tibia excluded   Transverse Section

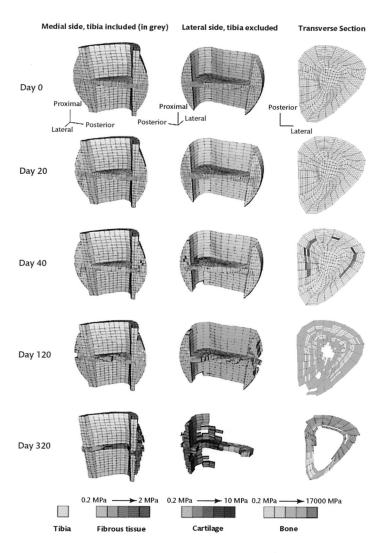

Day 0

Proximal
Lateral — Posterior
Lateral

Proximal
Posterior — Lateral

Posterior
Lateral

Day 20

Day 40

Day 120

Day 320

0.2 MPa ⟶ 2 MPa   0.2 MPa ⟶ 10 MPa   0.2 MPa ⟶ 17000 MPa

Tibia   Fibrous tissue   Cartilage   Bone

Figure 5.8 Cross-sectional views of a modelled fracture callus mesh over simulated time. The initial fibrous tissue gives way to cartilage in the fracture gap, while a bony bridge forms on the exterior of the callus. This reduces the stimulus sufficiently to promote more bone growth, and healing. The under-loaded bone elements resorb the deleted elements, restoring the bone marrow cavity to a physiological state. Image courtesy of Dr Julian Gunn, Senior Lecturer and Honorary Consultant Cardiologist, Department of Cardiovascular Science, University of Sheffield, UK.

how agent-based models can be powerful tools in exploring complex biological systems that have the behaviour of individual cells within cell populations as a key component. It is possible to use agent-based methods to produce complex tissue and organ behaviour from the simple algorithms that represent cell behaviour and that are formulated from information obtained both *in vivo* and *ex vivo*. Reducing the complexity of a model system to simple, repeatable algorithms based on the behaviour of single cells is one clear advantage of this approach.

These models often use algorithms that can be described as phenomenological; that is, they are based on empirical observations. The model described above, for example, did not explicitly model the mechanism by which differentiation occurs within the cell. These are often at least equal to the task of modelling a specific and clinically relevant process, as this example has shown. However, another benefit of agent-based models is that it is possible to slot new components into a defined modelling framework, so algorithms can be adapted, improved, and exchanged in a modular way.

5.6 Simulation Tools

5.6.1 Software

In the last section, we described two of the most commonly used methods for modelling cellular processes: differential equations and agent-based models. Different ranges of software packages and tools are used for these two methods.

When ODEs are used to model cell processes, the independent variable is normally time. This type of differential equation can normally be solved using conventional mathematical software, and a wide range of appropriate packages are available for this. These include:

- commercial software; for example Matlab, Mathematica, Mathcad, and Maplesim;
- open-source (free) software; for example R, Octave, Scilab, and JSim.

All these are generic packages that can be applied to many different types of scientific problem, and they work in similar ways. Researchers use the software to generate equations to fit the scenarios they are simulating and then solve them numerically to generate the simulation results. If the equations used are simple ones it may be possible to use some of these packages (such as the commercial ones named above, Scilab, and Simpy) to obtain analytical solutions to the equations.

Researchers engaged in systems biology and physiological modelling have already put a great deal of effort into generating computational tools that are more specifically adapted to the needs of their own modelling communities. Useful tools developed inside the community include the two main Extensible Markup Language (XML)-based markup languages for systems biology and cell-level modelling: SBML and CellML. These are discussed in more detail in the Appendix. These packages, which are essentially standardized languages for representing mathematical models, were developed to create a neutral format that modellers could use to exchange models and data. This had to be flexible enough to allow a wide range of data types to be stored in

sufficient detail and to be modified and used in a variety of ways. The markup language format was also designed to support meta-information, allowing the data and models stored to be fully comprehensible without requiring additional information. XML is a platform-independent, text-based markup language that has often been applied to this type of problem, and CellML and SBML are only two of a large number of specialized markup languages that are based on it. Others include MathML, which is widely used within languages such as CellML to specify mathematical concepts. All these markup languages take advantage of the general format of XML to construct representations of their specific domains of knowledge and to provide a generally accepted format for the exchange of data and models.

Markup languages are gaining wide acceptance in the biological modelling community. Well over 200 software packages that support SBML models are now available, and this number is likely to increase in the next few years (see trends illustrated in Figure 5.9). More information about modelling standards is also available in the Appendix.

5.6.2 Tools for Cell Modelling

We have heard that a wide range of specific modelling tools have been written for computational biomedicine, and that these are mainly based on the standards that have been set out in the markup languages SBML and CellML. We set out and briefly describe some of these in this section. They are all powerful enough for experienced modellers and user-friendly enough for beginners in the discipline.

Simulation Software

Table 5.5 summarizes some of the most widely used open-source software packages for biomedical

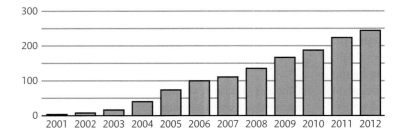

Figure 5.9 Total number of known SBML-compatible software packages from 2001 to 2012.
Source: http://sbml.org.

Table 5.5 Common open-source software packages for biomedical simulation

Software	Description
COPASI http://www.copasi.org/	A stand-alone program for the simulation and analysis of biochemical networks and their dynamics. It supports SBML models and can simulate their behaviour using ODEs or a stochastic simulation algorithm.
CellDesigner http://www.celldesigner.org/	An editor for drawing structured diagrams of gene-regulatory and biochemical networks. The networks are drawn based on the process diagram using a standard graphical notation system, and are stored using SBML.
JSim http://www.physiome.org	A versatile, Java-based simulation system for building quantitative models and testing their validity using experimental data. JSim's primary focus is in physiology and biomedicine. JSim models may intermix ODEs, PDEs, implicit equations, integrals, summations, discrete events, and procedural code as appropriate. Its model compiler can automatically check that the units in the equations balance, and the user may allow automatic unit conversion factors for compatible physical units (e.g. 60 s to 1 min) as well as detecting and rejecting unit unbalanced equations. Equations are stored using an XML-compatible variant of its equation language MM, and it also supports the other XML variant archival model formats SBML and CellML for import and export. It can use the commercial mathematical package Matlab as an engine, which is particularly useful for handling matrices. In summary, JSim is a powerful general package for the public use of reproducible modelling in data analysis that is particularly appropriate for biomedical applications.
JDesigner http://jdesigner.sf.net/	A stand-alone application with a graphical interface for drawing biochemical networks and generating SBML models. The designer has an interface to the Systems Biology Workbench (SBW) that allows it to be called from other SBW-compliant modules, for example Python. In addition, JDesigner is able to use Jarnac as a simulation server via SBW, which allows models to be run from within JDesigner; thus, it can be used as both a network design tool and a simulator.
NEURON http://www.neuron.yale.edu/	A simulation environment for modelling individual neurons and networks of neurons. NEURON provides tools for conveniently building, managing, and using models that are convenient, numerically sound, and computationally efficient. It is particularly well suited to problems that are closely linked to experimental data and to those that involve cells with complex anatomical and biophysical properties. GENESIS (http://genesis-sim.org) is an alternative platform for developing neural models and analysing data.

simulation. Most if not all of these are freely available to be downloaded from the web.

Pathway Databases

Databases that represent the data, for example in gene-regulation networks or metabolic pathways, are very useful as modelling tools. Computable representations of pathways can, for example, provide a basis for an analysis of network patterns, enable comparisons between different types of pathways and networks and correlations between networks and mRNA or protein abundances, combine data between pathways and to query by network connectivity, and link pathways into simulations. They can

therefore be very usefully built into and combined with cell-level models.

Some examples of network and pathway databases are shown in Table 5.6.

Databases of Biochemical Kinetic Data

Table 5.7 describes some databases of kinetic and biochemical information that can usefully be incorporated into cell-based models.

Model Databases

Cell-based (and, indeed, molecular- and system-based) models that have been developed by one group are often stored in databases for other modellers to

Table 5.6 Examples of network and pathway databases

Database	Description
KEGG (Kyoto Encyclopedia of Genes and Genomes) http://www.genome.jp/kegg/	A database resource for understanding high-level functions of a biological system such as a cell, organism, or ecosystem, using genomic and molecular-level information. It is a computer representation of the biological system, consisting of molecular building blocks of genes and proteins (genomic information) and chemical substances (chemical information) that are integrated into molecular wiring diagrams of interaction, reaction, and relation networks (systems information).
Reactome http://www.reactome.org/	An open-source, open-access, manually curated, and peer-reviewed pathway database. Pathway annotations are authored by expert biologists in collaboration with the Reactome editorial staff. They are cross-referenced to many bioinformatics databases including the NCBI Entrez Gene, Ensembl, and UniProt databases, the UCSC and HapMap Genome Browsers, the KEGG Compound and ChEBI small-molecule databases, PubMed, and Gene Ontology.
NetPath http://www.netpath.org/	A curated resource of signal transduction pathways in humans. It has been developed in collaboration between the PandeyLab at Johns Hopkins University and the European Bioinformatics Institute (EBI).

access and share. Some examples of these databases are given in Table 5.8.

5.7 Reproducible Cell Modelling

We have come back to the original question that we asked at the beginning of the chapter: how best can we model a cell? And, further to this question, is it possible for such a model to be reproducible?

We will explore these concepts further in the case study of hepatic clearance that is presented in Box 5.3, but before that we will highlight some fundamental issues about models to show the complexity of the problem. During this discussion you should remember, however, that no model is ever considered wholly 'final': all models are better thought of as stepping stones on the road to future models that will be more accurate, precise, and/or versatile.

Table 5.7 Databases of kinetic and biochemical information

Database	Description
BRENDA http://www. brenda-enzymes.info/	A web-based database and resource of molecular and biochemical information on enzymes as classified by the International Union of Biochemistry and Molecular Biology. Every enzyme is characterized with respect to the biochemical reaction that it catalyses. Its kinetic properties, substrates and products are described in detail. BRENDA provides a web-based user interface that allows a convenient and sophisticated access to the data. BRENDA was founded in 1987 at the former German National Research Centre for Biotechnology (now Helmholtz Centre for Infection Research) in Braunschweig, Germany, and the data was originally published as a series of books.
SABIO-RK (System for the Analysis of Biochemical Pathways - Reaction Kinetics) http://sabio.villa-bosch.de/	A web-based application based on the SABIO relational database that contains information about biochemical reactions, their kinetic equations with parameters, and the experimental conditions under which these parameters were measured. It aims to support modellers in setting up models of biochemical networks, but it is also useful for experimentalists or researchers with an interest in biochemical reactions and their kinetics. Information about reactions and their kinetics can be exported in the SBML format.

Table 5.8 Examples of model databases

Database	Description
JWS http://jjj.biochem.sun.ac.za/	JWS Online was one of the first resources to offer both curation of the models it distributes and online simulation. It is linked to a number of peer-reviewed journals including *Microbiology*, *FEBS Journal*, and *IEE Proceedings in Systems Biology*; paper authors deposit their models on submission of their manuscripts to make them available to reviewers. Models are now distributed in SBML and Pysces formats.
BioModels http://www.ebi.ac.uk/biomodels/	A repository of published, peer-reviewed computational models. All models stored are of biological interest, and they are primarily from the field of systems biology. This resource allows biologists to store, search, and retrieve published mathematical models. In addition, models in the database can be used to generate submodels, can be simulated online, and can be converted between different representational formats. This resource also features programmatic access via web services.
CellML Repository http://www.cellml.org/	This repository distributes curated models of biochemical and cellular systems, and it is possible to translate resources written in different languages and simulators to its main markup language, CellML.
Physiome Model Repository http://www.physiome.org/	This is a repository of physiological systems and transport models, most of which can be run over the web. It aids investigators by providing them with tutorials that guide them through a sequence of models towards advanced research models of physiological systems. Many of the models involve PDEs for modelling a wide range of physiological phenomena.

These principles are discussed here in relation to modelling cells, but many of them are also applicable to modelling other physiological systems.

5.7.1 Discussion: Most Published Models are Irreproducible

As a scientist, you should realize—very probably you will already have realized—that results published in the peer-reviewed literature are often irreproducible or even untrue. In particular, published models and simulations are often not reproducible. As recently reported [14]: 'There are a few hundred models in the CellML repository; I think there are maybe five or six of these that didn't have errors in the original publication'.

There are many factors behind this, including:

- errors in data,
- errors in software,
- errors in experimental and analytical methods,
- typographical errors,
- incompleteness.

Furthermore, even published models that are 'true' may not be useful for other researchers, for reasons including the following:

- the models are not publicly available,
- the data are not publicly available,
- there is insufficient documentation.

One particular problem is that the parameters used in models may be transferred from one model to another without further testing, even when the original parameters came from a different species from the one to be simulated. Parameters may also 'mutate' between models, as one model is adapted to become another, and then a third, from year to year.

This is clearly an unsatisfactory situation. Collaborative efforts in physiological modelling, such as those in the various Physiome Projects discussed in the course of this book, rely on the reproducibility of methodologies, models, data, and its analysis. Data are perhaps the most important of these. All models can be seen as 'snapshots', as they capture the modellers' evolving perception of the system modelled, but

Box 5.3 Case Study: A Reproducible 'Validated' Model of Hepatic Clearance Using ODEs

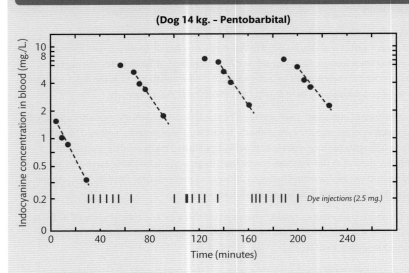

Figure 5.10 ICG injection and clearance. Blood concentrations in a 14 kg dog with 22 successive intravenous injections of 2.5 mg ICG (vertical pips). Dashed lines represent estimated single exponential decay after each series of injections. Data are from [15].

Indocyanine green (ICG) is a dye that is used for the estimation of blood flow. It has a narrow absorption peak at 800 nm, a spectral location at which oxygenated and deoxygenated haemoglobin have the same absorbance so the measurement of dye concentration is independent of blood oxygen saturation. It is bound to albumin and therefore does not escape from the bloodstream, except in the liver where it is taken up by selective transporters into hepatocytes and excreted into the bile. Some early experiments to test the reliability of estimates of cardiac output involved making a long series of injections of ICG into the circulation, taking blood samples periodically to measure its concentration, and using this data to calibrate the densitometers with indicator dilution curves (Figure 5.10).

In this case study, a model of the whole-body blood circulation was constructed, comprising a central blood volume V_1, from which flows the whole cardiac output, CO (Figure 5.11). Although modelling the blood transport between organs is known to require PDEs, this system involves the whole circulating blood and so it is possible to use a simpler compartmental representation (leading to ODEs). The cardiac output is distributed into three 'organs' of volume V_1 to V_3, that can be labelled 'kidney' (representing the kidney and myocardium), 'liver' (representing the liver and intestines), and 'muscle' for the rest of the body. The intravenous injections are essentially made into the central heart and lung compartment, but the observed concentrations result from recirculation of the injected amounts and clearance by the liver.

Each injection, C_{inj}, is defined when the model is run using a separate function generator. Clearance of the injected ICG

dye is really hepatic extraction via a saturable transporter followed by ATP-dependent excretion into the bile, but it is represented here by a passive first-order loss, $G3$. This is kinetically adequate only at low concentrations of ICG, when the transporter is mainly uncomplexed. This is the first stage of a model that will be composed of the circulation and a liver cell with a selective transporter and an ATP-powered excretory transporter into the bile.

The ODEs for the change of concentration over time in each compartment are given by:

$$\frac{dC_1}{dt} = \frac{(F_1(C_{recirc} - C_1) + Q_{injrate})}{V_1}$$

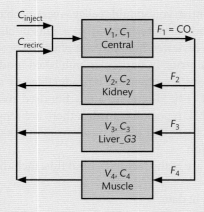

Figure 5.11 Compartmental circulatory model for the hepatic clearance of the dye ICG.

... continued

$$\frac{dC_2}{dt} = \frac{F_2(C_1 - C_2)}{V_2}$$

$$\frac{dC_3}{dt} = \frac{F_3(C_1 - C_3)}{V_4} - \frac{G_3 \cdot C_3}{V_3}$$

$$\frac{dC_4}{dt} = \frac{F_4(C_1 - C_4)}{V_4}$$

Here, C_{recirc} is the concentration recirculating to the vena cava flow entering the right atrium at the injection site, $Q_{injrate}$ (measured in mg/min) is the injection rate as a function of time, and is therefore finite during the injection itself and zero otherwise, and $G3$ (measured in mL/min) is a first-order approximation to the clearance rate from the hepatic sinusoids into the bile. The cardiac output was modelled to be 1.5 litres/min throughout the 4 h although variation was seen in the estimates made following each injection. Each V is the volume of a compartment and each C the concentration of dye in that compartment.

The extreme simplicity—over-simplicity, perhaps—of this circulation model has *not* compromised the results significantly, if at all. The reason for this is that the hepatic clearance is slow relative to the time required for the dye to become well mixed into the entire blood volume. The mean transit time through the circulation is V_{tot}/CO = 1.25 litres/(1.5 litres/min) = 0.83 min, and the clearance time constant is V_{tot}/G3 = 1.25 litres/(0.064 litres/min) = 19.6 min, which is more than 20 times slower. If the total circulation could be considered as a stirred tank and if the clearance is a first-order process, then the ICG removal should be described by a single exponential with rate constant $G3/V_{tot}$ (the red dashed curve in Figure 5.12). This approximation appears to fit both the basic model shown here and the experimental results fairly well.

An optimization of V_{tot} and $G3$ with the CO fixed at the average value of 1.5 litres/min, obtained from the 20 indicator dilution curves, gave the following results: V_{tot} = 1.254 ± 0.079 litres/min and $G3$ = 63.6 ± 1.3 ml/min. Keeping the number of free parameters low, or constrained by data, helps to narrow the confidence limits. Lowering CO to 1.35 litres/min gave V_{tot} = 1.256 ± 0.079 litres/min and $G3$ = 64.6 ± 1.35 litres/min, while raising it to 1.65 litres/min gave V_{tot} = 1.252 ± 0.079 litres/min and $G3$ = 62.8 ± 1.27 litres/min.

In contrast, when CO was unconstrained the results were V_{tot} = 1.24 ± 24 litres/min, $G3$ = 59.2 ± 7000 mL/min and CO = 3 ± 5000 litres/min; the reason for these enormous errors is that CO and V_{tot} are inversely correlated with a correlation coefficient of −1. The lesson that can be learned from this example is that it is necessary to constrain some parameters in order to evaluate the desired unknown. Data from anatomy, physical conditions, chemical thermodynamics, or direct observation should be used as constraints as far as possible.

Of course this model, however useful, is grossly incomplete. It is not mechanistic and allows for neither the saturation of the hepatic uptake process nor the biliary extrusion process. ICG clearance was developed as a liver function test and is still in common use today [16]. The next version of this model is expected to include many more mechanistic details and to be evaluated against clinical data. It is very common for complex models to be developed from simple ones, by adapting or even combining them; for example, see the Compartmental Tutorial at **http://www.physiome.org/Models/tutorial.html**. The fact that this model and all the modelling processes are reproducible, including the setup for the optimization and the confidence limit calculation, makes building complex models from this simple one an easier, and even a more valid, process.

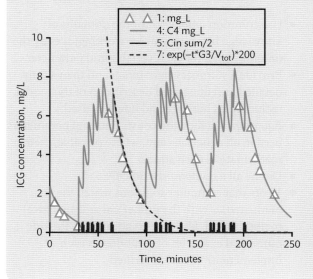

Figure 5.12 Model fit to ICG concentrations in a dog. The spikes on the x axis are the times of 2.5 mg injections. Triangles are concentrations of dye in the blood. The model solution shows that each injection gives rise to a sharp peak followed by decay; the RMS error was 0.31 mg/litre.

good data can be seen as permanent, and serve to provoke deeper explanations and more global inferences. In summary:

- models (whether descriptive, theoretical, or mechanistic) are designed to represent data or ideas about real systems;
- data and models should form complementary matched sets;
- good data should be able to be used as a permanent reference source.

It can be both tedious and difficult to make all the elements involved in a modelling project reproducible—data, models, modelling processes, and data analyses—but it is a worthwhile task and one that all scientists working in this field should strive towards. Reproducible models are building blocks and milestones of science that evolve through iterative refinements: the more data becomes available, the more comprehensive the resulting models can become, integrating information and reconciling contradictions.

In this context, it is useful to re-state the value of simplicity in modelling. Simple models help to define basic concepts. Einstein's admonition, 'Keep the model simple, but not too simple', is a good starting point, especially in models that are intended to be used even sometimes for teaching (see **http://www.physiome.org/Models/tutorial.html**). The more complex a model is, the more detailed the biology that can be described; however, more complex models will demand more computational resources and put higher demands on verification and validation measures. Integrative models define mechanisms and fit more datasets. They infer a higher level of validity and can be applied to a greater range of biological problems. Furthermore, the more complex the model, the more vital it is that the data used to generate it is kept with it and made equally accessible.

The more reproducible a model is, the more useful it becomes as an aid to experimental design, and the easier it becomes to go round the feedback loop, validate it, and improve it further. Reconciling disparities between experiment and model and broadening the factual basis improves the predictive capability of a model. A thoughtful phrase used

by T.H. Huxley in lecturing on Darwin's theory of evolution, 'active scepticism', can usefully be applied to the relationship between a modeller and his or her—and other researchers'—models: 'an *active scepticism* is that which unceasingly strives to overcome itself and by well directed research to attain to a kind of conditional certainty' (**http://aleph0.clarku.edu/huxley/guide2.html**).

The word active here can be assumed to mean 'well-directed research', inspiring researchers to actively seek to gain knowledge by testing and retesting hypotheses rather than waiting for others to supply answers. Testing models to their limits with the broadest range of input conditions and predictions will render the models more susceptible to disproof or invalidation. This, however, should not be thought of as a failure of the model but as an advancement of science. A reproducible model provides the stepping stone to the next model in an iterative process in which the validation experiment produces data that can be fed into the model to improve it.

5.7.2 Technical Requirements for Reproducibility

The processes that researchers should use to make sure that models of cells are, as far as possible, reproducible are basically technical ones. A minimal set for a model that is meant to be widely used can be assumed to include the following.

- Use common *ontologies* for the domains of knowledge in which you are working; specify the ontology used for each variable and parameter.
- Provide a summary *description* and then a fully detailed and illustrated description of the model, including its heritage and how it is used.
- *Equations*: include the constraints and parameters used in all equations and, where possible, their mathematical derivations.
- Define all *assumptions* explicitly.
- *Verify* that computer solutions match the mathematics with the numerical and analytical methods used. Use conservation tests. Provide numerical examples of solutions so each implementation can be verified.

- *Validate* the model by testing against suites of experimental data. *Fit* the models to the data directly, not just by shape analogy. Show data analyses for *multiple datasets*.
- *Assess* the model using (for example) residuals, covariance matrices, or Monte Carlo parameter distributions (see Chapter 3).
- *Interpret*: model, parameters, confidence limits, and the model's contributions.
- *Publish* all models in full in the peer-reviewed literature.
- *Archive everything*: data, models, solutions and verifications, validation suites, analyses, and their interpretations.
- Produce an *operation manual* to guide users.

5.8 Conclusions

We have seen in this chapter that the cell, situated as it is in the middle of the complexity scale that makes up the human 'system', between molecules and organs, is at the centre of the multi-scale modelling of human (or indeed of any organism's) physiology. Modelling the cell, however, has been seen to be multi-scale in itself. Cells have a wide range of functions including the regulation of gene transcription and communication within and between cells. Methods for modelling the biochemistry and physiology of cells are continuing to evolve and develop and are gaining in complexity, accuracy, and ease of use. However, they still need to be improved further, in particular to handle the vast and still increasing numbers of known interacting molecular and subcellular components that they are now known to contain.

A large number of research groups have generated simple and complex cell models in recent years, most of which either take a continuous approach and using ODEs or PDEs, or use agent-based methods. Many of these are written using widely used, standardized languages and stored in public databases or repositories, and these can therefore be readily adapted and extended to generate more complex models. Their use is often compromised, however, by a lack of reproducibility. Many models cannot be fully verified because the analytical

computational methods or, crucially, adequate, relevant experimental data is not available. New large repositories are needed for physiological and pharmacological data.

 Recommended Reading

Alberts, B., Johnson, A. et al. (2007) *Molecular Biology of the Cell*, 5th edn. New York: Garland Science.
Alberts, B., Bray, D. et al. (2009) *Essential Cell Biology*, 3rd edn. New York: Garland Science.
Beard, D.A. and Qian, H. (2008) *Chemical Biophysics. Quantitative Analysis of Cellular Systems*. Cambridge: Cambridge University Press. This is an introduction to the quantitative analysis of cellular systems.
Websites for cell-level models of biochemistry and cell biology are to be found at **http://www.physiome.org**, **http://sbml.org**, and **http://www.cellml.org** (see also the Appendix).

 References

1. Bassingthwaighte, J.B. (2000) Strategies for the Physiome Project. *Annals of Biomedical Engineering* 28, 1043–1058.
2. Patterson, S.W., Piper, H., and Starling, E.H. (1914) The regulation of the heart beat. *Journal of Physiology* 48, 465–513.
3. Bassingthwaighte, J.B., Noble, D., and Hunter, P.J. (2009) The Cardiac Physiome: perspectives for the future. *Experimental Physiology* 94(5), 597–605.
4. Ciechanover, A. (2012) Intracellular protein degradation: from a vague idea through the lysosome and the ubiquitin-proteasome system and onto human diseases and drug targeting. *Ramban Maimonides Medical Journal* 3, 1–20.
5. Reich, J.G. and Sel'kov, E.E. (1981) *Energy Metabolism of the Cell: A Theoretical Treatise*. London: Academic Press.
6. Goldbeter, A. (1996) *Biochemical Oscillations and Cellular Rhythms*. Cambridge: Cambridge University Press.
7. Pe'er, D. and Hacohen, N. (2011) Principles and strategies for developing network models in cancer. *Cell* 144, 864–873.
8. Díaz-Zuccarini, V. and Pichardo-Almarza, A. (2011) On the formalization of multi-scale and multi-science processes for integrative biology. *Interface Focus* 1(3), 426–437.

9. Alber, M., Kiskowski, M. et al. (2003) On cellular automaton approaches to modeling biological cells. In *Mathematical Systems Theory in Biology, Communications, Computation, and Finance*, IMA Volumes in Mathematics and its Applications, vol. 134 I, Rosenthal, J. and Gilliam, D.S. (eds), pp. 1–39. New York: Springer.

10. Moreira, J. and Deutsch, A. (2002) Cellular automaton models of tumor development: a critical review. *Advances in Complex Systems* 5, 247–267.

11. Pérez, M.A. and Prendergast, P.J. (2007) Random-walk models of cell dispersal included in mechanobiological simulations of tissue differentiation. *Journal of Biomechanics* 40(10), 2244–2253.

12. Byrne, D.P., Lacroix, D., and Prendergast, P.J. (2011) Simulation of fracture healing in the tibia: mechanoregulation of cell activity using a lattice modeling approach. *Journal of Orthopaedic Research*, 29, 1496–1503.

13. Checa, S. and Prendergast, P.J. (2009) A mechanobiological model for tissue differentiation that includes angiogenesis: a lattice-based modeling approach. *Annals of Biomedical Engineering* 37(1), 129–145.

14. Sainani, K. (2011) ERROR! What biomedical computing can learn from its mistakes. *Biomedical Computation Review* 7(2), 12–19.

15. Edwards, A.W.T., Bassingthwaighte, J.B. et al. (1960) Blood level of indocyanine green in the dog during multiple dye curves and its effect on instrumental calibration. *Mayo Clinic Proceeedings* 35, 747–751.

16. Kortgen, A., Paxian, M. et al. (2009) Prospective assessment of hepatic function and mechanisms of dysfunction in the critically ill. *Shock* 32(4), 358–365.

Chapter written by

James B. Bassingthwaighte, Department of Bioengineeering, University of Washington, Seattle, USA;

Colin Boyle, Department of Mechanical Engineering, University College London, London, UK;

Cesar Pichardo-Almarza, InScilico Ltd, London, UK;

Vanessa Díaz-Zuccarini, Department of Mechanical Engineering, University College London, London, UK.

6 Modelling Tissues and Organs

Learning Objectives

After reading this chapter, you should:

- understand the difference between modelling at the organ and tissue level and modelling at the cellular and molecular level, and appreciate some of the difficulties of organ and tissue modelling;
- know the implications of modelling at different scales of distance and time;
- understand the need for interoperability when mixing different scales in space and time and when using different types and formulations of data;
- know many of the main approaches that are currently used in tissue and organ modelling as applied to epithelial tissues and to the heart, kidney, and gastrointestinal (GI) tract, and the present and future prospects for clinical application.

6.1 Introduction

As explained elsewhere in this book, computational biomedicine is concerned with developing and using computational tools for modelling physiology; that is, for modelling the integrated function of the human body at every scale. It is self-evident that this includes modelling at the levels of whole tissues and organs. A tissue can be defined as an organized collection of different cell types, and an organ can similarly be thought of as an organized collection of tissues. You will remember from earlier chapters that it is not possible to simply deduce the properties of cells from the structures and functions of the molecules that constitute them. Similarly, it is not possible to derive the functions of organs, let alone tissues, from the knowledge of the characteristics of their constituent cells. Denis Noble describes

in detail in his acclaimed, and very accessibly written, little book *The Music of Life* (see Recommended Reading at the end of this chapter) how the action potential along a nerve and the oscillations of a cardiac cell depend on but cannot be deduced from the kinetics of individual ion channels. Similarly, the transport of solutes and water across epithelial barriers that separate tissues and organs depends on, but cannot be deduced from, knowledge of the various cell types making up the epithelium, and no cell type or gene alone can explain the regulation of blood pressure. Noble describes higher-level properties like these, which cannot be explained entirely from the sum of their parts, as emerging only at the 'systems level'.

This chapter is unlike many others in this book as it consists mainly of a series of case studies. Each case study describes the techniques that are currently used to model a particular tissue or organ. We therefore cover a variety of modelling techniques in the chapter, although just as we cannot cover every tissue and organ we cannot pretend to give an exhaustive coverage of modelling techniques. By reading through these case studies you should understand how smaller and simpler models can be fitted together, like jigsaw pieces, to represent higher-order physiological processes. It should become clear from this that there is no such thing as a definitive model that is appropriate for all organs and all scenarios. Different models may be appropriate even for the same organ in different circumstances, and the one (or ones) that you select will depend on your goal in setting up the modelling and what you expect to learn from the model that you will be constructing. In particular, not only the scale of the model—that is, whether you are principally modelling molecules, cells, tissues, or organs—but also its level of resolution (i.e. the detail at which the results will emerge) will depend on your aim.

Examples of common reasons for researchers setting up models include, among others:

- asking a research question, to explore alternative hypotheses about an unsolved problem;
- attempting to explain or represent current knowledge: a model can resemble a form of literature review, since it requires the modeller to make judicious choices among hypotheses and data types;
- developing a tool for other researchers to use, for example a platform for the exploration of drug effects *in silico*;
- developing and using a clinical tool to help with diagnosis or selecting a therapeutic strategy.

As all models in computational physiology are at least partly based on data, the scale of the models and the amount of detail that can be modelled will depend crucially on the quality of the data available. It is never possible to measure everything, and it is very likely that not all the data that could be used in setting up a particular model will be available. Furthermore, all experimental techniques have technical limitations, and all too often measurements that

could have been made were not carried out. When or if you come to design your own models, you should remember that experimentation and modelling should go hand in hand; try to have a say in the design of the experiments as well as the models that the experiments will inform.

Most strategies for modelling tissues and organs can be thought of as 'middle out' strategies. This means that, as Noble has put it, 'all levels can be the starting point for a causal chain', meaning that you can build a multilevel model of a tissue or organ starting from whichever level seems to be the most logical, and building 'up' (towards the organ and organism) and 'down' (towards the organelle and molecule) as seems appropriate. In this chapter, you will see examples of modelling strategies that start at different levels and that employ different formalisms and programming languages.

We will now move on to the case studies that illustrate typical approaches to modelling tissues and organs. We start with the modelling of a tissue, or rather of a group of tissues: the epithelial tissues that line all of our organs. Examples of two very different forms of modelling epithelia will be described: agent-based modelling of the dynamic renewal of the outermost layer of the skin (the epidermis) and the use of ordinary differential equations to model transport processes across the epithelia. We then describe some of the principal techniques used to model three organs: the heart, the GI tract, and the kidney. The chapter ends with a brief section on modelling a multiorgan process: the regulation of blood pressure.

6.2 Modelling Epithelia

The main organs of the body are lined with one or more levels of very tightly packed cells. These layers, which are termed epithelia, carry out several essential functions. They transport water and solutes from one body compartment to another and protect the organism from infection and from invasion by foreign bodies. In this section, we give an overview of the biology of epithelia and explain some methods of modelling their function.

6.2.1 Epithelial Biology

Epithelia, then, are the layers of tissue that form boundaries between 'inside' and 'outside'. They include both the outer layer of skin that covers the whole of the external surface of the body (the epidermis) and cell layers lining the surfaces of all body cavities. The epithelia lining different organs have different structures and different specialist functions: examples include the airway epithelium in the lungs, urothelium in the bladder, intestinal epithelia in the gut, and renal tubule epithelia in the kidney.

All epithelia, however, have similar general properties. They all control and/or carry out the transport of water and solutes across the interface between the relevant tissues and organs. Differences between the specialist function of different epithelia are reflected in more subtle differences in their structure. The skin and urothelium have tight junctions surrounding all the cells in their outermost layer and act primarily to *limit transport* across the boundary. In contrast, the primary function of intestinal, renal, and airway epithelia is to carry out *selective transport* of water and solutes across the interface.

Internal epithelia vary from one to a few cells in thickness (e.g. the intestinal, kidney, and bladder epithelia are a single cell layer thick, and the airway epithelium contains a few cell layers). The skin has more cell layers, and its thickness depends on its site; it is thin on the inner surface of the forearm and much thicker beneath the heel. Even the thickest epithelium, however, is normally not more than about 0.5 mm thick.

All epithelial cells have a common developmental origin, and share a set of common features that enable them to carry out their barrier function. They are strongly polarized, with clearly differentiated *apical* (meaning towards the outside), *basal* (towards the inside), and *lateral* (towards neighbours in the same layer) surfaces (Figure 6.1). They form *tight junctions* between cells within layers, so the cells form cohesive sheets with little space between the cells, and they adhere tightly to the collagen layer that separates them from the supporting tissue (which is known as the basement membrane). They therefore appear to be cuboidal or rectangular in cross-section, and resemble closely packed hexagons from above (see Figure 6.2). The specialized structure and function of epithelial cells, like that of all cells, derive from localized physical and chemical signalling.

Epithelial cells can detect strain. Mechanical transduction at the protein complexes that connect cells to the extracellular matrix (known as *focal adhesions*) and strain-sensitive ion channels are both common to many types of cell. In particular, urothelial cells have been shown to act as

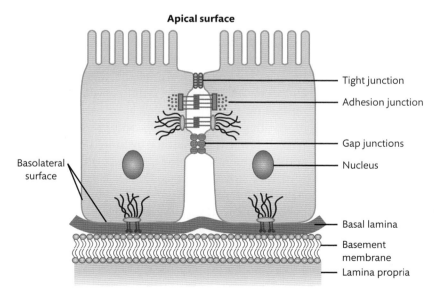

Figure 6.1 Epithelial cells. Schematic illustration of a typical layer of epithelial cells, showing the basement membrane, the tight junctions between the cells, and the clear differentiation between cell surfaces. Redrawn from Pocock and Richards (2009) *The Human Body: An introduction for the biomedical and health sciences.* Oxford University Press.

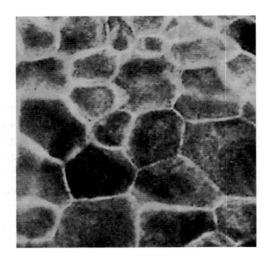

Figure 6.2 A surface view of normal urothelium, grown *in vitro* and showing the hexagonal appearance of the cells. The adhesion molecules forming the tight junctions between the cells have been labelled with a green fluorescent protein. Image © Professor Jenny Southgate, University of York.

strain sensors in the bladder by releasing signalling molecules that are detected by nerve endings in the bladder wall. They can also produce signals that affect the pain-sensing neurons or nociceptors, which gives rise to the pain that is common in many bladder diseases. It has also been suggested that a buckling collapse of the airways in an acute asthma attack can result in the release of cytokines from acutely stressed airway epithelial cells, which provokes an immune response that exacerbates the attack.

6.2.2 The Epidermis

We are all familiar—if not in ourselves then in others—with the effects of the ageing process on skin. However, the age-related reduction in skin elasticity that causes wrinkling is only visible over a timescale of years. Over shorter periods, at least superficially, there is no visible change in our skin. This, in fact, is quite remarkable because all epithelia including the skin are constantly being renewed; in the case of the skin this is over a period of about a month. This appearance of stasis despite constant renewal is a remarkable and little understood phenomenon that is central to a mechanistic understanding

of the behaviour of epithelia. Understanding the control mechanisms that ensure that the structure and function are maintained despite constant renewal is essential for an understanding of epithelia-related disease. One important example of this is the skin disease psoriasis, in which a failure of the homeostatic mechanism causes hyper-proliferation of the cells of the epidermis.

Although different epithelial tissues have slightly different structures, the renewal process appears to be similar in all epithelia. All epithelia contain a stem cell niche in which the stem cells from which epithelial cells are generated are localized. For example, the stem cell niche in the intestine is at the base of the intestinal crypts, and in the skin it is the bottom of the so-called *Rete ridge* (which is actually a conical depression extending the epidermis down into the dermis). Stem cells—and, in some epithelial tissue, progenitor cells as well—divide and differentiate in one or more stages and move away from the stem cell niche. The cells are finally shed into the lumen or (in the case of the skin) into the atmosphere.

We have already mentioned that the epithelium in an intestinal crypt is one cell thick. The stem cells at the base of the crypt divide continuously, and these dividing cells differentiate as they are pushed up the side of the crypt by further layers of dividing stem cells. They are eventually shed when they reach the luminal surface of the intestine.

The process in the skin is slightly different. Firstly, the stem cells divide to give rise either to more stem cells or to progenitor cells. If the progenitor cells remain attached to the basement membrane, they will continue to divide in response to signalling between the progenitor cells and that membrane. However, they will stop dividing when they are pushed up into the next cell layer by the next set of dividing cells. These cells will be continuously pushed towards the surface by more dividing cells at the basement membrane, and will differentiate as they approach the surface (Figure 6.3). In the final stage of differentiation, the cells (which are now called corneocytes) lose their nucleus, flatten, and are shed from the skin surface. Normally we do not notice this, but there are exceptions: dandruff occurs when the cells are shed as clumps that are large enough to be visible.

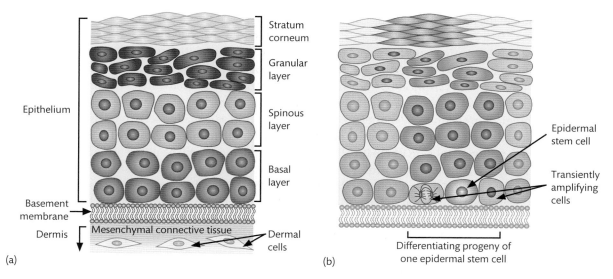

Figure 6.3 (a) Diagram showing skin epithelial histology. Cells of the basal layer attach to an underlying basement membrane. Basal cells are mitotically active, but they lose this potential when they detach from the basement membrane and move outwards toward the skin surface. As basal cells enter the spinous layer, they strengthen their cytoskeletal and intercellular connections, gaining resilience to mechanical stress. They then enter the granular layer, where they produce the epidermal barrier. As the cells enter the final phases of terminal differentiation, a flux of calcium activates the enzyme transglutaminase, leading to cell death through a complex mechanism. Cell death ensues, leaving dead, flattened squames at the skin surface, the end process of terminal differentiation. These eventually slough from the skin surface, to be replenished continually by inner layer cells moving outward. (b) Diagram of the epidermal proliferative unit. A putative slow-cycling epidermal stem cell occasionally divides, giving rise to a stem cell daughter and a transiently amplifying daughter. The transiently amplifying cell divides two to four times, and these progeny then leave the basal layer and execute a program of terminal differentiation. This model is based on retroviral transduction of a β-galactosidase gene into cultured keratinocytes, which were then used in engraftments onto nude mice to trace stem cell lineages. Both panels taken from Alonso and Fuchs [1].

Epithelial cells form sheets that can cover a very large surface area, and must therefore comprise a very large number of cells. For example, the skin of a normal adult human will contain, at any time, about 10^{11} epithelial cells, and their continuing renewal will require an even larger number of cell divisions (10^{12} per year in the skin). Therefore, even a very low probability of mutation in each cell division can give rise to significant numbers of mutated and perhaps pre-cancerous epithelial cells. Indeed, most carcinomas originate in epithelial tissues. It has been suggested that the presence of some pre-cancerous cells in epithelial tissues can be normal.

Modelling the Epidermis

The structure, function, properties, and behaviour of the epidermis all derive from the interactions between cells, the interactions between epidermal cells and the basement membrane, and physical and biochemical signalling within those cells. Influences from outside the epidermis that affect its function include the cells that synthesize the extracellular matrix, the fibroblasts, which are found in the dermis. These cells become activated by skin injury and form more extracellular matrix, which immediately begins to heal the wound. Furthermore, immune system components will migrate into the epidermis in response to an immune response that is triggered there.

Mathematical models of the epidermis and of all epithelia are, in general, based on the behaviour of individual cells; these are termed *agent-based* or *individual-based* models. As epithelial cells are often continually cycling, these models must include models of the cell cycle and the processes that trigger and control it [2].

The first stage in setting up an agent-based model, such as one of an epithelial tissue, is generally to put together a description of the various entities, relationships, and mechanisms that constitute the biological problem to be modelled. This is termed a *domain* or *discursive model*; the name *platform model* is given to a mathematical and computational representation of the problem that is described in such a model. For example, let us assume that the biological problem to be modelled concerns the growth, *in vitro*, of properly structured and functional human skin (Figure 6.4). Then, the types of information we might need to know include:

- the cell types present in the tissue and the successive stages of differentiation of each one: for example, the mature epidermal cells or keratinocytes that make up some 90% of human epidermis differentiate from stem cells through intermediate cell types that are termed transit amplifying cells and committed cells;

- cell–cell and cell–substrate adhesion, and its effect on cell behaviour;
- the signalling pathways that trigger processes in the skin, including mechanical transduction;
- determinants of cell differentiation.

This 'model' is actually a set of models, since the modelling paradigm that is best suited to each component of the full model will be different. Cells will generally be modelled using an agent-based or individual-based paradigm; signalling processes will be modelled using differential equations, and physical behaviour using finite element models. Any software framework that is used to develop a complete platform model of this tissue will need to be able to accommodate, and link, all these different modelling paradigms.

Validating the Models

We have already discussed briefly the importance of validating models, and this is surely important here. How can we be assured that a model is an adequate representation of a complex system? A common approach to validation is to start with a domain model of an *in vitro* or *ex vivo* system and see whether the

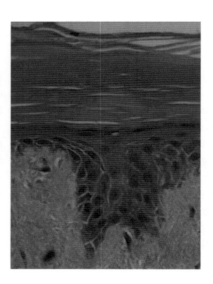

Figure 6.4 A comparison between the individual-based computational model of an epidermis (left) with a stained section of actual epidermis, grown *in vitro* (right). The histological section has been stained using haematoxylin and eosin (so-called H&E staining) to reveal individual cells. Stem cells and corneocytes are stained brown, progenitor cells white, stratum spinosum cells purple, and stratum granulosum cells pink. For simplicity, the model represents the cells as spheroids without the biological aspect ratio. Model image © Shannon Li, University of Sheffield; H&E section © Anthony Bullock, University of Sheffield.

behaviour predicted by the computational model is indeed similar to that observed in the laboratory. In this case, however, modellers will need to take care that the behaviour of the computational model is not peculiar to the very simplified laboratory system and can be generalized to a more physiologically appropriate one. In particular, most *in vitro* laboratory models of the epidermis make use of immortalized cell lines, which by definition are not normal cells and can have very different behaviour from normal cell lines.

As an example, researchers in the AirPROM project[1], who are modelling and predicting the progression of airway diseases such as asthma, use two *ex vivo* laboratory models for validating their cellular-level model of airway remodelling after exposure to allergens and irritants. The cells used in the *ex vivo* models are obtained from patients through airway brushing and bronchial biopsies. These provide populations of airway epithelial cells, fibroblasts, and smooth muscle cells from patients with different conditions.

6.2.3 Epithelial Transport of Solutes and Water

Modelling the transport of solutes and water across epithelia can be thought of as representing a crossroads between cell and organ models. Single-cell models are not sufficient to model transport, and the epithelium is often considered to be a simple and relatively insignificant part of an organ model. This is an important function in human physiology. The epithelial tissues that line our organs are responsible for the essential, and energetically demanding, transport of water and those solutes that are specific to the function of each organ across these boundaries. Some examples of this include the gut epithelia (see also Section 6.4) which reabsorb water and nutrients. The kidneys' role in homeostasis is carried out by secretion and reabsorption across the epithelia that form the successive segments of kidney tubules (the nephrons), with different cell types and different membrane transporters in each segment (see also Section 6.5). Specific transport functions are also carried out by many other types of epithelia

including the airway epithelia, salt glands, salivary glands, inner ear epithelium, and urinary bladder.

We mentioned at the beginning of this section that epithelia involved in transport are composed of asymmetric, or polarized, cells and with apical (or luminal), basal (or mucosal), and lateral cell membranes linked together into a continuous sheet by tight junctions (which may in fact be more or less tight); see also Figure 6.1. Epithelial transport involves the complex interaction of many specific transport processes carried out by membrane proteins that form specific ion channels or transporters and that are located within the epithelial cell membranes.

Many of these transport proteins are found in non-epithelial cells as well, but only accomplish transport across epithelia in one direction (net vectorial transport) due to the asymmetric arrangement of epithelial cell layers. These transport mechanisms may be relatively simple (such as electrodiffusion) or highly complex, such as active transport or co-transport. In addition, transport of a substance across an epithelium may be coupled either to ATP hydrolysis (primary active transport) or to the transport of other substances, either directly (secondary active transport) or indirectly because of secondary changes in electrochemical or osmotic driving forces (Figure 6.5). These active transport processes allow solutes to be transported against their own electrochemical driving force. This allows, for example, the virtually complete intestinal absorption of nutrients such as glucose and amino acids into the blood and their reabsorption in the kidney proximal tubule (see Section 6.5).

Many diseases are caused by defects in epithelial membrane transporters or of their regulation. Cystic fibrosis, which arises from a defect in a gene that codes for a protein that regulates the transport of chloride and sodium ions across epithelia, is one of the best-known examples of such diseases.

Modelling Epithelial Transport

Epithelial transport is usually modelled as a compartmental system using a set of ordinary differential equations and algebraic constraints. Conservation relations provide constraints that ensure that mass balance and electroneutrality are respected. The characteristics of each ion channel or coupled transporter are given using detailed kinetic equations, when

[1] http://www.vph-noe.eu/vph-projects/74-eu-fp7-vph-projects/496-airprom

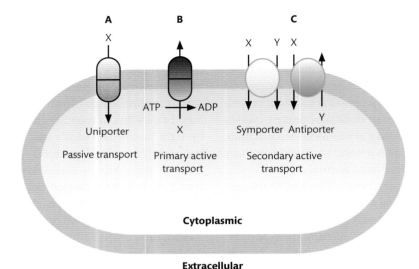

Figure 6.5 Types of transport across cell membranes. (A) Passive transport, or facilitated diffusion, in which a molecule (X) crosses the membrane through a channel or transport protein along its electrochemical gradient (i.e. from high to low electrochemical potential). No energy is involved. Passive-transport proteins can also be termed uniporters. (B) Primary active transport, in which a molecule crosses the membrane against its electrochemical gradient (i.e. from low to high electrochemical potential) with the direct expenditure of energy. The hydrolysis of the 'energy carrier' ATP into ADP shown here is common but is not the only way in which this energy is provided. (C) Secondary active transport, in which two molecules (X and Y) cross the membrane simultaneously. The flow of one molecule from high to low electrochemical potential creates energy that can be used to pump the other against its electrochemical driving force. If the molecules cross the membrane in the same direction, the protein is known as a symporter, whereas if they cross the membrane in opposite directions it is known as an antiporter.

these are available, or otherwise using phenomenological thermodynamic equations that respect energetic constraints. Which are chosen depends on the amount of information that is available about the transporters to be modelled. A typical illustration of a simple model of this type is shown in Figure 6.6.

A complete, generic explanation of how to set up and solve a mathematical model of epithelial transport is beyond the scope of this chapter. Readers who would like to take this forward are referred to [3]. We include in the Online Resource Centre for the book a listing of the code for an implementation of their generic model (which is known as a Koefoed-Johnson and Ussing model) that uses an excellent, easy-to-use ordinary differential equation solver, Berkeley Madonna®[2]. The free demo version of this program should be perfectly adequate for many fairly small-scale modelling projects.

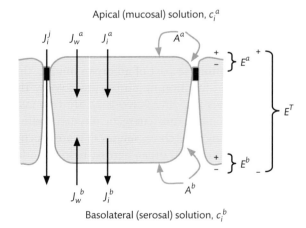

Figure 6.6 Schematic representation of a general model of transport across an epithelium. Symbols: flows, J; superscripts refer to apical (a) and basolateral (b) cell membranes and the tight junctions (j); subscripts refer to water (w) or a given solute (i); E is electrical potential difference across a given membrane, with orientation given by + and − signs (e.g. a positive value for E^T means that the mucosal solution is positive with respect to the serosal solution). The straight arrows denote direction of positive solute and water fluxes. Redrawn from [3].

---

[2] http://www.berkeleymadonna.com; the free demo version of Berkeley Madonna is perfectly adequate to solve this model by simply pasting the text of the listing supplied in the Online Resource Centre into the Equations window of the application.

It should prove possible to build up more complex models of specialist epithelia that are based on the principles in this model, which are outlined in [3]. The precision that is possible in these models will depend on the extent of the available experimental data that characterizes the overall transport processes across the epithelium concerned and its constituent channels and transporters. A wide range of methods have been used to study these transport processes over many decades, including isotopic tracers, intracellular and ion-selective microelectrodes, membrane vesicles, patch-clamp methods, and immunofluorescent labelling. Much of the data are available in the scientific literature.

Modelling studies have already been used to study many of the complex features of epithelia, including the expression of the channels and transporters in epithelial cells, their traffic to the correct membranes, and their insertion into and removal from those membranes.

It is also worth mentioning that some epithelia may transport large amounts of solutes and water. This can amount to many times the water and solute content of the cells themselves each hour. It is clear, therefore, that rates of inflow and outflow across membranes must be carefully coordinated to avoid the cells drying out or becoming bloated. This process, which has been called 'homocellular regulation' [4], requires some kind of co-ordination or cross-talk between transporters on opposite sides of the epithelium, and it has been studied using an approach that links modelling to experimentation.

It can be argued that one of the principal goals of computational physiology is to predict the behaviour of organ systems from physiological data. It is therefore clear that the more that is known about the physiology of an organ the easier it will be to develop accurate predictive models.

We now move on to discuss three case studies illustrating different techniques that are used to study the structures and functions of single organs in health and disease. We start with what may be the best-studied organ of all: the heart.

6.3 Cardiac Modelling

The human heart is a fist-sized muscular organ that weighs about 250–300 g in an adult and is located in the chest cavity between the lungs. It is divided into four cavities, the upper two of which are termed the right and left *atria* (singular: atrium) and the lower two the right and left *ventricles*. The walls of all the chambers are composed of thick, strong muscular tissue called myocardium. Unless the heart is seriously diseased, the left and right sides of the heart are completely separated from each other; the left atrium and ventricle carry oxygenated blood and the right atrium and ventricle carry deoxygenated blood. Each atrium is separated from its ventricle by a valve, and blood enters and leaves the heart through the great blood vessels (see Figure 6.7).

Students of biology learn before they reach high school that the heart is a pump. Its function is deceptively simple but also exquisitely controlled; a human heart will beat about 3 billion times in a 'typical' 70-year-plus lifespan. This heartbeat has four phases that can be described in very simple terms as follows:

- deoxygenated blood from the body enters the right atrium and passes into the right ventricle through the tricuspid valve;
- the right ventricle pumps the blood to the lungs, where it becomes oxygenated;
- oxygenated blood is returned to the left atrium, and passes through the bicuspid valve into the left ventricle;
- the left ventricle pumps the oxygenated blood round the body.

As the left ventricle has to pump blood a greater distance (and thus overcome a greater resistance) than the right one, its wall is noticeably the thicker and more muscular.

The regular pattern of heartbeats is controlled by a group of heart muscle cells in the right atrium called the *sinoatrial node* (SA node). This generates an electrochemical potential gradient that rapidly rises and falls and spreads through the atria, causing them to contract and empty into the ventricles. This electrochemical wave is then picked up by a second group of cells termed the *atrioventricular node* (AV node) causing the ventricles to contract and expel blood from the heart. Modelling cardiac function, therefore, involves characterizing a wide range of electrical and mechanical processes in the specialist muscle cells and tissues comprising the organ.

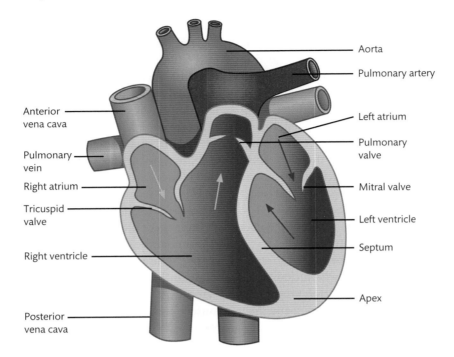

Figure 6.7 Anatomy of the human heart. The basic structure of a human heart, showing the four chambers (two atria and two ventricles), the four major blood vessels, and the principal valves. © Benjamin Bhattacharya-Ghosh.

6.3.1 Modelling the Heart

Currently, the heart, and the cardiovascular system more generally, is one of the best understood of all organs and systems and, therefore, arguably the one for which computational modelling is most advanced. There are three main reasons for this:

1. the social and epidemiological importance of heart disease as one of the most common and most often lethal pathologies in the developed world;
2. the availability of accurate and precise data on cardiac function at a range of scales; and
3. the perhaps rather surprising fundamental simplicity of this organ as, mechanically, a simple reciprocating electromechanical pump for the circulatory system.

These factors have motivated computational physiologists to develop increasingly complex and accurate mathematical models that assimilate the diverse data on cardiac physiology and function that have been made available from both the wet lab and the clinic. These models increasingly work at multiple spatial

and temporal scales and make use of a wide range of physical laws and assumptions (thus, they are multi-scale and multi-physics models; there will be much more about this concept in Chapter 7). Through this modelling work, frameworks are now emerging to provide new insights into cardiac behaviour and to predict the function of the heart under various physiological conditions.

A simple way of introducing the ideas behind cardiac modelling (and, thus, behind organ modelling more generally) is therefore to consider these two independent criteria: *scale* and *physics*. Firstly, established cardiac models span many dimensional scales, from subcellular models of transporters and ion channels, to whole-cell models constructed through combinations of these components, and then to models of the atria, ventricles, and whole heart developed through the integration of these cellular structural frameworks. Historically, whole-heart models have typically focused on capturing behaviour that is associated with a distinct subset of its physical processes. Specifically, these involve modelling the conduction of the wave of electrical current that passes through the heart muscle (the activation wave), the contraction and mechanical pumping mechanism of the heart that this electrochemical wave produces, and finally the blood flow within the cardiac chambers and the coronary vessels produced by this contraction. The next sections introduce these elements separately and then discuss the next steps that need to be taken in the development of cardiac modelling.

Electrical Activation and Cardiac Contraction

You will have seen in Chapter 5 how systems of ordinary differential equations have been set up to model heart cells, and how these have become more complex over time. Increasingly powerful computers have been needed to solve these systems of equations, but this has been more than compensated for by rapid improvements in high-performance computing. The next steps have typically involved extension from cell to tissue modelling: developing models of cardiac tissue in which models of individual heart muscle (myocyte) cells are combined with a continuous model of the general geometry of the tissue. These models have allowed the functional properties of electrical conductivity and mechanical stiffness to be predicted from detailed images of the cardiac tissue. Here the myocyte, fibroblast, and collagen microstructure has been used to determine parameters—numerical descriptions of (for example) the conductivity and stiffness of the tissue—for use in a continuous model of that tissue. Further, the spread of the electrical activation wave has been predicted in two- and three-dimensional tissue models (see Figure 6.8) by applying mono-domain or bi-domain equations.

Tissue deformation can also be predicted by solving equations for finite deformation using transient values for the tension in the tissue that have been calculated using the cellular models [5].

Linking short-term changes in calcium concentration (so-called calcium transients) calculated using a cellular model of electrophysiology to cellular tension generation provides the ability to couple electrical activation and muscle contraction. This is achieved at the tissue level by combining techniques for the solution of numerical equations; it is possible to do this in a way that preserves computational efficiency in most cases.

In computational simulations, complex three-dimensional objects are often covered with grids and parameter values obtained or calculated at each grid point; grids representing different properties may not have the same density, and in these cases interpolation is needed. Here, specifically, values for the cellular tension are interpolated from a high-density but low-order grid representing electrical activation to points on a low-resolution but high-order finite-element mesh representing the geometric structure of the heart. Currently, these tissue-level calculations require computational power that is not always available and there will be a trade-off between the volume of tissue to be modelled, the level of detail that can be included in the cellular model that is embedded in the tissue model, and the resolution of the grid, which is important to guarantee a correct solution. It has therefore often been possible to incorporate only relatively simple cell models into a model tissue that comprises a large number of such cells: this has limited the accuracy and precision to which the biophysical properties of the tissue can be modelled. However, recent improvements in both the algorithms and the way these have been implemented suggest that models on the scale of a whole human heart, at clinically relevant timescales, will soon become possible.

Models of Coronary and Ventricular Blood Flow in the Heart

Modelling the coronary circulation is important because this aspect of cardiac anatomy and physiology is relevant to many types of heart disease: both electrical disturbances that lead to irregularities in the heartbeat (arrhythmias) and simple mechanical dysfunction. Arrhythmias are often initiated by a

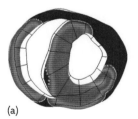
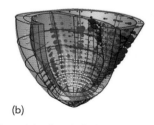

(a)   (b)

Figure 6.8 Electromechanical models of a whole human heart. (a) An anatomically based two-dimensional model of coupled excitation/contraction showing tissue deformation during electrical excitation. (b) Calculation of regional work distribution over a complete cardiac cycle.

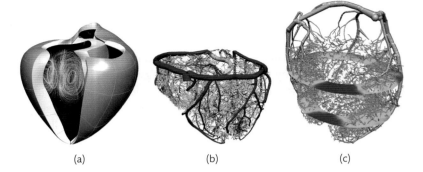

(a)   (b)   (c)

Figure 6.9 Integrated, fluid dynamic models of the heart: (a) computational fluid dynamic model of the left ventricle showing velocity vectors (© David Nordsletten), (b) coronary blood flow model, and (c) poro-mechanical model of coronary perfusion (© Eoin Hyde).

blood clot or thrombus that blocks a coronary artery, causing a local restriction in blood supply (ischaemia), extracellular accumulation of potassium ions, and a drop in intracellular pH. These changes also result in a loss of contractility in the ischaemic region, which can produce a pathological drop in the capability of the whole heart to act as a pump. This dysfunction arises from both the clot itself and the resulting changes to the activation of the electrical wave.

To understand fully the sequence of these events and the circumstances under which they produce significant disease, we need models of the blood flow to heart tissue. These blood flow models are currently being developed by solving equations for haemodynamics (blood flow) based on images of the network of cardiac blood vessels (Figure 6.9b) that are derived directly from imaging of the heart using computed tomography (CT) scanning and other techniques (see Chapter 4).

Applying specific types of flow models on these geometries decreases the computational cost by reducing the three-dimensional Navier–Stokes equations for fluid flow (see Box 6.1) to one dimension. To account for the non-Newtonian effects of reduced blood viscosity and differences in the concentration of different types of blood cell this framework has recently been extended to simulate blood flow in the microcirculation [6].

It is possible to simulate changes in the shape of blood vessels and the blood flow throughout the complete cardiac cycle by coupling intra-myocardial pressure to coronary vessel pressure (Figure 6.9c). The pressure is calculated by using finite-element-based methods to model deformations of the heart tissue during the cardiac cycle discussed above.

Fluid mechanics is another important influence on the mechanical function of the heart, through its coupling with the mechanics of the ventricular wall. The models that have been described above

---

Box 6.1 The Navier–Stokes Equations for Fluid Flow

The Navier–Stokes equations are a set of equations that describe the motion of continuous fluids in mathematical terms. They are derived from the application of Newton's second law (which, in its very simplest terms, states that force = mass × acceleration) to fluid flow. A general solution of the equations is always in terms of the velocity of the fluid at given space and time points, which is termed a flow field. In physiology, their main application is in the determination of blood flow.

The general form of a Navier–Stokes equation is shown below:

$$\rho\left(\frac{\partial v}{\partial t} + v.\nabla v\right) = -\nabla p + \nabla.T + f$$

Here v represents the flow velocity, ρ the fluid density, p the pressure, T one component of the total stress tensor, f the body forces acting on the fluid, and ∇ the del operator.

This equation can be represented in several ways, but all representations relate the rate of change of flow velocity of the fluid to its pressure and density and to the forces and stresses that are acting on it.

have been extended to include models of the three-dimensional fluid dynamics of the cardiac chambers. Specifically, this has been done through combining Cauchy's first law, which describes the conservation of linear momentum in any continuous medium (here, the blood volume within a cardiac chamber or the wall of the myocardium) with a form of the full Navier–Stokes equation (see Box 6.1). This forms part of the mathematical discipline of continuum mechanics; should you wish to learn more about this discipline you will find Spencer (2004) useful (listed under Recommended Reading).

6.3.2 Translational Heart Modelling

We have already mentioned the clear medical need for the development of computational models of the heart. Cardiovascular disease is the primary cause of morbidity and mortality in the Western world. The lifestyle that underpins this epidemic in the West is now spreading to developing nations, and cardiovascular disease is predicted to become the most common cause of death in many of these countries within a few years. Despite the significance of this disease, however, it can still be difficult to determine the exact clinical diagnosis and optimal treatment strategy for an individual cardiac patient. Whereas developments in clinical imaging have provided clinicians with the ability to measure the geometry, motion, perfusion, metabolic status, and other functions of an individual patient's heart, individual treatment decisions are still largely determined by relatively simple population-based metrics. These metrics are proving unsatisfactory for defining treatment options in some important patient populations. Indeed, a large number of patients are still undergoing unnecessarily invasive procedures, often with less than satisfactory results [7].

The use of multi-scale modelling, which is discussed further in Chapter 7, provides us with an exciting prospect for shifting treatment planning from a population-based to a patient-based paradigm. If it is to be successful, this will surely involve integrating multiple types of data into biophysically consistent mathematical models that can be considered as unique to individual patients. Furthermore, the development of these multi-scale cardiac models presents us with the ability to capture the complex and multifactorial cause-and-effect relationships that link underlying pathophysiological mechanisms. This, in turn, provides the opportunity to derive yet more parameters which are not directly observable, but which play a key mechanistic role in the disease process and to increase the precision and complexity of the models and further improve clinical decisions.

There are, however, important challenges if modelling techniques and frameworks are to be translated into a form that is appropriate for the clinic. It is essential for the paradigms that the models are based on to be robust (and, crucially, to be understood to be so by clinicians). It is also important that all the data that are built into the model as parameters should be obtained using minimally invasive techniques, particularly for seriously ill patients. Different strategies for data collection and parameterization have been proposed depending on the type of modelling to be undertaken (i.e. whether it is electromechanical or fluid mechanical) and on the sensitivity of the simulation results to the amount, type, and quality of the data available. Direct methods can be applied to cases in which small variations in clinical measurements imply large changes in the values of parameters to be used in the model. Techniques have been developed both to iteratively minimize the mismatch between equivalent observed and simulated values, and to correct mismatches at each time step of a simulation. A complementary approach involves the use of model databases that are being developed to characterize the variation in parameters used in cardiac modelling across populations. Using these, it should be possible to estimate parameter values that

Figure 6.10 Anatomically accurate human ventricular finite-element mesh fitted from magnetic resonance imaging data (© Pablo Lamata).

cannot be determined experimentally based on demographic characteristics of the patient such as age, gender, ethnicity, and medical history.

Several projects are under way to develop ICT tools and specify patient-specific data for cardiac modelling. One example of a recent project in cardiac modelling that incorporated such personalized data was euHeart[3]. This ambitious project set out to develop methods for the multi-scale modelling of diseased hearts based on patient-specific data and covering several common heart conditions: heart failure, coronary artery disease, and valvular and aortic diseases.

Some of the computational models generated within euHeart and similar initiatives are already being used in optimizing treatment planning, in selecting the most suitable patients for clinical trials, in clinical training, and in developing new devices and drugs. We hope and believe that these models will soon play a larger part in the practice of cardiology, and that the inclusion of parameters based on data from individual patients will allow treatment plans for those patients to be developed based on their specific physiology.

Similar computational modelling techniques are now being applied to many of these, however, and we will now go on to discuss two of them: first the GI tract and then the kidney.

6.4 GI Tract Modelling

6.4.1 A Brief Introduction to GI Tract Physiology

The GI tract is the system that is responsible for the digestion and absorption of food, generating energy, and for the elimination of some body wastes. Some people think of it as just comprising the stomach and intestines, while others include all organs from the mouth to the anus. In human adults the entire tract extends to about 5 m long (Figure 6.11).

The GI tract is also a prominent part of the immune system. The surface area of the tract in adult humans is estimated to be equivalent to the area of a football field. The immune system needs to prevent pathogens from entering the blood and lymph through this large surface area. Other factors in the

GI tract that help with immune function include some of the enzymes in saliva and bile. The intestine also contains antiporters, which transport molecules across membranes (see Figure 6.5), and some of the enzymes in the cytochrome P450 family. These enzymes catalyse the oxidation of molecules that are not normally encountered by the body (drugs and other xenobiotics) and their consequent elimination from the body.

The stomach is extremely acidic (pH 1–4) and this low pH is fatal for many of the micro-organisms that enter it. Other micro-organisms are neutralized by antibody-containing mucus. However, by no means all bacteria are harmful to human health, and not all are killed by the conditions found in the GI tract. Indeed, the intestine contains many trillions of health-enhancing bacteria which prevent the overgrowth of potentially harmful bacteria; the gut of an average human contains about 10 times as many bacteria as that human has cells.

Transit from the Stomach to the Small Intestine

Food enters the stomach through the oesophagus and stays there for about 2–4 h, and some digestion takes place there. The food mixture (now known as *chyme*) then leaves the stomach for the top segment of the small intestine, the duodenum. There are three sections to the small intestine (Figure 6.11): the duodenum, the jejunum, and the ileum (the last segment before the colon). Chyme takes about 3–10 h to pass through the small intestine, and then passes into the colon (large intestine). Only about 5% of undigested food products are broken down in the ileum, which explains why people who have had a small part of their intestine removed can still digest most foods with little problem.

Intestinal Absorption

The GI tract is folded, which provides a large surface area for absorption, and this area is lined with hairlike projections called villi; each of these includes many much smaller microvilli. The large total surface area of the intestines ('equivalent to the surface of a tennis court' [7]) greatly contributes to the efficiency of the process of nutrient absorption in the intestines.

On the molecular scale, the process of absorption involves the movement of molecules across the GI tract into the circulatory system. Vitamins, minerals, water, and most of the end products of the digestion

³ http://www.euheart.eu/index.php?id=26

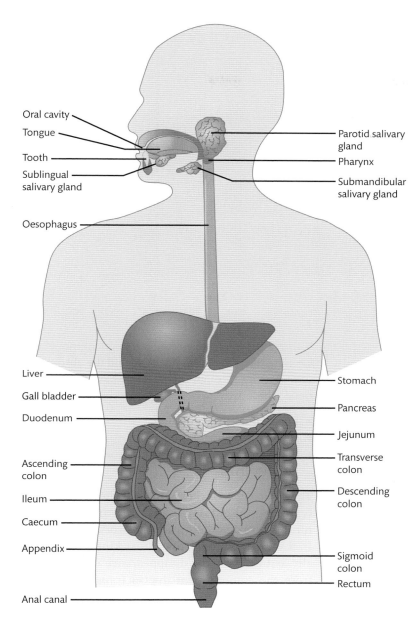

Oral cavity

Tongue

Tooth

Sublingual
salivary gland

Oesophagus

Liver

Gall bladder

Duodenum

Ascending
colon

Ileum

Caecum

Appendix

Anal canal

Parotid salivary
gland

Pharynx

Submandibular
salivary gland

Stomach

Pancreas

Jejunum

Transverse
colon

Descending
colon

Sigmoid
colon

Rectum

Figure 6.11 The human digestive tract. Largely schematic diagram of the human digestive or GI tract, which stretches from the mouth to the anus and includes all the organs necessary to digest food and process waste. Redrawn from Pocock, Richards and Richards (2013) *Human Physiology* 4th edition, Oxford University Press.

of proteins, carbohydrates, and lipids are absorbed from the small-intestinal lumen by the following four mechanisms:

1. *active transport* (which requires energy),
2. passive *diffusion*,
3. *endocytosis*,
4. facilitated diffusion.

The process of digestion involves breaking down complex molecules into smaller ones that can be absorbed through the intestinal wall (Figure 6.12).

Proteins are broken down by proteases into their constituent amino acids and into di- and tripeptides; all these can be absorbed from the intestine. Sugar molecules and larger carbohydrates (starches) are broken down by enzymes into the disaccharides sucrose, lactose, and maltose, and then finally into the monosaccharides glucose, fructose, and galactose; these are then absorbed mainly by secondary active transport. Lipids are broken down by lipase, an enzyme found in the pancreas and the small intestine, and bile from the liver, into fatty acids and

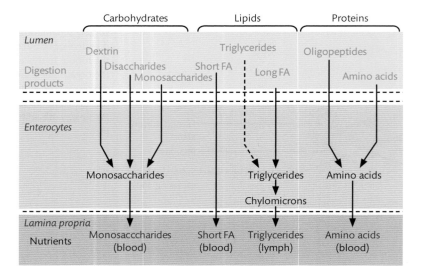

Figure 6.12 The main absorption processes of different products of digestion in the intestine. FA, fatty acids.

monoglycerides, and these end products are then absorbed through the villi as triglycerides.

Mathematical modelling has been used to study the transit of food and food products along the intestine and the absorption of the products of digestion and other molecules. In the next section, we will discuss a particular aspect of modelling intestinal absorption: the absorption of medicinal drugs.

6.4.2 Mathematical Modelling of Drug Absorption

Undoubtedly, the least-invasive, most popular, and cheapest method of delivering a drug into a patient's body is orally; that is, through the GI tract. Unless the physiological target of a drug is the surface of the intestinal tract, however, a drug that is delivered orally must be absorbed into the bloodstream where it can more easily reach its target organ. The most important organ in this process is the small intestine.

In general, drugs follow the movement of the intestinal fluids, which is faster at the beginning of the small intestine than in downstream segments and is slowest as the fluids approach the colon. Also, the flow is slower near the walls of the lumen and faster near the centre of the lumen, which leads to materials being distributed axially across the interior section, or lumen, of the intestine.

Three types of dynamic mathematical model describing the transit of food along and the absorption of molecules through the intestine have been developed. These are:

- one-compartment models,
- multicompartment models,
- dispersion models that include both transport and dispersion.

The *one-compartment* approach is the simplest; this represents the gut simply as a well-stirred compartment where dissolution and absorption take place. In this type of model, first-order kinetics is generally assumed for the disappearance of substances from the compartment due to transit out of the intestinal tract. This assumption implies that the rate of absorption depends on the concentration of only one substance; in this case, clearly, that of the drug or other substance that is being absorbed. This approach has been extended based on microscopic mass balance and considering time constraints on transit; that is, modelling absorption as ending after drugs and other molecules reach the end of the intestine. This approach has been used in the definition of the dissolution number (Dn) as a measure of the time taken for a drug to be released into solution [7] and the formulation of the *Biopharmaceutics Classification System* (BCS) is based on this.

Multicompartment transit models were introduced in the mid-1990s (see, for example, [8]). The model described in this paper is more precisely known as the *compartmental absorption transit* (CAT) model. Each compartment in the model represents one of seven specific intestinal regions, with first- (or zero-) order kinetic equations used to model the transit process from the stomach compartment to the first intestinal compartment. The one-compartment model was extended to incorporate the dissolution and release of drugs by considering drug concentrations corresponding to unreleased, released but undissolved, and dissolved molecules of the drug respectively.

The different intestinal segments are characterized by different transit times, which have an impact on drug absorption. In addition, the extent of drug absorption in each segment differs due to heterogeneity in the modelled physiology and structures of the compartments; the differences between the compartments include differences in the expression patterns and therefore concentrations of metabolizing enzymes and transporters. A later compartmental model based on these principles is known as the gastrointestinal-transport-absorption (GITA) model [9]. This divides the GI tract from the stomach to the large intestine into eight segments (stomach, duodenum, upper jejunum, lower jejunum, upper

ileum, lower ileum, caecum, and large intestine), and incorporates differences in the GI transit and absorption processes between each segment (Figure 6.13).

Dispersion models [10] can be seen as an alternative to the compartmental approach. In these models, the intestine is modelled as a pipe with the concentrations of drugs and other molecules described by spatially continuous profiles for the concentration of drugs (Figure 6.14). Partial differential equations are used to describe these concentration profiles in time and space:

$$\frac{\partial C(z,t)}{\partial t} = \alpha \frac{\partial^2 C(z,t)}{\partial z^2} - \beta \frac{\partial C(z,t)}{\partial z} - \gamma C(z,t) \quad (6.1)$$

Here α is the dispersion coefficient (which mainly arises from geometrical dispersion); β represents the velocity of the intestinal fluid; and γ is a constant representing the absorption rate of the drug. This rate constant is a function of the effective permeability, P_{eff}, and the radius of the intestine.

Other variants of this approach model concentrations for two species of molecule: the undissolved and the dissolved drug molecules are treated separately. Dispersion models may also be complicated by varying the dispersion coefficient α and the liquid velocity β over space to mimic, for example, the fact

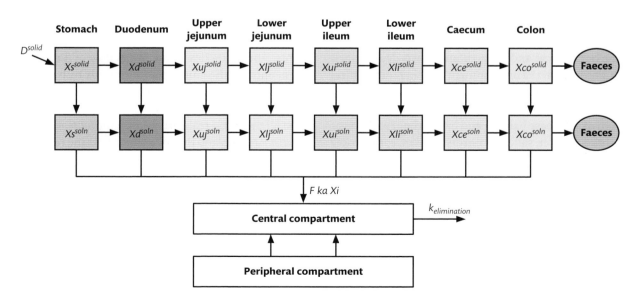

Figure 6.13 A compartmental model of drug transit through the GI tract. Image courtesy of Cesar Pichardo-Almarza.

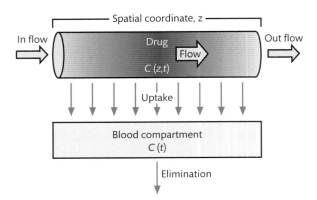

Figure 6.14 Dispersion model for the absorption of drugs from the intestine. Image courtesy of Cesar Pichardo-Almarza.

that the fluid slows down as it approaches the large intestine. A rather more complex model of this type, originally developed to model the rat intestine and extended to the human scale, has been incorporated into the PK-Sim® physiologically based pharmacokinetic model of the whole human body (Bayer Technology Services, **http://www.pk-sim.com**).

6.5 Modelling Kidney Function and Homeostasis

6.5.1 Human Kidney Structure and Function

Basic Renal Function

The kidneys are responsible for the maintenance of the body's fluid and solute homeostasis despite variations both in the environment (e.g. in temperature) and in diet (e.g. in the amount and type of liquids consumed). They thus play a major role in the regulation of acid/base balance and have primary responsibility for water, salt, and nitrogen balance; this triple role is unique to mammalian kidneys. Since the process of salt and water excretion through the kidneys is sensitive to arterial pressure, the kidneys also set a baseline for long-term control of blood pressure; this relationship between renal function and blood pressure has been termed the *pressure-natriuresis relationship* or the *renal function curve* [11].

Mathematical modelling has made a significant contribution to our understanding of kidney function in recent years. It has contributed to our understanding

of its basic mechanisms at many scales, from the kinetics of membrane-bound transporter proteins to models of exchange and recycling that involve the whole kidney. These have generally involved ordinary differential equations with constants and variables chosen to reflect different scales and aspects of kidney function. These might include, for example, kinetic models of the membrane-bound transporters and ion channels that are responsible for reabsorption and secretion in each segment of the nephron (see Figure 6.15 for a diagram of the anatomy of a human kidney), detailed models of epithelial transport along individual nephron segments, multinephron models of the medulla, or even reconstructions (see Figure 6.16) of the complete three-dimensional structure of a kidney from digitized sections of whole kidneys [12].

The Anatomy of the Kidney

The basic structure of a human kidney is shown in Figure 6.15. In the outer section of the kidney (the *cortex*) all the components of the blood other than cells and proteins are filtered in roughly spherical capillary beds called *glomeruli*. The filtered blood then enters the renal tubules or *nephrons*. A typical human kidney will contain approximately a million of these, each of which has a radius of about 10 μm. Nephrons begin at the glomerulus, in the cortex, and form a hairpin turn that extends to various depths within the medulla. They are surrounded by separate capillary networks in the three kidney regions, cortex, outer medulla, and inner medulla, that recover the fluid and solutes reabsorbed by the nephron segments. About 60–75% of the initial flow of fluid into the nephrons (the glomerular filtrate) is reabsorbed into the general circulation by the initial nephron segment, the *proximal convoluted tubule* (PCT); almost all the rest is reabsorbed from the *loop of Henle*, the *distal tubule*, and the *collecting ducts*. The remaining filtrate, which may be only about 1% of the initial filtrate, is excreted as urine from the papillary tip of the inner medulla into the pelvic space. This empties into the *ureter* and from there into the bladder.

The renal medulla is found in all higher vertebrates. It is formed from the hairpin turn sections of the nephrons (the descending and ascending limbs of the loop of Henle) and the blood vessels. Exchange processes within the medulla between the blood vessels (the *vasa recta*) and the loop of Henle allow the

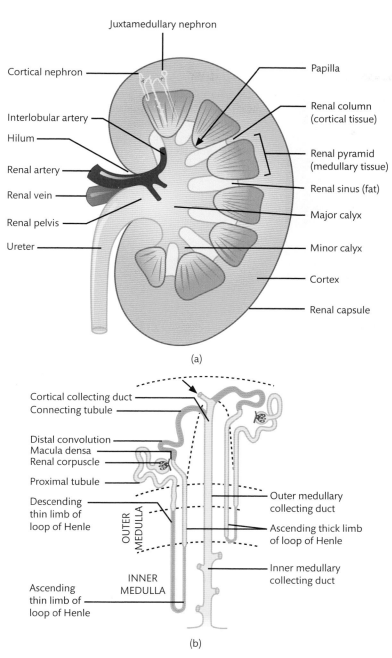

Juxtamedullary nephron

Cortical nephron

Interlobular artery

Hilum

Renal artery

Renal vein

Renal pelvis

Ureter

Papilla

Renal column (cortical tissue)

Renal pyramid (medullary tissue)

Renal sinus (fat)

Major calyx

Minor calyx

Cortex

Renal capsule

(a)

Cortical collecting duct
Connecting tubule

Distal convolution
Macula densa
Renal corpuscle

Proximal tubule

Descending thin limb of loop of Henle

OUTER MEDULLA

INNER MEDULLA

Ascending thin limb of loop of Henle

Outer medullary collecting duct

Ascending thick limb of loop of Henle

Inner medullary collecting duct

(b)

Figure 6.15 The structure of the human kidney. (a) Basic anatomy of a single human kidney, showing the cortex (outer layer), the medulla (inner layers), the ureter, and the main structures and blood vessels. Two nephrons, one short (cortical) and one long (juxtamedullary) are sketched towards the top of the kidney. (b) The structure of a short and a long nephron and their common collecting duct system, indicating the position of the loops of Henle. Redrawn from Pocock, Richards and Richards (2013) *Human Physiology* 4th edition, Oxford University Press; and Kritz and Bankir (1988) *American Journal of Physiology* 254, F1–F8.

generation of an osmotic gradient between the cortex and the medulla via a process that is not yet fully understood. This gradient, which is known as the corticopapillary osmotic gradient, can concentrate or dilute the urine that is formed within the kidneys, depending on the level of antidiuretic hormone (ADH). This, in turn, permits higher vertebrates to maintain a salt and nitrogen balance without excreting copious quantities of urine that would tend to dehydrate the body. The inner medulla, which has the highest osmotic gradients, is only found in mammals.

The blood supply to the kidneys is remarkable in a number of ways, some of which are described below.

- The blood flow through the kidney is higher when measured per gram of tissue (approximately 5 ml/min per g) than that through any other organ; for comparison, it is

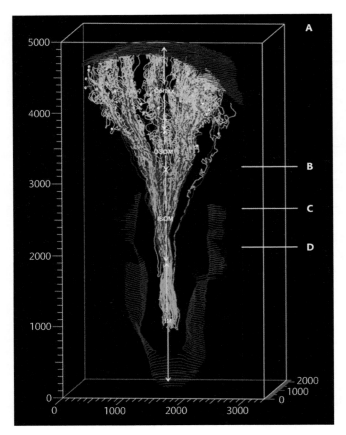

Figure 6.16 All short and long nephrons and collecting ducts traced within one mouse kidney. The red lines represent the renal capsule and the papilla. The white dots represent glomeruli (not actual size). Tubule segments are represented in different colours: proximal tubule, blue; thin limbs (TL), green; thick ascending limbs (TAL), red; distal convoluted tubule (DCT), purple; connecting tubules (CNT), orange; and collecting ducts (CD), brown. Scale = 50 μm. Reproduced with permission from [16].

about 0.5 and 0.9 ml/min per g in the brain and the heart respectively. This is not because the kidney needs more energy than any other organ, but because of its role in processing the blood to maintain homeostasis.

- The blood supplies to the three basic regions of the kidney (cortex, outer medulla, and inner medulla) are essentially independent. This is functionally very important but poses a number of technical problems for our understanding of kidney function and mechanism.

- The renal blood flow is autoregulated within certain limits, as it is in other organs. However, in addition, the rate of glomerular filtration (GFR) is also regulated by feedback so the flow into the distal nephron is maintained within strict limits. The loop of Henle has a hairpin structure (Figure 6.15b), and after it plunges into the medulla it returns to the vicinity of its own glomerulus, forming a structure called the *juxtaglomerular apparatus* (JGA). This consists of *macula densa*

cells in the distal tubule and *juxtaglomerular cells* in the arterioles. The macula densa sends signals to adjust the arteriolar resistances depending on the flow in the tubules and the salt concentration. This feedback is of course subject to a delay, due to the time it takes for filtrate to pass through the loop of Henle and reach the JGA; this leads to oscillations of pressure and flow that can be coupled among neighbouring nephrons.

- The inner medulla is constitutively hypoxic (i.e. it lacks oxygen), so most of the energy for the cells in this region must be provided by anaerobic glycolysis.

6.5.2 Modelling Kidney Physiology

Most models of renal physiology that have so far been developed have been steady-state ordinary differential equation models of internal renal mechanisms at appropriate scales and levels of resolution.

For example, modelling epithelial transport along nephron segments uses equations that are similar to those used generally to model epithelial transport, although integrated over the geometry of a tubular structure rather than a flat epithelium [13]. The model equations must respect the principles of mass conservation and, where appropriate, that of electroneutrality along each type of tube and vessel and among the structures that exchange water and/or solutes in each region. The latter principle states that an electrolyte is electrically neutral, with equal numbers of positive and negative charges.

Generally, the equations underlying these principles are mathematically relatively simple. Problems with solving them arise, when they do, not because of the complexity of the basic equations involved but because of a wide range of time constants or spatial scales (a property of models that is known as 'stiffness') or of problems related to the topology of the structures involved and differences in permeability and transport rates between the different regions of the kidney.

A more complete overview of the equations that have been used to model kidney structure and function than is presented here can be found in Keener and Sneyd (2009) (see Recommended Reading) and in [14,15]. Reference [15] in particular addresses a number of important topics including models for the mechanism of glomerular filtration, calcium signalling, and the selectivity of the glomerular filtration barrier.

In the next subsection we present an example of one type of kidney model in some detail, namely, a 'kinetic model' of water and solute transport (reabsorption and secretion) along a segment of the nephron. This example uses the proximal tubule (see Figure 6.15). The basic principles of transport are very similar in other nephron segments, but each segment has a different set of channels and transporters in its luminal and peritubular cell membranes corresponding to its specialized functions, and these need to be modelled differently.

Briefly, then:

- the blood plasma entering the glomeruli passes freely through the glomerular capillary barrier into the proximal tubules, driven by the net hydro-osmotic driving force (only the red and white blood cells and the plasma proteins are too large to pass the barrier);

  - the GFR is given by: $K_f \times (P_c - P_t - \pi_{COP})$ where K_f is the glomerular filtration coefficient, P_c and P_t are the hydrostatic pressures in the glomerular capillaries and in Bowman's space (just after the glomerular capillary barrier) respectively, and π_{COP} is the colloid osmotic pressure in the glomerular capillaries, due to the plasma proteins;
  - GFR is about 20% of total renal plasma flow;

- at the end of the proximal tubules, the fluid is still isosmotic to plasma, but its flow rate is reduced by two-thirds and virtually all of the nutrients (such as glucose and amino acids) and most of the bicarbonate have been reabsorbed;

- in the loops of Henle, with their hairpin turns at various depths within the medulla (Figure 6.15), the fluid is first concentrated in the descending limbs and then diluted in the water-impermeable ascending limbs, which rejoin their own glomerulus at the *macula densa* (part of the juxtaglomerular apparatus), thus allowing feedback regulation of the glomerular filtration rate (tubuloglomerular feedback) based on the flow rate and salt concentration;

- the remaining dilute fluid enters the distal tubules at a flow rate of about 10% of GFR. During its passage along the distal tubule, connecting tubules, and collecting ducts its contents are subjected to fine tuning under hormonal control of epithelial water permeability and transport rates through specific ion channels and transporters. This enables the overall balance of the body's water, salts, nitrogen wastes (i.e. urea), and its acid/base balance to be maintained;

- the fluid excreted from the end of the collecting ducts into the ureter (i.e. the urine) may thus be copious and dilute or very concentrated with a low flow rate, and its acidity and salt and urea concentrations may vary considerably. This depends on hormonally controlled adjustments of solute transport and water permeability along the distal parts of the nephron (i.e. the segments after the macula densa).

Virtually all aspects of this multifaceted process have profited from mathematical modelling studies. These have enabled researchers to interpret experimental data, to direct attention to relevant experiments, and even, in some cases, to stimulate the development of new experimental techniques to measure unknown parameters.

We describe here, briefly, only one example of this type of simulation: modelling solute and water transport by the proximal tubule.

Figure 6.17 shows solute and water reabsorption (from the tubule lumen towards the blood) and secretion (into the tubule lumen) across the epithelium of the PCT through some of the many channels and transporters that are specific to this segment of the nephron. In this initial portion of the nephron, more than two-thirds of the filtered fluid is reabsorbed along with virtually all of the filtered nutrients (such as glucose, amino acids) and most of the filtered bicarbonate (HCO_3^-). This massive reabsorption of water and solutes represents an important amount of work; it requires energy, which is supplied by ATP to drive the 'sodium pumps' situated in the basolateral membrane of the cells. These pumps transport three sodium ions (Na^+) out of the cell in exchange for two potassium ions (K^+). This process maintains an important electrochemical gradient for sodium entry into and potassium exit from the cell. This inward sodium gradient is then coupled to the reabsorption

of some solutes (e.g. glucose and amino acids) and to the secretion of others (e.g. H^+ ions). These transport processes are all secondary-active in that they move solutes against their net thermodynamic driving forces; the energy for this is indirectly provided by the ATP that drives the sodium pumps. Some chemical reactions, mainly for acid/base buffering, also occur. Water is reabsorbed (largely through aquaporin channels in the apical membranes of the cells; labelled AQP-1 in Figure 6.17) along with the solutes, thanks to small osmotic differences that are established by the transmembrane solute transport. To complete the picture, there is also passive paracellular movement of certain solutes and some water through the tight junctions in the PCT.

System Equations for Flow and Reabsorption along the Tubule

To formulate a mathematical model of this process we first develop a system of equations based on the basic conservation processes; namely, conservation of mass (both along the tubule and across the epithelial cells) and of macroscopic electroneutrality in each compartment. To be explicit, steady-state flow along the tubule is described by the relation for mass balance, assuming constant

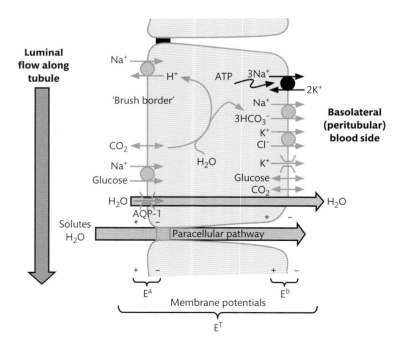

Figure 6.17 Transport processes involved in the reabsorption of water and solutes from the kidney tubule lumen into the blood, and secretion from the blood into the lumen.

luminal diameter and absence of radial gradients within the lumen. This is given using the following equations:

$$\frac{\partial F_v}{\partial x} = -J_v^{ap} - J_v^{tj} \qquad (6.2)$$

and

$$\frac{\partial F_i}{\partial x} = -J_i^{ap} - J_i^{tj} + S_i^{L} \qquad (6.3)$$

Here, F_v is the luminal volume flow, J_v^{ap} and J_v^{tj} are the total reabsorptive flux of water per millimetre of tubule length via the apical membranes and tight junctions, respectively, $F_i = F_v c_i$ is the luminal flow of solute species i, J_i^{ap} and J_i^{tj} are the total apical and tight-junction reabsorptive fluxes of species i per millimetre of tubule length, respectively, and S_i^{L} represents the net creation of i (per millimetre of tubule) due to chemical reactions. The relations describing transport across the apical and basolateral cell membranes and across the intercellular tight junction are given below.

In addition to these luminal conservation equations, the steady-state conservation of each species (i) within the cell can be described as follows:

$$\sum_i J_i^{ap} + S_i^{c} = \sum_i J_i^{bl} \qquad (6.4)$$

where each J represents a different type of transmembrane flux for a given solute i, and each S_i^{C} is a possible intracellular source of solute i (such as a particular chemical reaction). If the species are non-reacting, S_i^{L} and S_i^{C} will be zero.

Given the high concentrations of carbonic anhydrase both in the luminal cell membranes and within the cell, it is reasonable to assume equilibrium for the hydration of CO_2 to bicarbonate and protons. Therefore, using the Henderson–Hasselbalch relation for deriving pH as a function of pK_a:

$$pH = pK_a + \log_{10} \frac{\left(HCO_3^{-}\right)}{\left(0.03 \, pCO_2\right)} \qquad (6.5)$$

Here $pK = 6.1$, (HCO_3^{-}) is in millimolar, and pCO_2 is in millimetres of mercury (mmHg). We must also add to these equations mathematical relationships for enforcing macroscopic electroneutrality (total charge = 0) in each compartment, for conservation of current flow, and for constraining the total cell osmolarity to that of the peritubular solution, reflecting the high water permeability of cell membranes. A final equation reflects the assumption that all cells along the tubule can be expected to contain the same quantity of anions that cannot cross the cell membranes (impermeant anions).

The equations that we have set out and those we have mentioned here together give the global structure for a system of equations for fluid flow and reabsorption. Before any modelling can take place, however, it will be necessary to work out the detailed equations for the individual fluxes through all the membrane channels and transporters; that is, all the J_i terms that appear in equations 6.3 and 6.4. Equations have been defined to determine the diffusion of ions through channels in a membrane as a function of ion concentration, temperature, and charge; the flow of ions across tight junctions; and the kinetics of some types of specialist transport proteins. These are very complex, and will be of immediate interest only to researchers in this field (a representative number of them are described in more detail in the Online Resource Centre for the book).

To obtain a numerical solution for the flow of fluid and solutes along a tubule, a researcher will use the equations outlined above to formulate a set of ordinary differential equations and will set the initial conditions (which will include the volume filtration rate and the initial concentrations of solutes at the beginning of the tubule). These ordinary differential equations can then be solved using Newton's method, which is available in many generic mathematical software packages.

The equations set out here (and in the Online Resource Centre) have been specifically formulated to model transport along the PCT. Transport along other segments of the nephron has been similarly modelled using the kinetic properties of the specific transporters and channels embedded in the cell membranes in each segment to address their specific functions.

These mathematical formulations, which on the surface appear rather abstruse, should come to have important clinical applications. Defects in channels and transporters in kidney tubules are responsible for most renal disease and have been implicated in systemic pathologies such as hypertension and diabetes. Furthermore, this is only one relatively specific example of kidney modelling. We have not touched here on (for example) tubuloglomerular feedback or nephrovascular exchange in the renal medulla. Both these processes are of clinical importance, are amenable to mathematical modelling, and models of them are widely available in the research literature.

Future Developments in Kidney Modelling

Currently, there are no integrated models of the whole kidney that are equivalent in scale and complexity to the multi-physics models of the heart described in Section 6.3. However, several research groups are currently looking to produce such a model. Before an accurate and precise mathematical model of a whole kidney can be constructed, however, it will be necessary to find answers to a number of open questions about the fundamental mechanisms within the kidney. Some of these are now being addressed by new *in vivo* imaging techniques such as multiphoton microscopy [16]. This specialist form of optical light microscopy uses pulsed, focused laser beams to generate fluorescence, and it can be used to image living tissue up to a depth of about 1 mm. You should note, however, that the interior of the kidney must always be a difficult object to study experimentally, because virtually any experimental intervention will necessarily perturb the very processes that the researchers wish to study.

The enterprise of modelling a complete kidney can only be helped by the development of tools for enabling partial models of kidney structure and function to be linked together in similar ways to the more mature tools that have been developed to model the heart physiome. These renal physiome tools include the Quantitative Kidney Database (**http://physiome.ibisc.fr/qkdb/**) and an interface and grid portal (see Chapter 9) that is described in detail in [17]. We can hope that these tools will mature sufficiently for them to become useful not only to the modelling community but also to renal physiologists and clinical nephrologists more generally. However, there are still interesting unsolved questions in renal physiology that must be deciphered before this can

happen; skilled mathematical modellers will be required in this area for several decades to come.

6.6 General Homeostasis and Blood-Pressure Regulation

So far in this chapter we have described models of individual tissues and organs. (It is true that Section 6.4 concerned a system—the GI tract—but all the models discussed there were models of single organs within that system.) However, most physiological processes do not arise from the function of a single tissue or organ, but of multiple tissues and organs working together as systems. The concerted functioning of tissues and organs is regulated by hormones and through the nervous system. It is interesting that large integrated models of human physiology involving all these components—tissues and organs, plus regulatory feedbacks—were being developed long before the genome projects or modern computational physiology. The most famous of these were developed by Guyton and his colleagues, starting with the Guyton model of blood-pressure regulation which was first published in 1972 [18] and which is illustrated in Figure 6.18. This has continued to be developed, culminating in the HumMod model (**http://www.hummod.org** [19]). This includes over 5000 variables and has a sophisticated user interface that is oriented towards teaching physiology; there are a number of useful built-in clinical scenarios.

Guyton's original models of blood-pressure regulation have been extended in several ways. More recently, these models have been marked up in the CellML language (see Appendix and **http://www.cellml.org/**), principally for documentation. It has also been used as an example to demonstrate the concept of the 'core-model' approach applied to the integrated modelling of systems physiology. This involves a systems model being built up from low-resolution building blocks of the major organs and regulatory systems involved. These are based on validated experimental results and each individual component may at times be replaced with one or more higher-resolution detailed models. This enables the modellers to explore 'local' mechanisms, involving perhaps a single tissue, organ, or regulatory pathway, in detail while maintaining the global

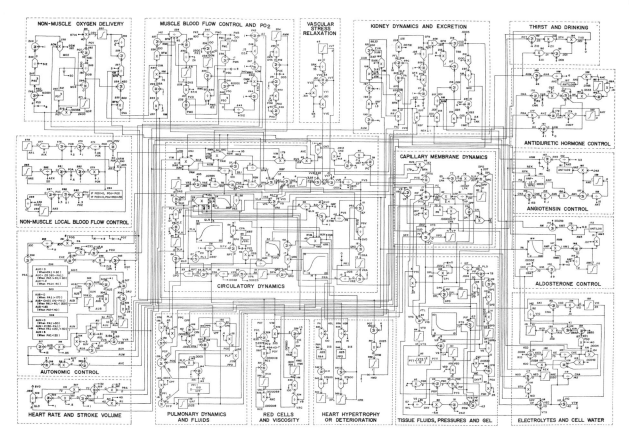

Figure 6.18 Systems analysis diagram of regulation of the circulation [18]. The model includes all major organs involved in blood-pressure regulation along with neural and endocrine regulatory loops.

regulatory feedback loops. In theory, the systems-level model may be as comprehensive as possible and as simplistic as needed, so even now it is possible to use this technique to model the whole of human physiology. The HumMod model, with its 5000-plus variables and parameters, is still probably the most sophisticated and complete model of human physiology to date, and it is now possible for users to modify its constituent equations and parameter files.

6.7 Conclusions

In this chapter, we have presented a series of case studies to illustrate the approaches that are currently taken to the computational modelling of human physiology at the tissue, organ, and, briefly, system

levels. We hope that you are now well aware that, given the complexity of the interlocking processes involved in physiology, modelling is already a necessary part of a physiologist's toolbox. It provides the means for evaluating the significance of experimental results, exploring the implications of hypotheses about the many questions that still remain open, and guides experimental design. Furthermore, the example of HumMod [19] has shown that comprehensive models are important in educating the next generation of physiologists, and this importance can only grow as virtual models of education and e-learning gain ground.

It cannot have escaped your notice that this chapter is far from comprehensive. We make no apology for this, as it would have been impossible to cover the whole of human physiology in the space that this book allows. The case studies presented here, however, do give a good overview of the range of mathematical techniques that are currently in common use in modelling human tissues and organs. Modelling projects are under way covering all the important human tissues and organs, and if your favourite organ is not among those covered in this chapter it will not take you long to find some relevant modelling papers in a database such as PubMed[4].

The challenge for the near future will be to bring some of these mathematical models of tissues, organs, and systems into the clinic; that is, to translate them from the bench to the bedside. Cardiac modellers have shown the way, and models of the heart are already being applied both directly in the clinic and in drug development. Much, however, remains to be done before models of other organs and systems—whether physiologically more complex, less well served historically, or simply less well funded—will be ready for clinical application. Maybe you will be among those who translate some of today's basic physiological models into useful bedside tools.

Recommended Reading

Advanced textbook(s) of human physiology, such as Silverthorn, Silverthorn, D.U. (2009) *Human Physiology: An Integrated Approach*. Harlow: Pearson.

Keener, J. and Sneyd, J. (2009) *Mathematical Physiology: Systems Physiology*, 2nd edn. Berlin: Springer. In particular, see pp. 821–850.

Noble, D. (2006) *The Music of Life: Biology Beyond the Genome*. Oxford: Oxford University Press.

Textbook(s) of continuum mechanics, such as Spencer, A.J.M. (2004) *Continuum Mechanics*. Dover Publications.

References

1. Alonso, L. and Fuchs, E. (2003) Stem cells of the skin epithelium. *Proceedings of the National Academy of Sciences USA* 100(Suppl 1), 11830–11835.
2. Smallwood, R. (2009) Computational modeling of epithelial tissues. *Wiley Interdisciplinary Reviews: Systems Biology and Medicine* 1(2), 191–201.

4 http://www.ncbi.nlm.nih.gov/pubmed

3. Latta, R., Clausen, C., and Moore, L.C. (1984) General method for the derivation and numerical solution of epithelial transport models. *Journal of Membrane Biology* 82(1), 67–82.
4. Schultz, S. (1981) Homocellular regulatory mechanisms in sodium-transporting epithelia: avoidance of extinction by flush-through. *American Journal of Physiology-Renal Physiology* 241, F579–F590.
5. Humphrey, J.D. and Yin, F.C. (1989) Constitutive relations and finite deformations of passive cardiac tissue II: stress analysis in the left ventricle. *Circulation Research* 65, 805–817.
6. Nordsletten, D.A., Hunter, P.J. et al. (2008) Conservative arbitrary Lagrangian-Eulerian forms for boundary drivenand ventricular flows. *International Journal for Numerical Methods in Fluids* 56(8), 1457–1463.
7. Yu, L.X., Crison, J.R., and Amidon, G.L. (1996) Compartmental transit and dispersion model analysis of small intestinal transit flow in humans. *International Journal of Pharmaceutics* 140, 111–118.
8. Agoram, B., Woltasz, W.S., and Bolger, M.B. (2001) Predicting the impact of physiological and biochemical processes on oral drug bioavailability. *Advanced Drug Delivery Reviews* 50, S41–S67.
9. Kadono, K., Yokoe, J. et al. (2002) Analysis and prediction of absorption behavior for theophylline orally administered as powders based on gastrointestinal-transit-absorption (Gita) model. *Drug Metabolism and Pharmacokinetics* 17(4), 307–315.
10. Ni, P.F., Ho, N.F.H. et al. (1980) Theoretical-model studies of intestinal drug absorption. 5. Non-steady-state fluid-flow and absorption. *International Journal of Pharmaceutics* 5, 33–47.
11. Guyton, A.C. (1991) Blood pressure control—special role of the kidneys and body fluids. *Science* 252, 1813–1816.
12. Zhai, X.Y., Birn, H. et al. (2003) Digital three-dimensional reconstruction and ultrastructure of the mouse proximal tubule. *Journal of the American Society of Nephrology* 14, 611–619.
13. Weinstein, A.M. (2003) Mathematical models of renal fluid and electrolyte transport: acknowledging our uncertainty. *American Journal of Physiology Renal Physiology* 284(5), F871–F884.
14. Thomas, S.R. (2009) Kidney modeling and systems physiology. *Wiley Interdisciplinary Reviews: Systems Biology and Medicine* 1, 172–190.
15. Edwards, A.I. (2010) Modeling transport in the kidney: investigating function and dysfunction. *American Journal of Physiology - Renal Physiology* 298(3), F475–F484.
16. Peti-Peterdi, J.N., Burford, J.L., and Hackl, M.J. (2012) The first decade of using multiphoton microscopy for high-power kidney imaging. *American Journal of Physiology - Renal Physiology* 302(2), F227–F233.
17. Harris, P.J., Buyya, R. et al. (2009) The virtual kidney: an eScience interfaceand Grid portal. *Philosophical*

Transactions Series A Mathematical, Physical, and Engineering Sciences 367(1896), 2141–2159.

18. Guyton, A.C., Coleman, T.G., and Granger, H.J. (1972) Circulation: overall regulation. *Annual Review of Physiology* 34, 13–46.

19. Hester, R.L., Brown, A.J. et al. (2011) HumMod: a modeling environment for the simulation of integrative human physiology. *Frontiers in Physiology* 2, 12.

Chapter written by

S. Randall Thomas, IR4M UMR8081 CNRS, Orsay & Villejuif, France and Université Paris-Sud XI, Orsay, France;

Rod Smallwood, University of Sheffield, Sheffield, UK;

Nic Smith, King's College, London, UK;

Cesar Pichardo-Almarza, InScilico Ltd, London, UK.

Multi-Scale Modelling and Simulation

Learning Objectives

After reading this chapter, you should:

- have an overview of how multi-scale modelling fits into computational biomedicine;
- understand the concept of scales and levels of biological organization;
- know an example of a generic framework for computational multi-scale modelling;
- understand some of the issues related to multi-scale computing;
- understand how to formulate a multi-scale model related to human physiology;
- know about some computer science methods that can help in multi-scale computing;
- understand some examples of multi-scale modelling applied to human physiology and medicine.

7.1 Introduction: Multi-Scale Modelling in Computational Physiology

If you have learned only one thing from reading this book so far, it may well be that the human body is a complex system. It comprises trillions of cells, each of which contains many thousands of different types of molecule: the cells work together in functional tissues and organs, just as the molecules work together to form functional cells. Furthermore, all humans live in a complex society comprising about 7 billion unique, interacting individuals who are influenced by many environmental agents, each of which can directly or indirectly impact their wellbeing. The complex systems that are human beings are not made of identical and undistinguishable components: rather each gene in a cell, each cell in a tissue, each tissue in an organ, and each individual has its own characteristic behaviour and provides a unique value and contribution to the systems in which it (or he or she) plays a part. Human physiology spans many different orders of magnitude and encompasses these scales in a continuous way, from the smallest microscopic scale up to the largest macroscopic one. We have seen that the human system or 'physiome' is influenced but not completely controlled by an individual's genome, proteome, and metabolome in health and disease, comprising a multi-scale, multi-science system [1].

In many cases it is possible (and, in fact, quite easy) to select the appropriate scale on which to study and to model a specific physiological feature, and the history of science has shown that this can be a very fruitful approach. We are now coming to realize, however, that many fundamental physiological processes are best modelled and simulated on many different scales. A heart model, for example, might include models of ion channels, heart muscle cells and tissues, and blood flow into and out of the heart. Many more examples are given in Kriete and Eils (2006) (see Recommended Reading at the end of this chapter).

Computational modelling or *in silico* experimentation on human physiology has only become common in the last couple of decades in response to the wealth of experimental and clinical data now available to test and validate these models. Now, however, modelling is frequently used in hypothesis testing, leading to the formulation of predictions that can be tested further using *in vitro* or *in vivo* studies. Many, but not quite all, of these published models demonstrate the need to go beyond modelling at a single scale. Typically, these propose a multi-scale approach to modelling physiology. This is often made explicit: for example, Friederike Gerhard and his coworkers state in a study of bone modelling [2] that '...in order to gain a deeper understanding of the whole process of bone modelling and remodelling, there is a need for hierarchical simulation models investigating different levels of resolution and incorporating these levels in a multi-scale approach'. Clearly, the more scales and levels that can be incorporated into such a model, the more comprehensive it may be. The current challenge, then, is not only to study fundamental processes on all these separate scales, but also to model the system as a whole, the mutual coupling through the scales in the overall system, and the resulting emergent structural and functional properties of the system to be modelled.

The previous chapters have described some popular methods for modelling human physiology at different individual scales. In this chapter we provide a brief overview of multi-scale modelling in the context of human physiology. We introduce a systematic approach to multi-scale modelling and computing that complements the classic dynamical systems approach, and discuss its application using one case study of an example system: the physiological process of in-stent restenosis in diseased blood vessels (Section 7.6).

7.1.1 The Scales of Interest

Human beings—in fact, all living organisms—can be thought of as inhabiting a four-dimensional world of space and time. Multi-scale modelling will therefore often involve modelling processes that act on a wide range of both spatial and temporal scales. Processes that are important in human physiology occur on spatial scales ranging from the molecular (most molecules are between 10^{-10} and 10^{-7} m in diameter) to the height of an adult human (up to 2 m); see Figure 7.1. Relevant timescales range from the time taken for a single intermolecular interaction (often about 10^{-6} s) to the human lifespan (10^9 s). These ranges encompass about 10 orders of magnitude in space and 15 in time.

This simple division of space and time into different numerical scales may not, however, be the most natural way of dividing up physiological processes into spatial and temporal ranges. A more abstract, but perhaps more logical, way of dividing up physiology is by levels of biological organization [3]. It is possible to identify up to 10 different such levels, ranging from the quantum level through the molecular, cellular, tissue, organ, and organism levels to the environment.

Different techniques are more or less appropriate for modelling the different levels; for example, on the molecular level ion channels can be modelled appropriately using molecular dynamics; on the cellular level models of cardiac cells can be derived using continuous ordinary differential equations; and on the environmental level an outbreak of an infectious viral disease such as SARS can be simulated using stochastic compartmental models. Irrespective of how the scale separation is expressed, however, such a coupling between scales must form part of the core of any approach to multi-scale modelling. Despite rapid progress during the past decade, it is still fair to say that the multi-scale modelling of complex biological systems including human physiology is a discipline that is barely out of its infancy.

To make sense of a complex, multi-scale system in order to model it, it will generally be necessary to disassemble it into tractable submodels. In practice, the number of submodels and the range of scales that can be represented will be determined by the availability of the supporting information and data that can be used to validate these models. We will describe later in this chapter some of the tools that are available to disassemble complex models into simpler ones that

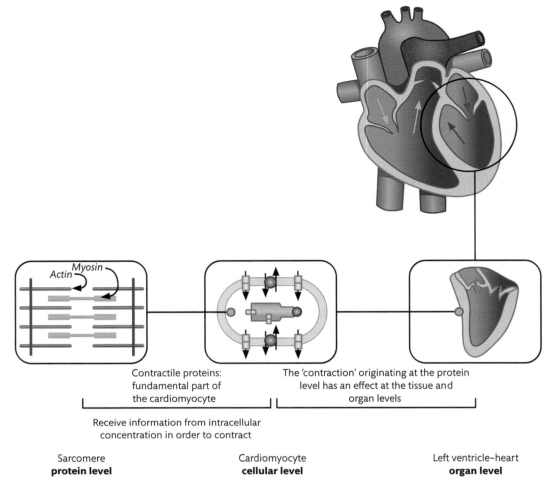

Figure 7.1 Processes that are important in human physiology take place over a wide range of spatial scales. This figure shows as an example how the function of the heart is built up from processes on the molecular, cellular, and organ levels. The contraction of the muscle proteins actin and myosin on the molecular level (10^{-10} to 10^{-7} m) drives the activity of the cardiac muscle cells or cardiomyocytes on the cellular level (10^{-6} to 10^{-5} m) and thus the pumping action of the whole organ (10^{-1} m). © Benjamin Bhattacharya-Ghosh.

can provide a common platform for discussion for both the biologist and modeller.

7.1.2 Examples from Biomedicine

It is clearly impossible to give a detailed overview of all applications of multi-scale modelling to the biomedical domain in a single chapter. However, to set the scene for a more formal discussion of a theoretical multi-scale framework that is appropriate for modelling human physiology, we list here a few representative examples that already have, or may soon have, clinical applications.

The discipline of multi-scale modelling as it is applied to human physiology is perhaps approaching a decade old. In 2005, Hunter and Nielsen described this process as 'the application of continuum field concepts and constitutive laws, whose parameters are derived from separate, finer-scale models' and identified it as 'the key to linking molecular systems biology (with its characterisation of molecular processes and pathways) to larger-scale systems physiology (with its characterisation of the integrated function of the body's organ systems)' [4].

Hunter and Nielsen assumed that the timescales are separated so that the lower-level, fine-scale processes are much faster than the higher-level, coarser ones. This implies that the lower-level processes are in quasi-equilibrium with the slower and larger-scale processes, and that they can be included at the higher level via, for example, constitutive equations or force fields. This type of multi-scale modelling, in which microscopic processes (small spatial scales, fast dynamics) are coupled to macroscopic processes (large spatial scales, slow dynamics) is the most common and features in most published studies; it is also exemplified by many of the examples that are listed below or described elsewhere in the book. However, other types of multi-scale coupling are also important; for example, the coupling of slow cell-based models on the tissue scale to fast macro-scale force fields such as those involved in modelling pulsating blood flow.

Multi-scale models of the heart are almost certainly more advanced than those of any other organ or system. Heart models are discussed in many other places in this book, and patient-specific heart models including multi-physics and multi-scale coupling are becoming precise and accurate enough to be used as a clinical decision-making tool in cardiac care [5].

Although no multi-scale models of any other organ system exist that are as sophisticated as those of the heart, multi-scale models of other organ systems are certainly emerging. Typically, models of smaller and simpler parts of an organ or system must be developed before they can be put together. These may themselves be multi-scale, but they will cover a smaller number and range of spatial and temporal scales than models of a whole organ will. Within the pulmonary system, for example, we already have detailed, anatomically based models of the lung, airway, and vascular trees; models of tissue mechanics and microcirculatory flow; and cellular-level models of erythrocyte gas kinetics. Scientists are only now beginning to integrate these into a multi-scale model of a virtual lung [6].

The musculoskeletal system has been addressed in a similar way. Multi-scale models of the human femur, for example, have already been produced, but there are still challenges involved in linking this and other detailed models into a full model of the system. Interestingly, this system can be said to demonstrate the 'middle out' approach introduced in Chapter 1 and touched on in other chapters. This implies that the researcher picks a particular biological level of interest as the starting point for the modelling exercise and then works up and down the scales as required. For example, models of bone remodelling following a fracture (an event that takes place at the organ level) also depend on both higher levels (the whole body) to provide bone loading conditions and lower levels (the tissue level) for constitutive equations and failure criteria and the cellular level for bone remodelling. The current state of the art in multi-scale modelling of the musculoskeletal system is described in detail in a recent book by Marco Viceconti (Viceconti, 2011; see Recommended Reading).

In the area of oncology, multi-scale, integrative frameworks have been set up for simulating tumour growth and how it is affected by treatment. In vascular medicine, multi-scale models have been produced for thrombus development. These couple a discrete model of cellular behaviour to continuous models of blood flow and biochemical reactions. They assume that the growth of a thrombus is a slow process. This assumption allows the transport equations for blood flow to be solved first and to provide boundary conditions for the model of thrombus growth. This is then iterated over many time steps of the slow model, which allows the growth of the thrombus to be studied in detail. Initially, activated platelets arrive upstream of the thrombus but, as this grows, the flow fields change and activated platelets and collections of blood cells are pushed back. This might explain the inhomogeneity and subsequent instability of the thrombus [7].

7.2 Why Multi-Scale Modelling?

A computer simulation of a system that spans many orders of magnitude in time and space is still simply out of reach, even for the fastest supercomputers. In any model, the minimum time step and the spatial resolution are constrained by the smallest scales of the problem. Suppose, for example, that the spatial scales of a model system span six orders of magnitude (from micrometres to metres) and that the temporal scales cover eight orders of magnitude (from seconds to decades). A fully resolved simulation of this model, in which space is represented by volumes (in three dimensions) and thus the whole model is four-dimensional, would require around 10^{26} calculations. Even if you assume that each of these operations requires one single floating point operation, such a simulation would still take years of central-processing-unit (CPU) time on an exa-flop machine (running at 10^{18} floating point operations per second), and no exa-flop machine has yet been developed.

Another problem with a fully resolved description of the system described above is that it would contain so much data that it would be almost impossible to extract anything of relevance from it. Therefore, this 'brute force' approach is currently impossible; it is probably also undesirable.

In order to confront these problems with the CPU and memory limits of currently available machines and with our limited capability to understand the output data, it is essential to devise a different theoretical approach. Multi-scale modelling techniques aim to provide a solution to this apparently intractable problem. Roughly speaking, these techniques assume that only a reduced number of scales play a significant role in the global behaviour of the system under consideration. Concentrating the computational resources on these relevant scales will usually make a numerical simulation tractable, although it will still require powerful machines. Care must be taken with this approach, however, as neglecting intermediate scales will always come at a cost in terms of the accuracy of the simulation and the quality of the predictions.

7.3 A Framework for Multi-Scale Modelling and Computing

We will now move away from discussing specific applications of multi-scale modelling that are relevant to human physiology, and, in fact, away from physiology and biomedicine altogether. This section introduces a theoretical and computational framework for multi-scale modelling that is designed for the implementation and execution of generic models in and for any scientific discipline. It is based on a computational approach that is known as complex automata [8].

The fundamental concept behind this modelling framework is that any multi-scale system can be broken down or decomposed into a number, say N, of single-scale models that interact together across the scales. Therefore, natural systems that span many orders of magnitude in space, time, or both can be described and simulated as a collection of simpler models, each of which acts on a specific scale, and each of which interacts (exchanges information) with the others. In computational physiology or biomedicine, each of these models will describe a different level of biological organization.

In this section we investigate systematically how these single-scale models can interact through the scales and discuss five different regions of time and space in which they might interact. We will then define a generic submodel execution loop (see Section 7.3.3) that itself can be used to define a small number of coupling templates. These then lead to a number of scale-bridging approaches that encapsulate many of the application-specific techniques that can be found in the literature.

7.3.1 The Scale Separation Map

Decomposition of multi-scale models can be achieved through building a scale separation map (SSM) on which each system can be represented as an area that describes its spatial and temporal scales. This easily defines the components of the multi-scale model as separate processes with well-separated scales. The selection of scales to use might be based on prior scientific knowledge or inferred from detailed studies of subsystems. We will not be concerned with this issue as such in this formal description of the SSM; we merely assume that this information is available.

Figure 7.2 shows an SSM in which the horizontal axis represents the temporal scales and the vertical axis the spatial scales. Panel (a) shows a system with the spatiotemporal domain A (Δx, Δt, L, T) as it is represented on the SSM. We assume that the overall process to be simulated is really multi-scale in the sense that it contains relevant subprocesses on a wide range of scales. In this case, it is not really feasible to run simulations that are based on the finest discretization of spatial and temporal scales (look back to Section 7.2 here). The formal approach in this case is to split the original multi-scale model into a number of simpler, single-scale models and to let these exchange information between them so that the

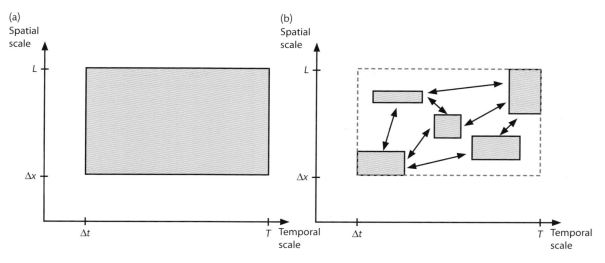

Figure 7.2 The SSM. (a) A single model spanning a wide range of scales; (b) a hypothetical multi-scale model of the same process that has been decomposed into five single-scale models.

dynamic behaviour of the multi-scale process can be represented as accurately as possible.

This separation is shown schematically in Figure 7.2b. This divides the single multi-scale model shown in panel (a) into five separate subsystems. The subsystem in the lower left-hand corner represents a model that operates on small spatial scales and short timescales, the one in the upper right operates at large scales of both space and time, and the other three operate at intermediate scales. These models might represent processes that operate on the micro-, meso-, and macro-scales.

The process of building an SSM starts by identifying all the different subsystems and placing them on the scale map, as shown above. The next step is to represent coupling between subsystems using edges on the map. An example of this is when one process can be coupled with another using a 'lumped parameter' deduced from a smaller-scale model. For instance, in modelling the vascular system, a model that deals explicitly with individual red blood cells might be used to compute a value for the local blood viscosity that is passed as a parameter to a simulation of blood flow on the macro scale. Another example is when models are coupled together through detailed, spatially and temporally resolved signals. In this case, both processes will typically share a boundary and exchange information synchronously; this occurs, for example, in fluid–structure interactions.

The distance between subsystems on the SSM indicates which model-embedding method should be used to simulate the overall system. In a worst-case scenario the modellers are forced to use the smallest scales throughout the whole of the model, which will probably result in a computationally intractable simulation. On the other hand, if the subsystems are well separated and the smallest-scale subsystems are in quasi-equilibrium, then they can be solved separately. Even in this case, however, there will probably be a need for feedback between the systems, although this may be infrequent and possibly driven by events occurring within the simulation.

7.3.2 Interaction Regions

Let us consider two processes, A and B, that occur on different spatial and temporal scales. Then process A can be represented as occurring on scales (Δx_A, Δt_A, L_A, T_A) and B on scales (Δx_B, Δt_B, L_B, T_B). Assume that process A has the largest spatial scale, or, if the spatial scales overlap, then A has the largest temporal scale. In other words ($L_B < \Delta x_A$) OR ($\Delta x_A < L_B < L_A$ AND $T_B < \Delta t_A$). We can now place process A on the scale map, and investigate various possibilities for placing B on the map relative to A. This leads to five types of multi-scale coupling, which are shown on Figure 7.3. Here, process A is shown in the upper right-hand region labelled 0. The type of multi-scale coupling will depend on the region of this diagram in which B is located.

If B is also found in region 0, then processes A and B overlap and there is no scale separation. This is the simplest case, and the result is a coupled single-scale model although processes A and B are still likely to be simulated using different processes (forming a multi-science or multi-physics model). If B occurs in region 1, there is a separation of timescales at the same spatial scale. Similarly, if B is found in region 2 there is a separation in spatial scales only, such as when coarse and fine structures function on the same timescale. If B is located in either of the regions labelled 3 then it is separated from process A in both the temporal and the spatial scales.

The case in which B is located in the region that is labelled 3.1 on the diagram is the well-known case of micro/macro coupling described earlier, in which fast processes acting on a small spatial scale are coupled to slower processes that act on a larger spatial scale. This type of multi-scale model has received much attention in the literature.

In region 3.2 we have the reverse situation in which a slow process on a small spatial scale is coupled to a fast process acting on a large spatial scale. This interaction region is often relevant to the coupling of biological processes on a cellular scale with physical processes on a larger scale. One biological example of this is the slow response of cells to a faster physical process on a larger scale, such as in blood flow in arteries.

It is important to realize that these are the only regions of the scale map that need to be considered in linking two processes. If process B occurs on the larger spatial and/or temporal scale, then the interaction regions shown on Figure 7.3 are simply reversed, and we fall back to one of the five cases identified above.

Consider again the two regions labelled 3, in which there is a separation in time and length scales. In region 3.1 we find that $L_B < \Delta x_A$ and $T_B < \Delta t_A$. As we have stated previously, this is the classical micro/macro coupling. Using the terminology that we are defining here, this means that the full spatiotemporal extent of process B (defined as $T_B \times L_B$) is smaller than one single spatiotemporal step $\Delta t_A \times \Delta x_A$ of process A. A large number of modelling and simulation paradigms have been developed for this type of multi-scale system [8].

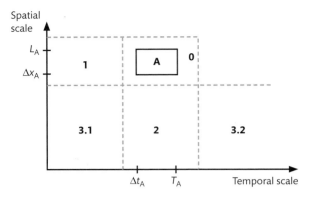

Figure 7.3 Interaction regions on the scale map.

A different situation involving separation of time and length scales is represented in region 3.2. Here, as in region 3.1, $L_B < \Delta x_A$ and so the spatial extent of process B is smaller than the grid spacing of process A. We also find, however, that $T_A < \Delta t_B$. In other words, the full timescale of process A is smaller than the time step in process B. Simulating a system that can be represented by two models that are coupled in this way will involve a different approach than that in the previous situation where B is located in region 3.1. In this case, the processes A and B will typically be coupled together through calculating time averages of the dynamics of the fast process, A.

7.3.3 The Submodel Execution Loop

Another equally important part of this modelling framework is the expression of each submodel using a generic, abstract instruction flow. Using this, we can express the multi-scale model formally and find regular patterns in the way that the models are coupled together that can be used to select a framework for the multi-scale modelling. This is also useful in selecting a computational environment that is appropriate for performing the calculations involved in an efficient way. The computational environments that are most useful for complex, multi-scale simulations and the methodologies and software that can be used to run them efficiently are discussed in Chapters 8 and 9.

Many, although not all, submodels can be described using the structure outlined in this section. This is a common workflow for subcomponents of multi-scale models that provides a way to identify generic coupling templates and develop a precise model for executing the simulation. In it, we represent the workflow using the following pseudocode abstraction. This is termed a *submodel execution loop* (SEL) [9].

```
t ← t_0
f ← f_init(t_0)
while t - t_0 > T do
O_i(f, t)
t ← t +Δt
f ← S(f, t)
f ← B(f, t)
end
O_f(f, t)
```

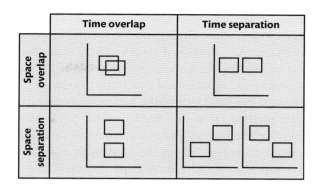

Figure 7.4 Interaction regions on an SSM.

In this pseudocode, f defines the current state of the submodel. This could be any number of things, examples being a simple variable such as a pressure; a complex field such as a strain field; or a discrete state space, representing for example the state of each cell in an agent-based cellular model (see Chapter 5). Time is denoted by t, and t_0 is used to represent the starting time for the submodel simulation. The five operators used in the pseudocode are shown in bold:

- f_{init} represents the initialization of the state and boundary conditions;
- **S** is the solution of one time step of the submodel (assuming a time-driven model);
- **B** represents re-establishing the boundary conditions;
- O_i and O_f are observations of the intermediate and final states of the submodel respectively.

Out of these operators, f_{init}, **S**, and **B** modify the state f. Each observation is accompanied by the time point of that observation.

7.3.4 Multi-Scale Coupling

Although multi-scale modelling is becoming more widely used, there is as yet no well-accepted generic methodology or nomenclature for this process. Several research groups have classified multi-scale models into different types and proposed strategies for solving each one, but none of these typologies has become widely used. The aim of these classifications is to illustrate how different strategies may be used to solve different types of problem, and help modellers select the coupling strategy that will enable a particular multi-scale problem to be solved with the greatest possible computational efficiency and accuracy.

However, it is not always possible, let alone easy, to find any correspondence between these different classification systems or between the terminology used or methodologies proposed. This indicates that there is still a significant lack of consensus in this field.

Following this caveat, however, we now return to our theoretical discussion of the SSM in Section 7.3.1 and the SEL in Section 7.3.3. There, we identified five different types of scale separation: where two perhaps very different processes operate on the same spatial and temporal scales; where the scales are separated in space but not in time, and vice versa; where one process operates at a smaller scale than the other in both space and time; and where a fast process on a small spatial scale is coupled to a slow process on a large one. These interaction regions are summarized again in Figure 7.4.

Multi-scale models can also be distinguished using a parameter known as the domain type. We can distinguish between *single-domain* (sD) and *multi-domain* (mD) types. In a sD model involving two processes, A and B, both processes can access the whole simulated domain and communication can occur everywhere. Alternatively, in a mD model each process is restricted to a different physical region and communication can only occur across an interface or a small overlap region.

7.3.5 Coupling Templates

Now we attempt to identify a method of multi-scale coupling that is appropriate for each combination of an interaction region and a domain type, basing our approach on the SEL discussed in Section 7.3.3, showing which operators from this model are coupled to each other. These can be known as *coupling templates*. Any published model can be described and analysed in these terms. For example, a generic method for computational modelling developed by Engquist et al. and termed the heterogeneous multiscale method [8] is described in these terms as a coupling template sD process. It would be found in interaction region 3.1 on the SSM, with a macroscopic process A and a microscopic process B. Then, the SEL for A and B can be defined as a coupling template $\mathbf{O}_f^B \to \mathbf{S}^A$; $\mathbf{O}_i^A \to \mathbf{f}_{init}^B$ (see also Figure 7.4). At each time step of the macroscopic process B a microscopic process A is initialized using information from the macroscopic model. It then runs to completion and sends its final information to the solver operator of the macroscopic process.

Our intention will often be to create a multiscale model from a collection of single-scale models. Coupling the submodels together is an essential part of this process, and, indeed, this coupling may be both as intrinsically interesting and as computationally expensive as the submodels themselves. Although each modelling scenario will involve a different set of multi-scale methods, the patterns of data flow between the models will be similar whatever the methods involved. Analysis of these patterns can give a useful insight into the dynamics of the multiscale model itself. A simple SSM for micro-/macrocoupling is shown in Figure 7.5a together with the associated coupling template in Figure 7.5b.

Coupling templates are defined as a possible data flow between the SEL operators of two instances of submodels, and they can provide a method for specifying these patterns. Looking back at the formulation of the SEL and the definition of its operators, the operators $\mathbf{O}_i$ and $\mathbf{O}_f$ only send data whereas the operators $\mathbf{f}_{init}$, $\mathbf{B}$, and $\mathbf{S}$ only receive data. This restriction renders coupling templates inherently unidirectional, as they are defined only between a pair of operators in which data are transferred from one to the other in only one direction. This means that there are six possible combinations of operators, each with its own interpretation. Four of these are listed in Table 7.1.

The operator $\mathbf{B}$ is not listed separately among the coupling templates in Table 7.1 but has a similar role as $\mathbf{S}$ in terms of the data flow. The role of these two operators can be distinguished using the concepts of mD and sD coupling. The general distinction between boundary operator $\mathbf{B}$ and solving operator $\mathbf{S}$ in coupling templates is that $\mathbf{B}$ is used in mD couplings while $\mathbf{S}$ is used in sD couplings with the same template. The motivation for this is that, in general, boundary conditions should deal with operations between domains, rather than with those that happen within a single domain.

It is possible to interpret these coupling templates in practical terms; that is, in terms of how they are applied to real simulations. First, if two single-scale models A and B overlap in their temporal scales, each will need to resolve multiple time steps of the other; that is, they will need to exchange information on their own temporal scales. For instance, if A operates with a time step of 0.5 s and B with a time step of 0.3 s, then A will run the inner loop of the SEL at

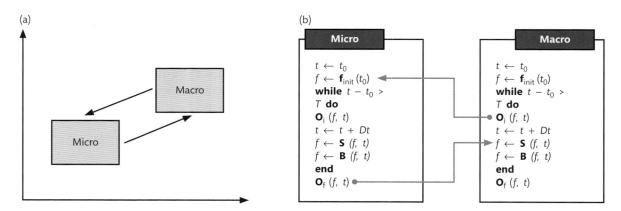

Figure 7.5 An example of micro/macro coupling. (a) SSM. (b) Coupling template.

Table 7.1 Four examples of coupling templates between submodels *A* and *B*, listing the temporal scale relation between the submodels and the most likely scenario in which each template is used

Coupling template	Temporal scale	Scenario
1. Interact $\mathbf{O}_i^A \rightarrow \mathbf{S}^B$	Overlap	Reciprocated by the same template
2. Call $\mathbf{O}_i^A \rightarrow \mathbf{f}_{init}^B$	Separation	Reciprocated by template 3
3. Release $\mathbf{O}_f^B \rightarrow \mathbf{S}^A$	Separation	Reciprocated by template 2
4. Dispatch $\mathbf{O}_f^A \rightarrow \mathbf{f}_{init}^B$	Any	Loosely coupled or state full

Note that more scenarios are possible.

timestamps 0, 0.5, 1.0, 1.5 s… whereas *B* will run at timestamps of 0, 0.3, 0.6, 0.9, 1.2, 1.5 s…. In this case, it is necessary to put a mechanism in place to resolve the coupling between *A* and *B* at these timestamps, which overlap but do not necessarily coincide. This is possible using the $\mathbf{O}_i$ and $\mathbf{S}$ (or $\mathbf{B}$) operators within the SEL, so the submodels will interact at each time step in both submodels. The details of the procedures that allow the results of one submodel (say *A*) to be interpolated to each timestamp of the other model (*B*) can be hidden in the detailed procedure for implementing the interaction. We discuss one way of doing this in the next section.

An alternative possibility is that the two single-scale models *A* and *B* are scale separated, so that *A* has a larger temporal scale than *B* and *B* may even need to be completely resolved once during each time step of *A*. Therefore, *A* may call *B* during each time step, and then *B* releases its observations when it finishes and sends them to *A*. When *B* is called, it gets the time of $\mathbf{O}_i$ as its starting time, and it must release its data to *A* before *A* makes its next step. When *A* (which may be an initialization or an *ab initio* state model) finishes, it may dispatch its state to *B*, which will continue computation at the time that *A* left off. Alternatively, *A* may just dispatch part of its state to the next instance of itself. In the example illustrated in Figure 7.3 the coupling template in region 3.1 can be described as a release/call pair.

7.3.6 Coupling Topology

The coupling templates indicate the type of data flow that can occur between a pair of submodels. However, if you need to interpret a complete multi-scale model

you should also consider its full topology. A coupling topology can be defined as a graph representation of a multi-scale model that has coupling templates as its edges. Each template is weighted by the number of times data is sent across it, and instances of the submodels form the nodes. This is similar to an SSM, although the maps contain submodels rather than submodel instances, and keep no knowledge of how often each coupling is used.

There are three properties of coupling topologies that are of key interest in multi-scale computational models. These properties can be summarized in the following questions, which are also visualized in Figure 7.6.

1. Is the model loosely or tightly coupled, or is it partially tightly coupled?
2. Are there multiple instances of certain submodels, and, if so, can the number of these instances be determined when the model is compiled?
3. Can the number of synchronization points be determined when the model is compiled?

The first question, about whether the model is loosely or tightly coupled, can be answered by constructing a directed graph of the coupling of the model. This is a graph with submodel instances—that is, individual points when a specific submodel code is run—as its nodes and coupling templates as described above as its edges. A model is loosely coupled if and only if that graph is acyclic. If a submodel is loosely coupled, computation becomes quite straightforward as the submodels can then be ordered and executed consecutively in a simple workflow.

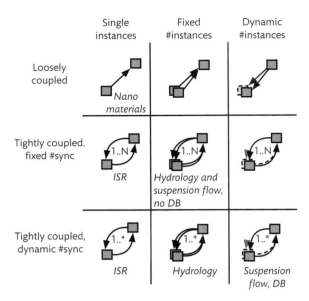

Figure 7.6 Graphical overview of different coupling topology properties, such as being loosely or tightly coupled, having a fixed or dynamic number of submodel instances, and having a fixed or dynamic number of synchronization points. All the examples shown here use two submodels, but this can be generalized to any number. #, Number of; DB, database; ISR, in-stent restenosis.

If this is not the case, a more complicated execution model such as dynamic execution or prediction of data flows will have to be used. When the *interact* coupling template is used and reciprocated, the resulting multi-scale model is tightly coupled. Cycles can also be formed using any of the other coupling templates, and these can include any number of submodels.

The second property, multiplicity, provides an indication of the number of instances of each submodel that are needed. If there is more than one instance of any submodel, these instances need to be addressed separately and it might also be necessary to provide a means of forking and joining data. If the multiplicity is determined during the runtime of the model, there must be a mechanism through which the model can spawn new instances of submodels, increasing the complexity of the model and the framework that is needed to run it.

One property of a multi-scale model that can lead to a need for multiple instances of a submodel is spatial scale separation. Imagine two models, again called A and B, and with the spatial scale of A being

the larger of the two. Then, the domain of A may contain many subdomains modelled by B, and many instances of B may be needed to model them all. Furthermore, if the phenomenon that is modelled by A is not regular in either time or space then it might not be possible to predict the number of instances of B that are needed for a particular model. In some cases this might even change over time.

The third property listed above—and the third question to be answered—concerns the number of times in which each submodel will transfer data and do a computation. If this is static, such as when the temporal scales of all submodels are regular, data transfers and submodel execution times can be predicted and thus scheduled. However, if this is not the case, and if model communication is organized only within the model, then the amount of time needed for the model to run is more difficult to estimate. Alternatively, if communication is scheduled at particular points during the simulation run, the scheduler must be informed of the number of communications and calculations that are due to take place.

The three properties defined here have a major impact on the complexity of software that is needed to compute the multi-scale model; if a property is more dynamic then more complex software will be needed. A loosely coupled topology with a single or fixed number of instances of each submodel can easily be expressed as a workflow, and multi-scale models like this can easily benefit from existing solutions. Tightly coupled topologies with fixed instances and, perhaps, unknown numbers of instances of certain submodels are more complex and dedicated coupling software will very often be needed.

7.4 Scale Bridging

So far in this chapter we have described the models and their coupling templates in abstract terms, and have not been concerned with what this means in practice. Obviously, when you are developing a multi-scale model of a biomedical or physiological problem, elucidating how the different single-scale models are coupled together forms a major part of the process. If you intend to capture the multi-scale character of a soft tissue, for example, the models that you need to couple together will include a macroscopic structural mechanics model of the tissue and a smaller-scale model of the cells that make it up. Developing a framework for bridging these models together across the scales is exactly where the hard work starts. What we have aimed to do so far in this chapter is to define the issues involved and build a framework for multi-scale modelling that will make a problem like this more tractable.

Most of the literature that has been published specifically on multi-scale modelling concerns this type of scale bridging, often translating cell-level models (such as the dynamics of suspended red blood cells) into the macroscopic scale (in this instance, the macroscopic rheology of whole blood).

We cannot possibly, in this chapter, discuss the process of scale bridging in detail for a representative number of types of submodel. Instead, we take a different approach. Again, we step away from the domain of specific applications, and instead try to observe common strategies for bridging scales within models and to connect them to the theoretical framework that we have introduced in this chapter so far.

A key question when dealing with a multi-scale system is how to decompose it into several coupled single-scale subprocesses. This decomposition is certainly not unique and may well require an in-depth knowledge of the system concerned. Once the subprocesses are chosen, however, they specify the relationship between the computational domains and the interaction regions on the SSM. The expected coupling templates can then be determined using the classification scheme described in the previous sections.

Below we describe several strategies for dealing with systems that are inherently multi-scale and for simplifying them by reducing the area that they occupy on the SSM.

7.4.1 Time Splitting

This approach is appropriate if you are considering two processes that act at different timescales, such as the dynamics of individual erythrocytes (red blood cells) suspended in plasma and the fluid flow of whole blood. Let us assume that we have a sD problem that involves two processes and that can

be described using a solver operator $\mathbf{S}$ that is itself a product of two operators:

$$\mathbf{S}_{\Delta t} = \mathbf{S}^{(1)}_{\Delta t}\mathbf{S}^{(2)}_{\Delta t}$$

Here, Δt specifies the finer scale of the process. Then, if $\mathbf{S}^{(1)}_{\Delta t}$ acts at a longer timescale than $\mathbf{S}^{(2)}_{\Delta t}$ we can approximate M iterations of the dynamics as:

$$[\mathbf{S}_{\Delta t}]^M \approx \mathbf{S}^{(1)}_{M\Delta t}[\mathbf{S}^{(2)}_{\Delta t}]^M$$

This means that we evaluate the faster process (2) every time step but we only evaluate the slower process (1) every M time steps.

7.4.2 Coarse Graining

The goal of coarse graining is to express the dynamics of a given multi-scale system at the larger temporal and/or spatial scale only in that part of the computational domain where less accuracy is needed. It will result in a new process that can be specified by new $\mathbf{S}$ operators and that occupies a reduced area on the SSM. For example, coarse graining in space and time with a factor of 2 can be expressed as:

$$[\mathbf{S}_{\Delta x}]^n \approx \Gamma^{-1}[\mathbf{S}_{2\Delta x}]^{n/2}\,\Gamma$$

Here, Γ is a projection operator, Δx the fine scale, and n is the number of iterations that are needed to simulate the problem.

7.4.3 Amplification

This strategy can be used to reduce the larger timescale of a process. Consider, for instance, a process that acts with low intensity but for a long time, in a time-periodic environment. One example of a process of this type which is of relevance to biomedicine is cellular growth and proliferation of the interior wall of an artery, which can be thought of as a growth and remodelling process within a pulsatile flow. Using our formalism, let us consider two coupled (mD) processes that are iterated n times, where $n >> 1$.

$$\left[\mathbf{S}^{(1)}\right]^n \text{ and } \left[\mathbf{S}^{(2)}(k)\right]^n$$

In this formulation, k expresses the intensity of the coupling of process 1 to process 2. So, if $\mathbf{S}^{(1)}$ is periodic with period $m \ll n$, we can approximate the above evolution as:

$$\left[\mathbf{S}^{(1)}\right]^{m} \text{ and } \left[\mathbf{S}^{(2)}(k')\right]^{m}$$

with k' the new effective intensity of the coupling. For a linear coupling we would have $k' = (n/m)k$.

7.5 Multi-Scale Computing

So far in this chapter we have described a theoretical framework for multi-scale modelling in some detail. This section is rather more practical. In it, we describe a framework for actually implementing a model of this type and running a simulation on a computer. The computational systems that can be used for this type of modelling range between desktop workstations and advanced, high-performance, distributed resources. These resources will be described in much more detail in Chapter 8.

7.5.1 The Multi-Scale Modelling Language

In order to bridge the gap between the theoretical framework that is discussed above and computational implementations of such a framework, it is necessary to use well-defined, standard terminology for the modelling processes. A markup language termed the multi-scale modelling language, or MML, has been developed to introduce just such a methodology. This language can be used to describe, verify, analyse, and execute multi-scale models.

When multi-scale models are described using MML, the coupling topology must be specified explicitly, but the properties of this topology may be deduced. The language, however, does not specify how execution software should communicate with a multi-scale model; this, rather, is left to the way in which the software is implemented. Although the coupling topology will give rise to a few interesting properties, a more precise specification will be necessary to formulate a well-defined model that is ready to run. Within MML it is possible to specify submodels and their instances, and also their scale, their computational requirements, and how they should

be implemented. Couplings are made explicit using the concept of conduits that bind to specific ports of submodels. Submodels should not be aware of other submodels; instead, data messages between submodels are manipulated inside the conduit using conduit filters. So-called fan-out and fan-in mappers are used for distributing and collecting messages. Therefore, MML includes most of the features that are necessary for a full description of a model at the level of software architecture.

This description mainly concerns the formulation of MML that allows computation to run. Human modellers and operators will generally interact with models written in MML via its graphical representation, which is known as gMML. An example of this is shown in Figure 7.7. It includes all the elements listed above but does not contain any implementation details or information on scales. It is particularly useful for composing or communicating the architecture of a multi-scale model.

The gMML language was inspired by Unified Modelling Language (UML); it is a graphical representation of MML that includes some of the UML icons. A submodel instance is shown as a rectangle containing a name label. Mappers are drawn as hexagons, again with name labels, and a conduit filter is a rounded square with a name label inside or close to it. If there are a number of submodels with a sD coupling this can be emphasized by drawing a dotted rectangle around the submodels involved, optionally including the name of the object. Conduits are represented as edges between elements, each of which may be annotated with the type of data that is sent over that conduit. The head and tail decoration of each edge depends on the SEL operator that a conduit is sending to and from respectively. The

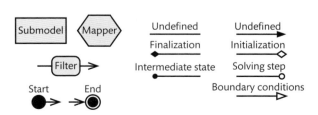

Figure 7.7 The elements that make up gMML. On the left the computational elements are shown; on the right are the edge terminals corresponding to the current operation in the submodel.

operators $\mathbf{f}_{\text{init}}$ and $\mathbf{O}_f$ are denoted by a diamond; the operators $\mathbf{O}_i$ and $\mathbf{S}$ are shown as circles, being called within the loop; and the operator $\mathbf{B}$ is shown as a triangle. The shapes of the sending operators $\mathbf{O}_i$ and $\mathbf{O}_f$ are filled with black, while the others are filled with white. Since a mapper does not have an SEL, it has a simple arrowhead coming to it and no arrow tail going from it.

Furthermore, the submodel instances that are initially instantiated by the model will have an edge from a filled black circle, and, conversely, the submodel instances that determine the ending of the model have an edge to a filled black circle with a black round border. These circle icons correspond to the start and end states of a UML state diagram.

The largest advantage of gMML is that it provides a visual way of communicating the computational architecture of a multi-scale model. It does, however, also have some limitations. These include a lack of detail, and the fact that there is no means within the language of representing or deducing the scales involved.

Obviously enough, computers interpret MML in a different way from human modellers. The Extensible Markup Language (XML) format xMML captures the features of MML in a way that is easily interpreted by machines. This can also capture a wide range of metadata, including scale information, possible parameter settings, a system of data types, binding ports of submodels and mappers, details that are needed for the implementation of the model, such as the number of cores that each submodel will need, and also descriptive and documentation facilities. In contrast to gMML, xMML can be automatically processed and it can act as an exchange format between models.

xMML aims to be both a descriptive and a prescriptive language. It is descriptive in the sense that it describes, in detail, all the submodels that make up a multi-scale model; their computational requirements; and the couplings between them. It is prescriptive in the sense that the scales that are specified by the xMML are expected to be maintained during execution and the conduits between the submodels are only meant to be used for the coupling templates for which they are defined. The descriptive part enables an execution framework to run and schedule a multi-scale model simply by reading its xMML specification. The prescriptive part allows

a stricter reasoning about the runtime behaviour of the multi-scale model than would otherwise be possible. The descriptive and prescriptive aspects of xMML should together enable the correct execution of a multi-scale model.

An example of a macro/micro model described using gMML is shown in Figure 7.8. This shows a two-dimensional macro submodel A and a set of 10 one-dimensional micro submodels, B.

Figure 7.8 illustrates a macro submodel that starts executing and then transfers an observation in the form of a grid to a fan-out mapper, which distributes different values to the micro submodels. Following this, the reverse direction is executed using a fan-in mapper after a filter is applied to reduce the resulting one-dimensional array to a single value. This is all specified in MML within the `<topology>` tag. The combination is an example of micro-/macro-scale separation that uses the coupling templates of call $(\mathbf{O}_i^A \rightarrow \mathbf{f}_{\text{init}}^B)$ and release $(\mathbf{O}_f^B \rightarrow \mathbf{S}^A)$. It implements this coupling within a single domain, since the micro submodels use a subdomain of the macro submodel.

More detailed specifications are available in the MML using the `<definitions>` tag. Here, the scales of the submodels are specified using basic SI units (metres etc. in space and hours, minutes, and seconds in time). If no SI unit is given, the default SI unit (seconds for time and metres for space) is presumed. In the example shown in Figure 7.8, all scales are regular except the second spatial dimension of the macro submodel, which uses the complete scale specification. Inputs and outputs of the submodel to which conduits may attach are defined using the `<ports>` tag. A SEL operator is assigned to these to associate conduits with a coupling template, and to predict how often the port will be used. A data type is also assigned to the ports to ensure that both the

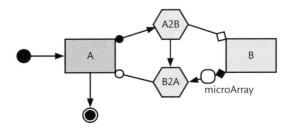

Figure 7.8 A simple micro/macro coupling shown in gMML.

sender and receiver use the same type of data. The ports and function of the fan-in and fan-out mappers can be specified using the `<mapper>` tags, and the `<filter>` tag allows a data type converter, scale reduction, or interpolation to be specified. Moreover, a model and its submodels may be versioned, described, and named, and all implementable computational elements may be given an executable class name. This enables all aspects of the multi-scale model to be defined using MML.

The computational aspects as well as the mathematical format of a multi-scale model can be specified using xMML. In the example above, the size estimate can be included in the data type and used to estimate the size of any message that is passed between the submodels involved and that depends on their scales. Message sizes may either be implied or explicitly listed for each coupling. Within this size estimate, submodel instances may be placed closer together in a distributed environment based on the amount of data that is transferred between them. An `<imple-mentation>` tag may be added to submodels, mappers, and filters to specify the estimated memory and CPU requirements of the submodel, as well as stating library or platform dependencies. When the implementation descriptions are combined, they will, for example, allow scheduling software to make more informed decisions about how submodels should best be distributed.

In summary, once a model has been implemented and fully described using MML (and optionally also gMML and xMML) it is possible for software to verify, analyse, and execute the model [9].

7.5.2 A Multi-Scale Coupling Library

So far, we have described some of the conceptual ideas that underpin a general framework for multi-scale modelling. These include decomposition into single-scale models, restriction to a common instruction flow, and the specification of a finite number of coupling templates. These ideas have been used to develop a software environment in which multi-scale models can be implemented naturally. This is known as the *Multiscale Coupling Library and Environment* (MUSCLE)[1]. Within this coupling library, both the single-scale models (or kernels) and the multi-scale coupling (or the conduits) are software agents.

These kernels and conduits (which are conceptually central to the modelling language) communicate with each other using two communication primitives:

- a non-blocking `send` operation, to send data from a kernel to a conduit entrance. This returns as soon as the data is sent to the conduit;
- a blocking `receive` operation, to allow a kernel to receive data from the exit of a conduit. The receiving kernel waits until the data is available before computations are resumed.

In this framework, the single-scale models that make up the multi-scale model do not need to be aware of each other and all information on the coupling and global setup is held by the framework. This allows the implementation of complex interfaces where multi-scale couplings can be performed using smart conduits. Furthermore, the structure of the coupling library allows complete independence from native codes. These can be replaced with a different source, provided that the interface, with respect to the framework, remains the same.

In the next section we move on to describe a particular example of multi-scale modelling that is relevant to biomedicine. This multi-scale model of in-stent restenosis involves three single-scale models that have been implemented using different programming languages (Fortran 90, C++, and Java). These are wrapped as agents and connected together to form a multi-scale model using the MUSCLE framework.

7.6 Case Study of a Multi-Scale Model: In-Stent Restenosis in Coronary Arteries

In this section we introduce and then describe in detail an example of a multi-scale model from computational biomedicine. This model has been developed within the Multiscale Modelling and Simulation framework. The problem that we discuss is that of in-stent restenosis (ISR), which is a pathogenic physiological response to a common method of treating coronary artery disease. Should you wish to learn

---

[1] http://apps.man.poznan.pl/trac/muscle

more about this case study, you will find more information in [10,11].

7.6.1 In-Stent Restenosis

In vascular medicine, the word *stenosis* is used to refer to a narrowing of the lumen or central cavity of a blood vessel. Stenoses arise in the arteries of patients with atherosclerosis due to a build-up of fatty deposits such as cholesterol and triglyceride (together known as atherosclerotic plaques). These can be corrected using a process known as balloon angioplasty, which is commonly carried out in conjunction with a metal mesh or stent, which is inserted into the blood vessel to prevent it from collapsing. The insertion of the balloon and stent in the blood vessel will always injure the blood vessel wall, and this can cause an abnormal proliferation of smooth muscle cells (SMCs) leading to growth of the blood vessel wall ('neointimal growth'). This maladaptive biological response can produce a new narrowing of the blood vessel, which is termed in-stent restenosis (ISR).

In this process, there is a recognized interaction between the biochemical pathways involved in cell growth and the physical processes in the smooth muscle and endothelial cells of the vessel wall in response to the mechanical stimuli involved in this process. These stimuli include the initial injury to the vessel, the strain that is imposed by the stent, and the shear that is imposed on the artery by normal, pulsative blood flow. This process can therefore be described using a multi-scale and multi-science model [11].

The design and geometry of the stent that is employed will influence the biological events occurring in the vessel following its deployment. The thickness, number, cross-sectional shape and arrangement, and length of the struts all influence the haemodynamics, degree of injury, and stretch that is observed within the stented segment. These, in turn, are critical determinants of the response of the arterial tissue, and thus the degree of ISR that is observed. Additionally, stents may be coated with active compounds (drugs) that are targeted to the biological processes that are responsible for driving the progression of restenosis. These drugs will be eluted locally at the stented site and can prevent the proliferation of SMCs and neointimal growth.

A multi-scale *in silico* model of the process of ISR has been developed. This model is designed to be able to test the influence of both stent geometry and drug elution on the process of restenosis. Its development has been motivated by the need for a better understanding of the dynamics that regulate this process. Such a model would provide a powerful tool for improved understanding of its biology, and assist in improving the design of devices and the deployment of drugs.

The process of ISR is like many other biological systems in that its dynamics span many orders of magnitude, from the smallest microscopic scales to the largest macroscopic ones. It is therefore an ideal process for modelling using the multi-scale approach that is described in this chapter.

Restenosis can be loosely described as a 'loss of gain'; that is, the vessel lumen returns, after restenosis, to a size that is similar to that observed before medical intervention (see Figure 7.9). It has, historically, been considered to be an over-reaction of the

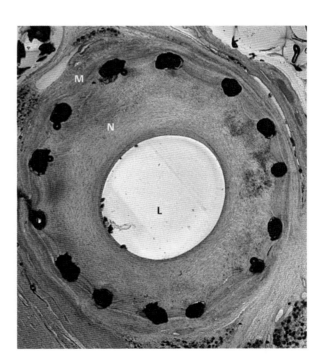

Figure 7.9 A stented coronary artery showing a significant restenosis due to neointimal growth (N) into the lumen (L). Stent struts are clearly visible and penetrate into the original vessel wall (M). Image courtesy of Dr Julian Gunn, Senior Lecturer and Honorary Consultant Cardiologist, Department of Cardiovascular Science, University of Sheffield, UK.

general wound-healing response in vascular tissue. From a biological standpoint, the injury caused by stent deployment and during inflation of the balloon is thought to trigger a cascade of inflammatory events that ultimately results in the development of new tissue, the neointima, within the lumen.

Again historically, most researchers studying this phenomenon have considered the biological and physical processes independently. This is a simplification as there is, in fact, a complex interplay between the two. The blood flow over the stent, its geometry, biological processes such as inflammation, drug elution and diffusion all affect the overall response of an artery wall to the deployment of a stent. We describe here a modelling process that aims to improve the understanding of this complex biological phenomenon by considering it explicitly as a multi-scale system.

7.6.2 An SSM for ISR

The development of the model that we discuss here started with an in-depth literature review to define the processes that are key to the regulation of restenosis. The temporal and spatial scales of these processes were then defined, with coupling considered in terms of the interactions between these processes. The next step was the generation of a complex SSM. This contained all the submodels that were thought necessary to capture the behaviour of this complex system, and depicted the coupling between them as the flow of information between models. The result of this was a detailed SSM that defined the process and its submodels. This was seen to be a useful way of summarizing and organizing the many and varied processes that are known to play a role in ISR, and was used to build a systematic, quantitative model of ISR that could be used as a vehicle for discussion between modellers and biologists. This shows that even before the first computational model is constructed, an SSM can play a useful role in aiding discussion between inter-disciplinary groups of researchers who will not necessarily find collaboration easy. Some other examples of SSMs defining multi-scale models that are relevant to biomedicine are described in reference [12].

This section, however, presents a simplified version of the complete ISR model that focuses on the behaviour of SMCs in interacting with the blood flow and the drug that is eluted from the stent. This can be represented by a simplified version of the SSM, and this is shown completely in Figure 7.11.

In this model, the insertion of the stent is modelled as a separate process to provide an initial condition. After the stent has been inserted, the SMCs of the artery involved start to proliferate in response to the mechanical insult. The proliferation rate will depend on the blood flow [defined, specifically, by two parameters, the artery wall shear stress (WSS) and the oscillatory stress index (OSI)], on the number of neighbouring SMCs, and, if a drug-eluting stent has been inserted, on the local concentration of the drug. In turn, the blood flow in the artery will depend on the geometry of its lumen (and thus changes as the muscle cells proliferate) and drug concentration also depends on muscle cell proliferation. The model described here relies on the assumption that scale separation between the three single-scale models considered is confined to the temporal scale. It is, however, worth noting that spatial scale separation exists within the model of SMCs. The SMC model itself can be subdivided into processes that occur on the cellular level and those that occur on the level of the tissue. Incorporating this level of complexity into the overall model would give a hierarchical multi-scale model.

In the simpler representation discussed here, however, the slowest process is SMC proliferation, which is dictated by the cell cycle. Blood flow is a fast process that is dictated by the length of the cardiac cycle (heartbeat). Diffusion processes can occur on different temporal scales, but in this example, the diffusion coefficients and the spatial dimensions of the arteries involved result in diffusion on a timescale that is between that of blood flow and that of SMC proliferation. More complex

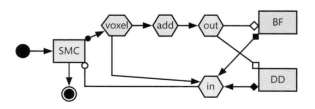

Figure 7.10 *A gMML representation of the ISR model described in this section.*

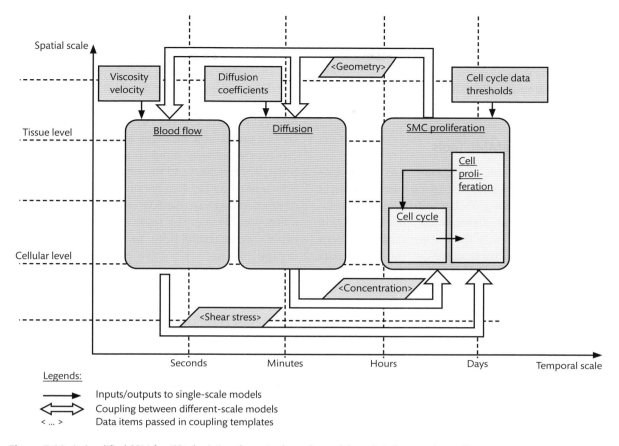

Figure 7.11 A simplified SSM for ISR, depicting three single-scale models and their mutual coupling.

models might also explicitly consider separation on spatial scales.

7.6.3 The Main Characteristics of the Model

The model that we are describing here is a simplified, multi-scale model for ISR (Figure 7.11). According to the framework discussed in Section 7.3.6, this is a tightly coupled multi-scale model with three submodels, describing SMC proliferation, blood flow, and drug diffusion respectively. The SMC submodel is coupled to both the blood flow (BF) submodel and to the drug diffusion (DD) submodel in a tightly coupled fashion. However, the latter two are not connected together. The coupling between the SMC and BF submodels is mD because cell proliferation, obviously enough, occurs in the blood vessel tissue whereas blood flows in the lumen. The interaction between these processes, therefore, occurs at the

boundaries of the system. In contrast, the coupling between the DD and SMC models is sD, because the DD model simulates the diffusion of drug in the vascular tissue.

This method employs the time-splitting method of scale bridging, which is one of those described in Section 7.4. Both drug diffusion and blood flow are much faster processes than cell proliferation, and the DD and BF models can be assumed to be in quasi-equilibrium with the SMC model. Therefore, after each time step of the SMC solver, both the DD and the BF models are called (see Table 7.1) by sending tissue and lumen geometries respectively. These models run to completion and are then released (Table 7.1).

The BF model sends its computed values for WSS and the DD sends values for drug concentrations in the tissue to the SMC model. This allows the SMC model to take its next step, and the complete process then iterates until a stopping condition is reached.

It is important to realize that the BF and DD models are independent of each other and can therefore be executed concurrently. This feature of the model, which affects the way it can run, can be detected and exploited automatically using the xMML specification for the model.

7.6.4 Using the gMML Tools

The full specification of the simplified multi-scale model of ISR that we are discussing here is shown in Figure 7.10 using the gMML specification. The specification of this three-dimensional model (ISR3D) in MML contains a few of the mappers that have been mentioned above. These carry out the simple data transformations that are necessary to ensure that the different single-scale submodels are not aware of the other submodels, or their internal representations, and scales. In practice, these mappers themselves will have a computational cost that should be considered when the model is implemented.

The submodels used in ISR3D are heterogeneous. Each is implemented using a different programming language (Java, C++, and Fortran), and they differ in whether they are serial or embarrassingly parallel (see Chapter 9 for an explanation of the difference between these types of parallelism) and whether they use particle or grid-based domains. Each of the submodels and mappers used in the multi-scale model is custom-made with the exception of the BF model, which uses the Palabos Lattice Boltzman simulator[2].

In order to create a software application from the MML specification of this model, and therefore to implement it on one or more computer systems, it was necessary to integrate the model with a software framework that can run tightly coupled multi-scale models. In order to do this, each of the submodels and mappers that make up ISR3D was wrapped as a MUSCLE component (see Section 7.5.2). The BF model, for example, acted as a MUSCLE controller that executed a Palabos simulation.

7.6.5 Simulating ISR

We now describe an example of the use of the multi-scale ISR model in which a two-dimensional

version of the model is used to simulate the overall dynamic response of the system. In this, the effect of strut thickness on the development of restenosis was studied by simulating the insertion of two bare-metal stents with different strut thickness. The depth of stent deployment was varied, and the number of new cells inside the vasculature (neointimal cells) was recorded as a function of time. The resulting growth curves were fitted to a logistic (sigmoid or S-shaped) function and the simulated ISR was characterized using just two parameters. The percentage of proliferating vascular SMCs was also measured as a function of time. The results were compared with experimental results from an *in vivo* model of ISR in pigs.

We now consider running this model using a segment of blood vessel that is 1.5 mm long and 1.24 mm wide, with a wall thickness of 120 μm. The strut thicknesses D considered were 90 and 180 μm, and the stents were deployed at four different depths in the artery wall (70, 90, 110, and 130 μm). Each simulation was repeated five times to obtain a mean and a standard deviation for the neointimal growth measurement.

The number of simulated SMCs (N) was measured in each simulation as a function of the time t after the stent was deployed. Some of the results obtained in this example are shown in Figure 7.12. Panel (a) of this figure shows the growth in cell number with a thick strut ($D = 180$ μm) deployed at a depth of 90 μm. The dynamics of this simulated growth does indeed resemble a logistic (sigmoidal) growth curve in that the initial exponential growth phase is followed by a slowing of growth and eventually by a plateau in which the number of neointimal cells reaches its maximum, N_m. In this case, all the simulations gave a similar pattern of cell growth, whatever values were used for the strut thickness and depth, although the maximum cell number N_m varied. The logistic growth of a population can be modelled using the simple equation:

$$dN / dt = rN\left(1 - N / N_m\right)$$

Here r is the growth rate, and as the initial number of cells N_0 is known in each case this can be solved analytically.

The next step in analysis of this model was to perform non-linear fits of the logistic growth dynamics

---

[2] http://www.palabos.org

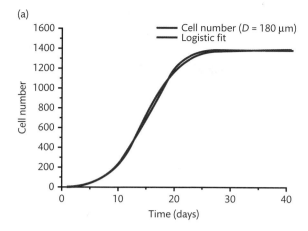

(a)

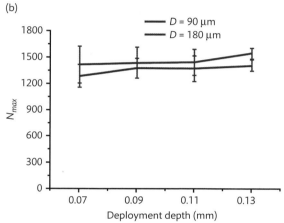

(b)

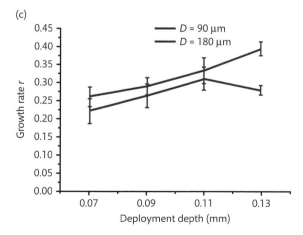

(c)

to the measured values of neointimal growth to obtain fitted values for the model parameters N_0, N_m, and r. As expected, these parameters were seen to vary with both strut size and deployment depth, and some of the results are summarized in Figure 7.12b and c.

The results summarized in Figure 7.12b demonstrate that this model predicts little change in the outcome of the model between the strut sizes. Interestingly, the model has predicted that the number of neointimal cells will be slightly higher if the implanted struts are thin. Furthermore, the cell growth rate increases when the strut is deployed further into the vessel wall. The only exception to this was observed for the thin strut deployed at the maximum depth (130 μm).

These simulated results were compared with experimental results obtained from pig arteries. Sections of stented pig tissue were taken from a database and the degree of injury imposed on the artery wall by each stent was noted. These data were used to generate a database in which this degree of injury could be related to the thickness of the stent, the growth of neointimal cells, and the length of time after stenting.

These data are summarized in Figure 7.13, which shows mean values and standard deviations from the

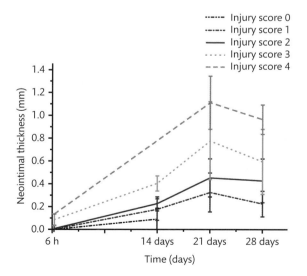

Figure 7.12 (a) Non-linear logistic curve fitting of neointimal cell number from a thick strut (D = 180 μm) deployed at a depth of 90 μm. Curve-fitting parameters N_{max} (b) and growth rate r (c) are plotted with respect to different deployment depths for both strut sizes.

Figure 7.13 Neointimal thickness as a function of time after stent deployment. The mean and standard deviation of the neointimal measurements are plotted at 6 h and 14, 21, and 28 days post-stent deployment. The data show a positive correlation between neointimal thickness and the Gunn injury scores.

full set of histological measurements available. Note that the measured neointimal thickness after 6 h does not correspond to an extremely rapid proliferation of SMCs, but to the formation of a thrombus immediately after stent deployment.

Analysis of these *in vivo* data shows that the extent of injury correlates with increases in the thickness of the layer of neointimal cells. A decrease in this thickness can be seen in almost all cases 28 days after stent insertion. This might be related to negative vessel remodelling, in which the neointima becomes reorganized. This process is well beyond the scope of the model that is described in this section. There do seem to be some similarities between these experimental data and the simulated human data. Comparing the results completely is not at all straightforward, however, since the quantities measured are different and the model and *in vivo* data are thus not truly comparable.

This model shows that SMC proliferation can be inhibited when the shear stress at the vessel wall (WSS) exceeds a given threshold. As the blood flux through the vessel is kept constant in this model, the decrease in the diameter of the vessel lumen caused by ISR results in a gradual increase of WSS. The threshold value of shear stress for inhibition of cell proliferation is reached at a lumen diameter that is more or less constant, independent of both strut size and deployment depth. Neither clinical data nor the data from pigs that is summarized in Figure 7.13 shows this type of end point. This, therefore, indicates that there is a flaw, or at least a simplification, inherent in the model that has been described here. In practice, an observation of this type of error will generally lead to a modification of the model.

Another useful finding of this model, however, concerns the effect of the deep tissue injury that is caused by the stent on the development of neointimal growth. Compare Figure 7.12c with Figure 7.13 and you will see that (ignoring the limitation of the final end point, as discussed above) the computational simulation follows the same basic pattern as the *in vivo* data. The degree of SMC proliferation increases with the degree of injury caused by the stent, which can be assumed to be higher the deeper that the stent is inserted. Differences arise only when you consider what happens 28 days after stent insertion; the decrease in neointimal thickness recorded *in vivo*

at this point, which may be related to negative remodelling, is not simulated in the model described here. It is, moreover, not possible to try to correlate this decrease with the output of the simulation because of the end-point limitation described above. The only solution to this discretion would come from increasing the complexity of the model to include these features.

Even in this simplified model, however, it has proved possible to fit the simulated data to a logistic growth function curve and thus to describe the dynamic response of the muscle cells to injury in terms of a growth rate. Although it is unreasonable to claim just from this that ISR is a logistic growth process, it is clearly possible to characterize it in terms of the logistic growth parameters r and N_{max}. This may prove useful in the preclinical testing of stent designs and perhaps, eventually, it may lead to the development of a clinically useful tool for predicting stent design that can be optimized for each patient with arteriosclerosis.

The model presented here is a simplified version of one that is still being developed and has not yet been fully tested. Further development of this model, and of the submodels that it incorporates, will lead to models that are more realistic in three dimensions and that will include further single-scale models representing, for instance, thrombosis, cell signalling, and vessel remodelling. This is relatively simple to do, at least in theory, using the framework described in Section 7.3. Furthermore, even a model as simple as the one described here provides a useful proof of principle that a complex biological response such as ISR can be usefully characterized with just a small set of parameters.

7.7 Conclusions

This book describes one of the central challenges in science and technology today: namely, how to understand, describe, and model complex systems that arise from the interaction of different processes. Many of these modelling scenarios also involve the interaction of different disciplines in biology, physics, chemistry, and medicine. These processes usually span a very large range of spatial and temporal scales, which makes a fully resolved approach computationally intractable if not completely impossible. It is therefore necessary to devise advanced multi-scale and multi-science methods to describe, model, and simulate these complex phenomena.

The contribution of this chapter to the broad discipline of computational simulation as it relates to biomedicine is the description of a theoretical framework to underpin multi-scale modelling and simulation, which can be applied to many different complex physiological and pathological processes. This framework is based on the separation of the complete multi-scale model of a phenomenon into a set of coupled submodels, each of which acts on a limited range of scales. We have described a few concepts that are fundamental to this understanding, including the SSM, the SEL and its operators, and the templates that are used to couple the different models together.

This framework therefore offers a general and coherent description of a multi-scale problem. These concepts can be exploited using a conceptual language, of which MML is a quite widely used example. This language provides scientists with a method for expressing the architecture of a complex multi-scale application and for communicating effectively with colleagues from different scientific disciplines. It also disentangles the 'scale-bridging' strategies from the specificity of each submodel, making it easier to reuse model components. Finally, a representation of a multi-scale model in MML will contain sufficient information to enable that model to be executed efficiently on an appropriate computational resource. Coupling consistency and scheduling constraints can be inferred automatically, allowing modellers to concentrate on the application side of the problem, rather than on the computational infrastructure.

We ended this chapter with a long case study that illustrates the application of this approach to the problem of ISR. This case study, although a simplified one, showed that it is possible to produce new, relevant scientific results with at least potential clinical applications using this type of simulation.

Multi-scale modelling remains a very active field of research with many challenges. One of them is undoubtedly that of validating multi-scale models. This may be more of a problem in biomedicine than in some other disciplines, as we have to deal with experimental datasets that are noisy or incomplete, with heterogeneous population data, and with data derived from experimental animals that may have more or less validity when applied to the human

system. The issue of the robustness and validation of multi-scale models of human physiology is therefore still a wide-open question. Another important challenge is that of timescale. Many simulations of the type we have discussed in this chapter are computationally intensive and it may be difficult, if not impossible, to obtain results on a meaningful clinical timescale, even using high-performance computational resources. These challenges and others are discussed, and some solutions offered, in the later chapters of this book.

 ## Recommended Reading

Kriete, A. and Eils, R. (2006) *Computational Systems Biology*. Burlington, MA: Elsevier Academic Press.

Sloot, P.M.A. and Hoekstra, A.G. (2010) Multi-scale modelling in computational biomedicine. *Briefings in Bioinformatics* 11(1), 142–152.

Viceconti, M. (2011) *Multiscale Modeling of the Skeletal System*. Cambridge: Cambridge University Press.

 ## References

1. Noble, D. (2002) Modeling the heart–from genes to cells to the whole organ. *Science* 295(5560), 1678–1682.

2. Gerhard, F.A., Webster, D.J. et al. (2009) In silico biology of bone modelling and remodelling: adaptation. *Philosophical Transactions of the Royal Society A: Mathematical, Physical and Engineering Sciences* 367(1895), 2011–2030.

3. Southern, J., Pitt-Francis, J. et al. (2008) Multi-scale computational modelling in biology and physiology. *Progress in Biophysics and Molecular Biology* 96, 60–89.

4. Hunter, P. and Nielsen, P. (2005) A strategy for integrative computational physiology. *Physiology* 20(5), 316–325.

5. Smith, N., de Vecchi, A. et al. (2011) euHeart: personalized and integrated cardiac care using patient-specific cardiovascular modelling. *Interface Focus* 1(3), 349–364.

6. Burrowes, K.S., Swan, A.J. et al. (2008) Towards a virtual lung: multi-scale, multi-physics modelling of the pulmonary system. *Philosophical Transactions of the Royal Society A: Mathematical, Physical and Engineering Sciences* 366(1879), 3247–3263.

7. Xu, Z., Chen, N. et al. (2008) A multiscale model of thrombus development. *Journal of the Royal Society Interface* 5(24), 705–722.

8. Engquist, B.W.E., Li, X. et al. (2007) Heterogeneous multiscale methods: a review. *Computer Physics Communications* 2, 367–450.

9. Borgdorff, J., Falcone, J.-L. et al. (2013) Foundations of distributed multiscale computing: formalization, specification, and analysis. *Journal of Parallel and Distributed Computing* 73(4), 465–483.

10. Tahir, H., Hoekstra, A.G. et al. (2011) Multi-scale simulations of the dynamics of in-stent restenosis: impact of stent deployment and design. *Interface Focus* 1(3), 365–373.

11. Evans, D.J.W., Lawford, P.V. et al. (2008) The application of multiscale modelling to the process of development and prevention of stenosis in a stented coronary artery. *Philosophical Transactions of the Royal Society A: Mathematical, Physical and Engineering Sciences* 366, 3343–3360.

12. Caiazzo, A., Evans, D. et al. (2011) A Complex Automata approach for in-stent restenosis: Two-dimensional multiscale modelling and simulations. *Journal of Computational Science* 2(1), 9–17.

Chapter written by
Alfons Hoekstra, Computational Science, Faculty of Science, University of Amsterdam, The Netherlands;
Bastien Chopard, Computer Science, University of Geneva, Switzerland;
Pat Lawford, Medical Physics Group, Department of Cardiovascular Science, University of Sheffield, UK.

Workflows: Principles, Tools, and Clinical Applications

Learning Objectives

After reading this chapter, you should:

- understand the general meaning of the term workflow when applied in the context of computational biomedicine;
- be aware of the theory underpinning this concept;
- know the difference between the two main classes of workflow implementation;
- be aware of some of the current open-source systems available for executing scientific workflows;
- understand the concepts of provenance and where they might be necessary;
- appreciate a number of practical workflow implementations from real research projects;
- understand the main considerations involved in selecting a workflow system for a specific project in computational biomedicine.

8.1 Introduction

With this chapter, we move from considering biological and biomedical systems and the different techniques that are used to model then, to some more practical considerations of how to implement them. You should have begun to appreciate just how complex an accurate model of any aspect of human physiology needs to be. This implies that many complexities are involved in how these models are set up and implemented on a computer system, or, as you will see, very often on more than one computer system.

8.1.1 What is a Workflow?

The dictionary definition of the term *workflow* is quite simple. It is: 'the sequence of industrial, administrative, or other processes through which a piece of work passes from initiation to completion'[1]. This rather vague statement illustrates why the term ends up meaning different things to different people. It can clearly be applied in many different contexts, some of which have nothing at all to do with scientific computing. We begin by discussing some of the background to the theory on which almost all computational workflows are based in order to clarify what this term means in this specific context. We will look at some of the domains in which workflows are used, and identify some relevant issues that are associated with scientific workflows. We will then give some concrete examples of workflows taken from the biomedical engineering community.

[1] http://oxforddictionaries.com/definition/english/workflow

The main focus of this chapter will be on workflows within the basic biomedical research community, with an emphasis on biomedical engineering. However, we will also compare and contrast these with the rather different requirements for a clinical workflow. You should already have appreciated that the end point for much research in computational biomedicine is expected to be the provision of clinical tools that will need to be fully integrated into a primary or secondary healthcare environment.

We will begin by discussing and highlighting some of the underlying theories involved in setting up computational workflows, with their advantages and limitations. These discussions are put into context by outlining how some of the computational workflow engines are most commonly used in a scientific environment to perform these operations. We will concentrate on three open-source systems: Taverna, Triana, and Pegasus. These are by no means the only such solutions, but they are ones that are currently most commonly found in applications across the whole of the biomedical and bioinformatics domain.

We will next discuss the importance of *workflow provenance*, which is a crucial aspect of any workflow execution in both the scientific and clinical domains. Finally we will take some case studies from computational biomedicine, looking at their requirements for executing workflows and how the solutions that have been chosen meet these requirements.

There are some topics, however, that will not be considered in this chapter, as they are covered in detail elsewhere in this book. In particular, the important issues of security and data protection, and in particular how they relate to the use of data derived from individual patients, will be covered extensively in Chapter 10. Here we will merely note briefly some of the issues involved in using data and tools from third parties.

Scientists and engineers designing workflows will often find challenges involved in the management of access to tools and, more often, data that have come from outside sources (third parties). The main reason for this is that the system will be set up so that the execution of the workflow is handed over to a central server. This server will need to 'impersonate' users who have authorized access to each tool or resource within the workflow. Many solutions have been found to this problem, but none of these can be completely satisfactory for all stakeholders, both users and suppliers. Generally, the level of security provided will be directly proportional to the level of inconvenience that must be experienced by the researchers who use the system. People who are involved with selecting a workflow will need to take this into account and select the most appropriate level of security; this will be greater if clinical data and, in particular, patient-specific data need to be processed.

8.2 Computational Workflows

The basic definition of a workflow, then, is just a set of events or tasks that are arranged to be executed in a particular order. Anyone who has studied logic or computer science, even at the most basic level, will be aware that there is a similarity between this and tool-chains and flowcharts. However, the concept of a computational workflow differs subtly from the other two concepts. If a flowchart is a way of representing a particular procedure—perhaps logical thinking, process control, or data manipulation—the computational workflows that are the subject of this chapter are simply implementations of these flowcharts in a programming language. Workflows consist of a set of *states* (which encapsulate sets of actions), *state transactions* (which define transitions between states), and *rules* (which control actions or transaction logic based on process data).

There are two main types of workflow implementation, and these have been termed *sequential* and *state machine*. Most of the commonly used scientific workflows are initially defined as sequential but we will see later how this often changes as the workflow gets closer to a formal implementation.

Workflow-oriented programming can often be more efficient as it enables users to manage several jobs simultaneously, separate logic implementation from code writing, and rerun common procedures automatically.

Workflows, therefore, are not intended to run as stand-alone programs [whether console- or graphical-user-interface (GUI)-based] but as a set of 'standard' instructions that can be run constantly or periodically or repeated as required. This introduces the concepts of the workflow *type* and workflow *instance*. The former term is given to the set of flow rules that are defined at workflow level, and the latter to a particular instance or 'run' of these rules.

A *workflow engine* controls the running of individual workflow instances; this is analogous to processor multithreading although it is implemented at a much higher level. Individual instances of a workflow are therefore accessed through this main engine, which is free to queue calls to instances; this is analogous to a switch or router. Higher resource efficiency can be achieved if individual workflow instances are able to persist in a database when they are idle without using the computer's working memory. This is analogous to the well-known procedure of suspending or hibernating an idle job in an operating system such as Unix.

The concept of *abstraction*, as applied to workflows, refers to the need for end users to be able to manipulate the behaviour of the workflow even if they have no knowledge of programming languages [or even just of the language(s) that the workflow is written in]. This is built into the workflow by providing a user interface that allows the modification of workflow logic. Abstracting the workflow logic and processes from flow control (think of a simple flowchart) has the added benefit that workflow definitions will not need to be recompiled when particular activities or services are changed, or when changes are made to the logic or rules.

If you are working on a big interdisciplinary modelling project that involves a complex workflow, you are very likely to find that the project to design the workflow is divided into a number of separate specialist areas. These might involve, for example, programming the workflow environment, coding specialist activities, the design of subsets of the workflow, and managing its day-to-day running. Workflows may be designed so that specific users or types of user are given or denied access to particular workflow tasks or routines.

8.2.1 Petri Nets

The mathematical concept of a Petri net provides a formal definition and a graphical representation for the execution of distributed processes. It can, therefore, be a useful way of representing the formation of a workflow. Petri nets consist of *places*, *transition nodes*, *directed arcs*, and *tokens*. In this formulation, places represent systems, transition nodes represent rules for changes in place to happen, directed arcs represent actions by connecting places with transitions, and each token represents a current location or active process. Progress through the process (which might, but need not, be a workflow) can therefore be represented by the consumption of tokens.

Figure 8.1a is an example of a Petri net that shows the consumption of tokens by an event firing (yellow). There are two possible ways arcs could branch, at places or transitions. Figure 8.1b shows branching at transitions. Since there is no conflict, concurrency is possible. However, the specific state of the system is ambiguous. This representation in which the pipeline is pre-set and depends on the release of tokens by the places may be used to describe a data-driven system. Figure 8.1c shows branching at places. There is a clear situation of conflict, so we must suppress the ability for concurrency. The plus side of this is having

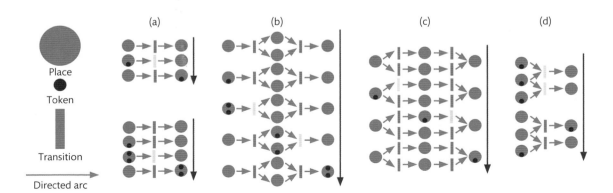

Figure 8.1 Basic illustration of a Petri net. See text for details.

the system represented at a specific state at any point in time. This representation, in which the pipeline is not known until completely executed and in which it depends on the particular events that are triggered, represents an event-driven system. With concurrency we lose the certainty of the state of a system.

Conflict is the situation in which token transitions are not certain. If concurrency and conflict occur at the same time it is called confusion. Figure 8.1d shows a state of confusion in which both transitions are ready to fire but the actions of one cancel the other.

We may therefore use Petri net notation to represent a workflow graphically by mapping each Petri net concept onto the components of the workflow as follows:

- *places* represent states or activities (circles);
- *transition nodes* are state change rules (bars);
- *directed arcs* are processes (arrows);
- *tokens* mark the current state of the system (dots).

If a workflow is to be valid, the presence of a token must allow for *deterministic execution*; this means that the execution of that task is fully described by the workflow rules, the initial conditions and formal decision points within the workflow logic. If the situation is ambiguous—that is, if a transition does not define a single, unique, and available place—then that transition is said to be in conflict. This might occur when a place is connected with more than one transition. If there are tokens in two different places at the same time, the workflow is said to be concurrent, with two processes being executed at the same time.

During the execution of a workflow we may need to know the state of the workflow at any given point in time. This means knowing where tokens will be allowed to transition next. If this is not clear, the workflow cannot execute and we say that the workflow is in a state of confusion. This happens when there is both concurrency and conflict, creating ambiguity during interpretation and the implementation of logic. It is an unacceptable situation for a computational implementation.

To avoid this state, the Petri net that defines a workflow may be constrained to allow only one phenomenon. How this is achieved depends on the type of workflow (see Figure 8.2). In a sequential workflow, each place is limited to one input and one output which ensures that there can be no conflict in logic flow, but concurrency is allowed. In contrast, in a state machine each transition is limited to one input and one output, which ensures that there is no concurrency.

The use of Petri net notation allows us to understand the limitations of workflow engines. However, while the execution order of a flowchart might be obvious to a human, the workflow itself will be interpreted by a machine. We must therefore be able to translate such a flowchart into a discrete set of rules that a machine can process. This provides a framework behind the design of workflows. Also, we often require the workflow engine to do many different tasks, including checking data types, validating the workflow itself, and executing complex sub-workflows. One of the most basic points to consider when implementing a workflow system is the selection of the most appropriate model for the workflow: the major paradigms being sequential and state machine-driven.

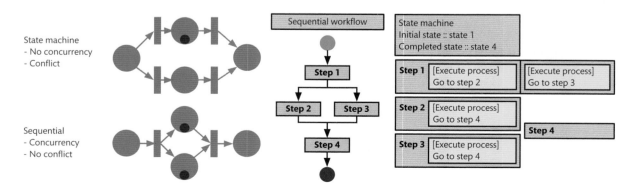

Figure 8.2 Purpose, automation, control, and design of workflows shown using Petri net notation. The left-hand panels illustrate the difference between a state machine and a sequential workflow.

Many workflows are implemented using distributed computer systems, in which more than one action—often many actions—can be automatically performed at the same time. This means that a flowchart procedure that needs to be heavily automated should be implemented as a sequential workflow. In contrast, a state machine is designed to exist in a state where it waits for an event to be triggered by another part of the workflow or by human intervention before a transition can take place [1,2]. Translating Petri net concepts into workflow terminology, a place refers to an activity (which could be a task and/or state the workflow is in) and flowchart-style arrows replace transitions and directed arcs. The interpretation of this control logic is dependent on the workflow engine.

A sequential workflow may be easier to conceptualize in that it represents a sequence of actions or activities that are dependent on each other and that must be executed in a particular order. This assured execution order gives confidence in data linkage and the implementation of logic. However, if you are designing a sequential workflow it is important that you work out the sequential order of all possible actions at the beginning of the process. This means that it is difficult to use this type of workflow to model behaviour that is highly conditional.

State machine workflows are those in which the workflow waits for events to take place in order to trigger workflow activity. Even though events may commonly occur in a given order, we cannot rely on this. The workflow logic must ensure that any interdependencies between workflow activities are met under all possible combinations of external events. This flexibility makes checking errors in logic and data in state machines a harder task.

In practical state machine implementations, activity interdependency may exist due to shared data; quite commonly an activity will work on a file that was created by another activity. Another common type of activity interdependency arises when a task depends on multiple decisions being made beforehand. The user must ensure that the logic is able to handle all combinations of external events.

It is possible to translate any formal Petri net graph into a general workflow design. A formal representation like this is generic and may seem a little abstract, but it is important to realize that these are the logical foundation on which all workflow engines are constructed. If you understand them completely you will be able to evaluate critically the limitations of different workflow implementations.

It is also possible to define a given scientific workflow using several different paradigms. One common task for a workflow involves data processing and mining. This clearly shows the importance of checking data types, the interdependence of events during the workflow (as each output serves as the input for the next process), and the importance of defining the sequence of processes. A given situation might require the execution of several tasks in parallel, or it might involve the execution of several extremely computationally intense programs or sub-workflows. A different situation arises when the motive of the workflow is a time-dependent simulation (an *in silico* experiment). This type of workflow is commonly used in chemical processing; one example of a scientific workflow of this type involving modelling at the molecular level is a molecular dynamics simulation (or a sequence of such simulations). These examples suggest that the workflow solution to a specific problem will need to be individually tailored to the specific situation.

8.2.2 Clinical Workflows

There is a clear distinction between a workflow that is designed to implement a series of tasks in scientific computing and one that is designed to be used in the clinic. A clinical workflow is likely to be closer to a business workflow than to a scientific one: essentially, it can be thought of as a customized implementation of a business workflow. This type of workflow will be driven by fairly static business logic and may fragment into many sub-workflows and processes. It is likely to be written in a language that has been customized for business use. The Business Process Execution Language (BPEL) is probably the most commonly used example of a language that is used to represent sequential workflows in a business environment. In addition, business workflows often require human interaction, and this can be difficult to manage in this system. The introduction of an extension to BPEL, called BPEL4People, demonstrates the specific difficulties that this type of workflow originally had when dealing with humans instead of machines.

Table 8.1 Differences between clinical and scientific workflow paradigms

	Clinical workflows	**Scientific workflows**
Purpose	Long-running patient monitoring	Computational intensive data manipulation
Automation	Human interaction	Highly automated
Control	Flow control	Data flow
Design	Stable design	Evolving design

The differences between clinical and scientific workflow designs (Table 8.1) are very useful, in fact essential, ideas to understand given the differences between the tasks that they are used to automate. However, these differences make the integration of scientific and clinical workflows within the same framework particularly difficult. Scientific workflows that are intended for clinical use are data-flow-oriented, and are created to automate and expedite the manipulation or analysis of data, particularly where this involves tedious repetitive processes. Individual functions are arranged into a pipeline of tools. Each of these tools will manipulate a data object by adding or extracting information and then present it to the user (or, of course, to another process). In effect, the tools can be thought of as a single process with one input and one output.

Most clinical workflows in current use are designed mainly to automate important non-scientific procedures, such as patient tracking and health centre management. No workflows have yet been produced that can effectively carry out diagnosis and clinical decision making, particularly as the condition of patients entering a hospital, for instance, cannot be assumed. The most pressing need is to integrate workflows into larger and more open organizations (such as a group of hospitals). Data will need to be shared between these organizations, although the tools used to analyse it may differ; in addition, data formats will differ depending on whether the data is meant to be analysed by humans or machines, and there might still be good reasons why the access to some data will need to be strictly restricted. (This is discussed in more detail elsewhere in the book.)

Clinical workflows are generally more static than scientific workflows, which will often need to change

as the scientists involved in the modelling explore new and better methods. Both types of workflow will need to incorporate many different levels of user; in the scientific environment, for example principal researchers, computer technicians, postdoctoral researchers, and students, are bound to have different access needs. In a health centre, the users of a workflow are likely to include clinicians, IT administrators, and software engineers.

Scientific workflows need to be flexible and might change in order to explore new and better methods. In contrast, clinical protocols are often long established making the workflow design static. Added considerations include the workflow design approval process; this indeed reflects the difference between the different user levels. In a scientific environment, users are more likely to be involved or require access to different levels of the environment than in a health centre, where users range from clinicians to IT administrators to software engineers.

Figures 8.3, 8.4, and 8.5 illustrate typical scientific and clinical workflow scenarios. The illustrations of clinical workflows (Figures 8.4 and 8.5) show that it is not always easier to represent all processes as sequential and that state machines provide a much simpler representation for some types of process.

Scientific and clinical workflows also differ in their scalability. While hospitals need to keep audits for long periods and thus host workflows that (at least ideally) will run for a very long time without stopping, a typical scientific workflow will have a much shorter timescale. A clinic or hospital will typically run a large number of workflows, each with a relatively low demand for computational resources, and these will need to keep track of a large quantity of patient data as they accumulate over time.

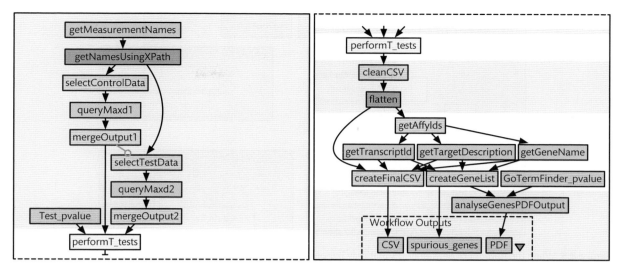

Figure 8.3 A sample workflow used in a scientific laboratory. It performs some data manipulation, executes some tests, and queries a database to analyse results. Figure reproduced from Lo (2008) *Identification of differential genes using the LIMMABioconductor package within R*, http://www.myexperiment.org/casestudy_microarray_workflow.png [3].

Scientific workflows tend to be involved with the manipulation of data rather than with making judgements, which is important in clinical workflows. These are conceptually similar, but if decision making is to be involved the workflow implementation will need to be modified to take interaction with humans into account. Scientists will tend to use workflows to reproduce or model processes while clinicians and managers of hospitals will have a stronger requirement to audit them. However, this is not necessarily a clear-cut definition, and the most powerful workflow software will of course provide the functionality for both.

In recent years, the research environment—particularly the applied research environment—is becoming more pressurized, with industry always pressing to shorten the time between a discovery and its exploitation. Medical researchers suffer from a lack of real clinical data that is both available and appropriate for their research. Clinicians, even research clinicians, quite reasonably expect higher levels of data confidentiality than scientists, as they will

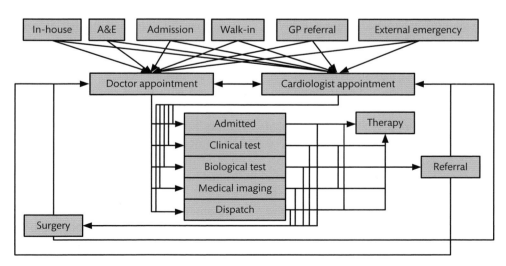

Figure 8.4 Workflows for a single clinical department can be troublesome to model sequentially.

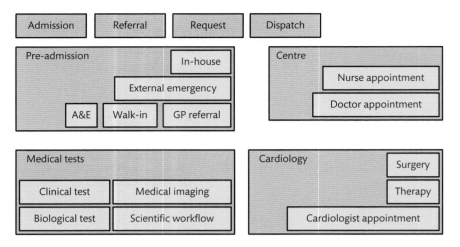

Figure 8.5 When the number of permutations that may happen is high, as in a hospital, the state machine paradigm makes more sense.

often be working with data from individual patients (see also Chapter 10). It is becoming increasingly obvious that researchers and clinicians will need to work even more closely together—and, therefore, to understand each other's mind-sets—if the results of that research are to be applied successfully in a real clinical environment.

There are two clear advantages to adopting a workflow system in the clinic: patient monitoring and the integration of clinical systems. A typical clinic will be full of specialist ad hoc solutions for obtaining, storing, and manipulating clinical data; a comprehensive clinical workflow system will need to be able to access all these. The key benefit of bringing all these solutions together under one common framework is that while the independent systems might have had minimal ability to communicate between themselves, the workflow system will take centralized control of the whole system.

If a system for monitoring patients is incorporated into a workflow, this will produce data and information that will be extremely valuable for future analysis. It will produce, for example, a labelled timeline of each patient's condition and how he or she has been treated: an idealized, computer-based form of a patient's medical record. The staff responsible for the day-to-day running of a clinic will be more able to plan and balance their often hectic workloads, and it will be easier to monitor and control the capacity of the clinic. The resulting data will be valuable for treatment planning, optimizing working protocols and improving the pathways through which patients

with a particular condition will be treated. There are many more possible uses for the data that can be obtained and stored by clinical workflows, and these may be financial—such as auditing—or involve the communication of data and protocols outside the hospital, in, for example, research papers or publications aimed at patients or the public.

Ideally, therefore, the outcome of incorporating workflows into a clinical environment will have direct benefits for that institution's patients. These are not all clinical: they may include, for example, reductions in the time spent on treatment as bureaucracy is reduced and the whole environment becomes more efficient.

The benefits of introducing workflows into the clinic may be extended further if research-based or engineering workflows (including workflows set up for modelling projects along the lines of those discussed in Chapters 5–7 of this book) are fully integrated with those clinical workflows. The most immediately obvious of these may be further opportunities for collaboration between basic biomedical and clinical researchers, and the parallel development of these fields. There are, however, many additional benefits, and these may include:

- automation of computational tasks in the clinic;
- integration of computationally intensive processes into long-running workflows;
- improving the prospects for collaboration between clinical centres and departments;

- translating engineering and computing specialties into the hospital environment;
- improving the speed and ease of the adoption of new technology in the clinic.

If you are selecting a workflow paradigm to solve a particular problem, whether in basic biomedical or clinical research, it will be important for you to take the human element into account. The following statement may be a useful one to note in this context: 'Humans can just be considered as a special type of long-running and unreliable task'. One value of this statement is that it helps us to remember to assess processes involving humans rather than machines alongside the other computational tasks in the workflow, instead of treating them separately.

However, the real power of this approach lies in the realization that complex computational tasks share some properties with humans. This might sound odd, but it can be explained using an example. Computational flow dynamics (or CFD) simulations can be unreliable in much the same way that humans can. They take weeks or even months to run, and their output can turn out to be nonsense, or the code can simply stop executing at some point if the problem becomes unstable. These are not uncommon attributes for complex scientific codes, but can lead to problems if the workflow in which they are embedded is defined in a sequential way. In many cases this problem is addressed by setting one or more points in the workflow at which the set-up is validated by a human operator. This step, which might even be thought of as obvious, builds a number of human interaction tasks into the majority of workflows. While there are often ways to 'coerce' sequential workflow engines to deal with breaks in the process at points when human interaction is needed, these do tend to limit the flexibility of the computational set-up. Dealing with this type of problem is even more complex in a state machine-type workflow, so there seems to be no general solution that will be optimal for all the workflows that are likely to be encountered in biomedical and clinical research.

It is also important to realize that if you decide to use one of the well-established methods for defining a workflow process, such as the BPEL, the output will not be tied explicitly to a single workflow application. It will, therefore, be possible to change approaches even during the course of a project. It often makes sense to use a highly flexible system, such as Taverna, during the initial development phases of the process. However, once testing has been completed and the workflow can move into a production-type environment, the definition of the process can be imported into another more suitable framework. This strategy can work very well, but only if it is well prepared and the researchers have ensured that the short-term development environment is compatible with the long-term production system. The main constraint is that there should be semantic compatibility between the two systems being used; that is, that they should share a common representation of control flow and logic. If this is true then often an automated transformation process can be applied to achieve syntactic compatibility, reducing the amount of work required and the errors that often accompany such complex manual transcriptions.

8.3 Implementing Workflows

The procedures involved with setting up and running a workflow can be divided into three main stages. These are roughly sequential, and represent different levels of abstraction between the user and the components or tools involved. They are:

- firstly, the design or composition of the basic shape and logic of the workflow: ideally, this is the only level at which a scientific researcher is involved;
- secondly, the lower-level planning of the workflow execution: this stage defines, for example, where—that is, on which hardware—each component of the workflow will run. There might be cost implications in this: some commercial systems will charge for any use, and different levels and priorities of use might attract different costs in some models;
- the final stage simply involves executing each stage of the code on its specified resource, checking its progress, and starting the next stage of the workflow when it is complete.

We will now describe each of these stages briefly in turn, focusing on the different tools and techniques that are available for scientists interested in workflow design, development, and execution.

8.3.1 **Workflow Design and Composition**

The process of developing a scientific workflow generally starts with gathering the requirements for the proposed workflow from the researchers concerned. This leads to the development of a specification for the desired functionality of the workflow. Once this is complete, a draft of the actual workflow can be assembled.

Workflow development differs from general computer programming in many ways. The most important difference is probably in the number of pre-existing components that are available. A special-purpose workflow will generally be assembled and configured from more general-purpose components, sub-workflows, and services. If you have done any script- or shell-based programming you will recognize this approach. During the composition of a workflow the user (who might be either a scientist or workflow developer) may create a new workflow by modifying an existing one, or compose it from scratch using components and sub-workflows obtained from a repository. In contrast to the business workflow world, where standards have been developed over the years (e.g. WS-BPEL [4]), scientific workflow systems tend to use their own internal languages and exchange formats. These include Simple Conceptual Unified Flow Language (SCUFL) [5], GPEL [6], and MOML [7]. This diverse range of systems reflects the wide range of computational models used in scientific workflows, and the emphasis, during the initial development of these tools, on improving the range of functionality available for scientists rather than standardization.

Most workflow systems that have been developed for scientific applications (e-Science) include a graphical tool for composing workflows. For example, Kepler and Triana have sophisticated graphical composition tools for building workflows using a graph or block diagram metaphor. Here, nodes in the workflow graph represent tasks and edges represent dataflow dependencies or the precedence of tasks. This is intended to simplify the task of describing workflows for the scientists involved, who are generally not trained in computer science theory. Other task-based systems, such as Pegasus, focus on the mapping and execution capabilities and leave the higher-level composition tasks to other tools.

There is a fundamental difference between workflow systems that operate on a task level and those that operate on a service level. Task-level workflow systems focus on resource-level functionality and fault tolerance, while service-level systems provide interfaces to services for the management and composition of the workflow. These can both be equally appropriate for scientific workflows: scientists will be most interested in whether the tools and services available through these systems provide sufficient variety and functionality for them to create the applications they need.

Once scientists have selected a workflow system that provides services and tools that are adequate for their purposes they will be able to start composing their system. Different systems provide different graphical interfaces and tools to help with this. Pegasus, for example, works with a workflow description in a form of a Directed Acyclic Graph in XML format (DAX). This can be generated using a Java application programming interface (API), any scripting language, or semantic technologies. In some scientific applications, users prefer an interface that simply supports metadata queries while hiding the details of how the underlying systems work. One example of this is in astronomy, for example, where many users will simply want to retrieve images of an area of the sky that is of interest to them. In a case like this, Pegasus will usually be integrated into a portal environment, and the user merely presented with a web form for entering the attributes of the metadata that they are interested in. Behind the portal a specific instance of the workflow will then be generated automatically based on the user's input. This will be given to Pegasus for mapping and then passed to another workflow-execution engine for execution.

Taverna, in contrast, provides a GUI-based desktop application that uses semantic annotations that are associated with services; a screenshot is shown in Figure 8.6. Taverna employs semantics-enabled helper functions and reasoning techniques to infer service annotations. It includes over 800 services described using ontologies that have been expertly annotated by a full-time curator. These are used by client software such as Find-O-Matic, its discovery tool, or Feta, which is only available as a plug-in from the Taverna Workflow Workbench. The BioCatalogue project incorporates the experiences of the

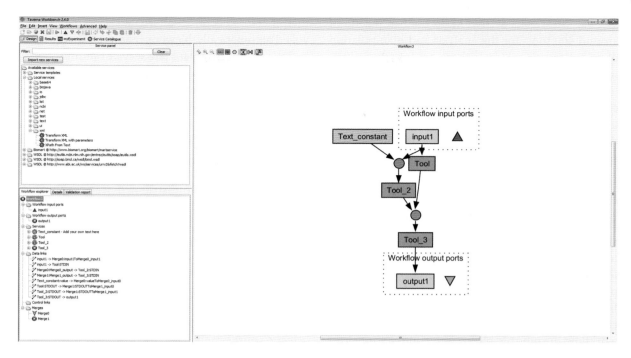

Figure 8.6 Taverna screenshot.

Taverna registry and myExperiment to build, richly annotate, and manage a catalogue of web services in the life sciences. The descriptions of the services in this catalogue are curated by experts with help from the community through social networks; operational information, such as the quality and popularity of each service, is automatically obtained through monitoring the use of each service.

The BioCatalogue is a free-standing component with its own APIs that can be embedded within and accessed from third-party applications. Workflow developers can incorporate new services simply and can, for instance, load a pre-existing workflow as a service definition. This can then be used as a service instance within the current workflow. Taverna uses an Extensible Markup Language (XML)-based directed acyclic graph format called SCUFL as its internal representation. The Taverna workflow model is based on data flow but allows both data and flow control, and is able to switch from one model to the other according to the particular conditions. The appearance of the graphical representation of workflows can also be configured within Taverna, for example enabling only higher-level processes to be viewed.

One of the most powerful aspects of Triana is its GUI. This has evolved in its Java form for over 10 years. It contains a number of powerful editing capabilities, wizards for the on-demand creation of tools, and GUI builders for creating user interfaces. The editing capabilities of Triana include, for example, multilevel grouping for simplifying workflows, cut/copy/paste/undo facilities, the ability to edit input/output nodes (to make copies of data and add parameter dependencies, remote controls, or plug-ins), zoom functions, different cabling types, optional inputs, and type checking. The basic workflow element is called a 'unit', and this can either be chosen from a palette of already available units or created specifically by the user. These units can be dragged and connected in the workspace in order to design the required workflow. Triana workflows are modelled as a dataflow system with data flowing from one neighbouring unit to another. Conditionals and loops, which are otherwise missing from this workspace, can be simulated, respectively, by outputting data on different branches or by looping back an output into an input. It represents workflows using an internal directed acyclic graph format (DAG) but can import and export the workflow in other formats as well. The Triana system has a built-in palette of around 400 tools for the analysis and manipulation of one-dimensional data; these are mostly written in Java with some C extensions. Other extensive toolkits have been added recently, incorporating facilities for audio analysis, image processing, text editing, and medical workflows. There are also tools for data mining, which are based on configurable web services to aid in the composition process.

8.3.2 Workflow Resource Planning

There are several necessary stages between the design and description of a scientific workflow and its implementation. These include validating the workflow, allocating resources, scheduling, optimization, parameter binding, and staging data. The term workflow mapping is sometimes used to describe the decisions made about optimizing and scheduling the workflow during this phase. The computational resources that are to be used for the execution of the workflow are generally not chosen until well after the workflow has been designed. Workflow mapping then refers to the process of generating an executable workflow from a basic description that is abstract and resource-independent [8]. In some cases, the intended user of the workflow performs this mapping directly by selecting appropriate resources; in others, those resources will be selected automatically by the workflow system. In these latter cases, the initial workflow is constructed at a more abstract level that will allow it to be executed using a range of different resources.

The approach that is taken to the mapping processes will depend on the particular stand-alone applications and workflow services that have been incorporated into the initial abstract workflow. With service-based workflows, this mapping consists of finding and binding the workflow to suitable services. If the workflow is composed of stand-alone applications, however, the mapping will involve not only finding sufficient resources for executing these applications, but also, perhaps, optimizing or even modifying the original workflow.

Some systems, including Taverna, rely on the user to make the choice of resources or services. A Taverna user is able to provide a set of services that each match a particular workflow component, so that, if errors occur, an alternate service can be invoked automatically. Newer versions of Taverna will include the capability for late service binding.

Triana is able to interface with a variety of workflow execution environments using the Grid Application Toolkit for task-based workflows and the Grid Application Prototype for service-based workflows. Users of service-based workflows can provide the location of, and information about, each of the services that the workflow might invoke. A user can also map part of the workflow on to distributed services through the use of one of the internal scripts available (either parallel or pipeline). In this mode, Triana distributes workflows by using distributed services that can accept Triana task-graphs as input. In contrast, the user of a task-based workflow can designate portions of the workflow as compute-intensive and Triana will send these tasks to appropriate, high-performance resources for execution. It can also use a broker to select resources only when the workflow is run.

It is often possible to specify a workflow using built-in scripts that can map from a simple specification (for example, of a loop) to the selection of multiple distributed resources in order to simplify the process of task distribution. These scripts can map sub-workflows onto available resources by using one of several service-oriented bindings available, including the Web Services Resource Framework (WSRF), web, and peer-to-peer (P2P) services using built-in deployment services for each binding.

Pegasus is able to map the entire workflow, portions of the workflow, or individual tasks onto the resources that are available. In the simplest case it will choose the sources of the input data and the locations where each task is to be executed. Pegasus provides an interface to a user-defined metadata catalogue and a scheduler that allows users to add their own defined metadata to each file, and its scheduler includes a number of different algorithms. As with many other scheduling algorithms, the quality of the schedule derived will depend on the amount and quality of the information provided about the data, tasks, and resources to be incorporated in the workflow (including the length of time that each task is expected to take to execute). In addition to its basic mapping algorithm, Pegasus is able to optimize the workflow through task clustering, data reuse, data clean-up, and partitioning.

Using Pegasus, a workflow can be partitioned into any number of sub-workflows before it is mapped. Each of these sub-workflows will then be mapped separately, with the order of the mapping dictated by how the sub-workflows depend on each other. In some cases they can be mapped and executed in parallel. In a dynamic environment in which the resources allocated to tasks may be changed quickly, partitioning a workflow into sub-workflows containing small numbers of tasks is preferable in that only a small number of tasks will be bound and committed to resources at any one time. In contrast, there are many situations (such as a dedicated execution environment) in which it will be possible to map an entire workflow at once. Pegasus can also reuse intermediate data products that are available, and this can possibly reduce the amount of computation to be performed. It also adds data clean-up nodes that remove the data at execution sites when it is no longer needed to the workflow, resulting in a reduction in the workflow's data footprint. Pegasus can also cluster tasks, treating a set of tasks as one in order to schedule them to run in a remote location. The tasks in the cluster will then be executed at the remote site in either a sequential or a parallel fashion. Task clustering can be beneficial for workflows that contain a large number of small tasks. Pegasus has also been used with resource provisioning techniques to improve the overall performance of workflows.

8.3.3 Workflow Execution

Once a workflow has been mapped and the data have been selected and made available to the workflow (in a process that is known as staging) the workflow can finally be executed. Workflows are usually set up so that, during execution, they will monitor their own performance, record each process, and provide functionality that will rescue the workflow if a component fails. It may record, for example, the steps invoked during workflow execution, the data that is used and produced in each step, data dependencies (which state which data was used to derive which other data), and/or parameter settings. These properties are together termed the provenance of the workflow, and this concept will be discussed in more length in the next section. A workflow may change during execution, for example if the resources available to the workflow

change. If this happens these changes will be recorded and may be used to develop the workflow further.

Each workflow system handles the execution of workflows in a different way. Pegasus, for example, does not include its own internal engine to execute workflows, but it is interfaced to DAGMan/Condor for workflow execution. It will map workflows onto a range of target resources including those managed by Portable Batch System, Load Sharing Facility, Condor, and individual machines. The authority of a workflow or user to make use of remote computational resources (authentication) is carried out via the Grid Security Infrastructure. During workflow execution, Pegasus captures provenance information about the executed tasks which includes a variety of information including, for example, the hosts where tasks executed, task run times, and environment variables.

Triana supports job-level execution through Gridlab GAT integration, which makes use of external components for submitting jobs. It also supports service-level execution through the GAP bindings to web, WSRF, and P2P services. During execution, Triana will identify component failures and provide feedback to the user if this happens. It does not, however, contain fail-safe mechanisms within the system, such as the facility to retry a service if it fails initially.

Taverna uses multithreaded parallelization that is specified by the user to speed up iterations. Fault tolerance such as repeating a step a set number of times can be specified by the user during workflow composition. The job execution panel displays the current state of the job and allows the user to browse and store intermediate and final results of the workflow. The workflow shown in Figure 8.3 was composed in Taverna.

Comparing these different platforms and strategies for executing workflows makes it clear that there is no apparent 'one size fits all' solution. Here, as in so many areas of computational biomedicine and mathematical modelling, different solutions will fit different designs. For example, the workflow execution engine within Pegasus, DAGMan/Condor, is a very mature and reliable platform with some built-in fault tolerance. It is optimized, however, for job-oriented scientific workflows that can be expressed as acyclic task-dependency graphs. Some types of scientific workflow, however, such as those that involve remote data streams, will require other models of computation that deal more effectively with loops and parallelization.

Taverna uses a heavily modified version of the FreeFlo workflow execution engine that supports multithreaded parallelization, and remote tasks are explicitly performed via web services. Triana, on the other hand, is a service-oriented system that supports a large variety of different Grid and service-oriented execution environments. It is therefore essential to consider all your requirements and the constraints of your application when deciding which workflow execution model or system to choose.

8.4 Provenance

The term *provenance* is generally used to refer to the context and particularly to the history of an object. In computer science, the word can be used to refer to the cause (and ownership) of an event, process, or data item. It therefore relates to the consumption or creation of data, rather than the content of the data itself. It is essentially a historical track of how, why, and when a particular piece of data was created and who or what were involved in the process [9]. It changes its meaning subtly depending on the subject matter: the provenance of a database entry, for example, will include who inserted the data, when, and possibly by what means, whereas the provenance of a data measurement (such as the amount of a chemical) might contain information on the method and equipment used and the conditions under which the measurement was taken. It might also include the name of the technician who took the measurement and the time at which it was taken.

Provenance is a crucial part of scientific workflow design and is required to meet clinical research medical standards. The provenance of a scientific workflow can be divided into two categories, as follows.

Process provenance refers to the execution of a single process (which might be a client process or server-client interaction) and to the logical trace of the workflow (i.e. which steps of the workflow were taken and for what reasons). This two-scale level allows the tracking of processes within the scope of the workflow and interactions between the workflow and data and events held outside it. This is also referred to as *coarse provenance*.

The term *data provenance* is used to refer to the derivation route of a particular data product which might be processed entirely within the workflow or serve as an input to the workflow or the invocation of an external code. The *depth of data provenance* refers to whether this information has an immediate relationship with the data or is a recursive property. This varies from application to application and is also known as *fine-grain provenance*.

It has been suggested that the term 'provenance' can be over-used [10]. It is therefore useful to consider which aspects of workflow design are best suited to the provenance approach.

We consider that workflow provenance should address the follow aspects.

- *Data quality*: accurate determination of the quality of data is already very important for determining the reliability of data sources and how data should be used. This is increasingly so with the current rapid increases in the amount of data (whether this is experimental results, raw data, or published data) that is available to modellers and with the integration of computer modelling with experimental science.
- *Data audit trails and attribution*: closer links between academic research and the business and clinical communities mean that academics increasingly need to think along the same lines as industrial scientists and clinicians in providing legally valid audit trails for their data. Scientific workflows, in particular, cannot be exempt from this requirement as they are typically extremely complex, involving large numbers of processes and operators, and generating vast volumes of data. The use of provenance provides a solution that addresses this need in programming terms. Data attribution, which refers simply to giving official credit for work done, assessing individual contribution to a project, or measuring work performance, is also now hugely important in many academic and business situations.
- *Reusability, reproducibility, and repeatability of measurements and data*: repetition of experiments has always been part of the scientific method. If workflows can provide a means of quickly and efficiently repeating tasks (which may be trivial or complex, immediate, or long-running) this will be an exemplary use of technology in the development of the scientific process. Since

provenance can be broken down to process-independent blocks, it may also be used to reuse parts of experiments in different contexts and to demonstrate the reproducibility of measurements or new evidence for a hypothesis (e.g. peer review).

- *Revision*: if the experimental parameters used to obtain a particular result can be retrieved, this will offer steps in post-processing and reviewing each stage of an experiment or model. Similarly, recovering the logical steps followed or processes involved in a workflow are also important during the process of optimizing procedures and improving a workflow. This may also include determining sources of error or reasons why processes fail. Furthermore, provenance can be used to compare similar workflows that produce differing or diverging data in a more meaningful way than if the only information available came from the final results of the workflow.
- *Annotation*: annotating data with associated metadata not only expands the meaning of the original data but may also add value to it that can be used for other purposes. Metadata may be designed either mainly for human interpretation (often as free text) or for computers (in a structured format).

The recording of provenance during a workflow operation can give rise to other secondary considerations that may still have an impact on the design of the provenance system. These include the following.

- *Migration and storage*: this refers to the place where the provenance information is stored, which may or may not be the same location as that of the data to which it refers. Links between the data and the description of its provenance need to remain when that data is moved or consumed. It becomes an architectural problem when data is moved across domains, shared between different platforms, or stored in different databases. Issues with specific data standards can arise if provenance information is added to the metadata of each file. Consider, for example, an imaging file in DICOM format (see Chapter 4). It is not clear where in this file type it is possible to store a formal provenance trail, and although

it is possible to add such information in private fields this can lead to problems if the data is later anonymized. The process of anonymization, which is designed to remove any information that links the data to an individual, will explicitly remove all data marked as private.

- *Querying data*: the most useful data is data that can be queried in a variety of ways, with the output presented either graphically or in text form. In a good provenance model, the provenance information that is attached to the data should aid the formation of complex and useful queries and improve response speed.
- *Data tracking*: the ability to query data (and its status) from within a workflow without having to load it into memory can improve both the speed and the overall performance of data queries. A variety of efficient workflow tracking services are available, including the standard Windows Workflow Foundation persistence and tracking services. Similar services are available in Taverna.

Figure 8.7 illustrates the taxonomy of provenance and highlights the issues that should be considered when selecting a provenance model for a workflow.

Just as it is hard to reach a consensus on a workflow package that is perfect for all possible workflow designs, or even for all those that are appropriate in computational biomedicine, it is hard to reach consensus on the perfect provenance model. The choice of a provenance model depends on factors such as the computer standards to be adopted, the requirements of the amount of data and data types to be collected, and optimization. For instance, provenance models and workflow models that are strongly coupled together will use fewer computing overheads, be more automated, and may make better use of the available data. However, these models may have less flexibility, scalability, or variability than a more loosely coupled approach.

In the most general terms, an abstract provenance system [12] is expected to:

- provide an open and interoperable interface to collect provenance information;
- track the workflow and data in virtual organizations that are independent of workflow model and data format;

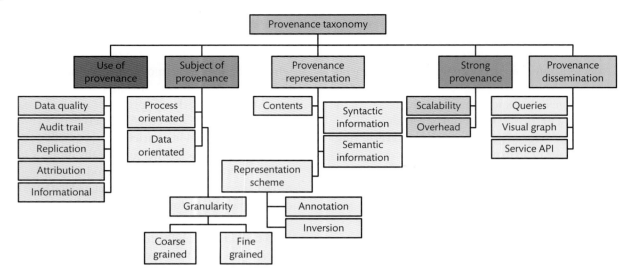

Figure 8.7 The taxonomy of provenance (adapted from [11]) describes points to consider when defining a provenance model.

- minimize the performance overhead and necessary modifications to the workflow components;
- provide an open and interoperable interface to query provenance information.

Despite attempts to work towards a common workflow provenance model that can be applied within all disciplines, most projects still implement their own provenance model. This might well be due to the conundrum of which comes first, workflow design or provenance model? It should be possible to develop a standard common provenance model that is independent of any specific workflow, but this will require a much greater consensus for any such standard to be developed. Currently, many provenance models are considered to be incomplete, unreliable, and non-portable [13].

The Open Provenance Model (OPM) has been developed as a standard interoperable model that can understand, collect, and represent provenance. This defines the concepts of agents (actors), processes (actions), artefacts (data), and p-assertions (provenance-related connections between these three). The connections between all these are described in OPM graphs, which can be considered as similar to database table schemas. In addition, the OPM specifies a standard XML-based schema for representing provenance and Java-based libraries to convert XML documents to and from OPM graphs; Figure 8.8 shows the concepts used in this schema [14,15]. An important issue in this model is the lack of control of the provenance information concerning data objects before they enter the workflow domain. A similar issue arises with the versioning of external web services. The system, in fact, relies on good coding practice, and it can be argued that if this practice were always followed there would be less need for provenance information. It will always be easiest if all data and processes can be managed within the hosting environment. Taverna has recently been extended to support the OPM, but it has not yet been fully integrated into the workflow system.

An excellent example of an implementation of the OPM is given by the Karma project[2], which was developed at Indiana University. This takes the form of a stand-alone Java-based service to decouple workflow execution and tracking. Karma captures streaming information and defines two abstract layers, termed the *execution layer* and the *registry layer*. Streams are then further abstracted into data products, workflow types, services, and methods, and these can be linked to individual workflow instances or runs. Karma provenance data are stored using MySQL with an extension for extracting specific workflow provenance in XML format.

---

[2] http://d2i.indiana.edu/provenance_karma

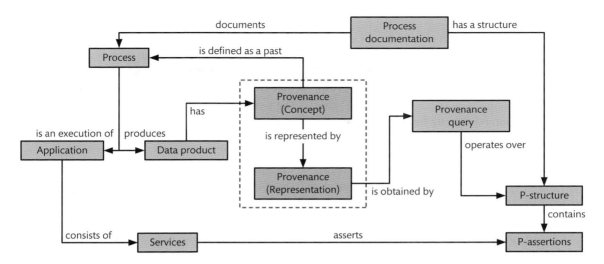

Figure 8.8 The provenance concept map (adapted from [14]) shows the different ways provenance is used and how it links together different data and process management concepts.

The subject of workflow provenance is an extensive one, and it has only been possible to touch briefly on it here. Further open-source alternatives to the OPM include operating-system-based provenance tracking systems that track program execution at a low level. Other issues that are worth considering, although rather beyond the scope of this chapter, are the efficiency of provenance querying, further privacy and security issues associated with provenance access, and the use of web semantics for provenance annotation [16,17].

It is also worth noting that for an effective provenance model all processes, and commonly also all executables, must be unambiguously identifiable; this includes the versions of each code that are to be executed. A detailed discussion of this is beyond the scope of this book; many tools including *git* and *subversion* are available to assist with this management process.

8.5 Examples of Scientific Workflows

The computational biomedicine community in general aspires to foster collaboration, data sharing, and integration of workflows between laboratories and research groups. More ambitiously, and with its aim of solving pressing clinical problems firmly in sight, it aims to collaborate also with the clinical and industrial research communities. With careful design, workflows can be implemented in this type of set-up to facilitate just such interdisciplinary and collaborative endeavours.

In this section we will briefly describe a few representative workflows that are currently used in large, interdisciplinary projects in computational biomedicine. We will use these to highlight where more automated workflow execution might help, and how this might be achieved given the current state of the computational infrastructures in each of the projects. It is important to note that none of the current workflow engines can be considered a one-stop shop for this type of scientific workflow. Once the researchers involved in a project have chosen a workflow engine it will almost always require more development work to ensure that it integrates well into the existing project's infrastructure and roadmap.

8.5.1 @neurIST

The @neurIST project, which ran from 2006 to 2010, was a European Union-funded integrated project to develop computational tools for the assessment of the risk of rupture where there is a localized weakness and dilation of a blood vessel surrounding the brain (a cerebral aneurysm). It was hoped that this would aid treatment decisions relating to aneurysms that are discovered incidentally (i.e. in the course of a routine examination or one for a different condition). It is thought that up to 1 in 20 people have a cerebral aneurysm, and that in a very large percentage of these cases the person will live a full lifespan without rupture and haemorrhage; he or she may even be completely unaware the condition exists. However, if an aneurysm is discovered incidentally, in the course of some other clinical investigation relating to, for example, a motor vehicle accident, the clinician and patient are left to decide if surgical intervention is appropriate. Since there is no intervention to treat cerebral aneurysms that is without risk, any information that can be derived about how likely this particular aneurysm is to rupture can strongly influence the decision. The philosophy underpinning the project is that the risk of rupture must be related, at least partly, to the physical characteristics of the aneurysm structure and the associated characteristics of the blood flow through it.

This project aimed to produce simulations of cerebral aneurysms, personalized with the characteristics of the individual patient and his or her aneurysm. These were then expected to produce indices related to physical processes such as stress on the walls of the blood vessels that could be used as indicators of rupture risk. It is therefore an example of a common class of problem in computational biomedicine in which risk assessment in a disease state is to be based on complex, physics-based models derived from the anatomy (and, in some cases, the physiology) of individual patients. There is therefore a common need to develop workflows, ideally generalized ones, which can be applied to a range of projects of this type.

No computational workflow was initially built into the @neurIST project as during the initial development stages it made more sense to process the data for each step of the pipeline manually. By the end of the project, however, this manual pipeline had reached a point of maturity at which it would make sense to automate many of the tasks into a formal workflow engine, illustrated in Figure 8.9. This work is now being undertaken in the VPH-Share project, which we discuss in the next section.

We have already mentioned one of the main issues complicating workflow design in computational biomedicine, which is that many of the tasks require human validation or interaction at various stages, and these are rarely represented in the process diagrams. We discussed four workflow engines earlier in this chapter, and it is not always obvious how

any of these—or, indeed, any others—can be persuaded to support processes that need this level of human interaction. Take Taverna as an example. If a researcher using Taverna needs to use graphics to validate the quality of a finite-element mesh before allowing a costly simulation to run, this can only be done using file-system triggers. This means that a file must be created in a specific location to trigger a workflow to start. In this example the graphical process used for validation of the mesh would also need to create this 'magic' file. This process is relatively simple to implement on a small system but becomes very problematic if you are dealing with remote services where the workflow engine itself is not on the same physical system, or even the same network, as the client software carrying out the validation. The complexity of this relatively simple process is an indication that this type of workflow engine is not designed to support these types of human interactions easily. It can generally be done, but rarely easily and some interactive processes are impossible using some workflow engines. This will often be a critical factor in deciding which workflow solution to use for a particular application.

8.5.2 VPH-Share

In contrast to @neurIST—and to the BrainSpan project discussed in the next section—VPH-Share is a general infrastructure project. The researchers in the large VPH-Share consortium aim to develop an environment that can manage and execute the types of workflows that are common in computational biomedicine projects including that one. Like the other projects discussed in this section, this—as its name implies—was funded through the European Commission's Virtual Physiological Human (VPH) initiative with the aim of developing workflows that can be used and shared by other VPH-funded projects. However, there is nothing to stop the environment that is being developed within VPH-Share from being used by researchers, consortia, and projects in computational biomedicine from outside this initiative.

The project goals include, in particular, to build on the current generation of cloud-based technologies and to address some of the issues and limitations facing the current workflow systems, in particular as they are applied in computational biomedicine.

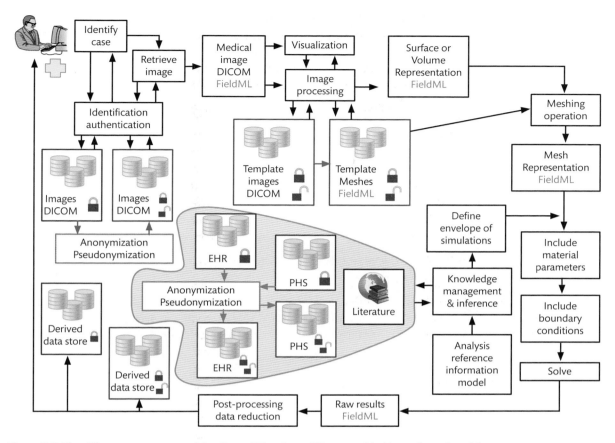

Figure 8.9 The different components of the @neurIST project will be assembled into a formal workflow.

Many of these issues have been discussed earlier in this chapter. Another part of the project involves the annotation of tools and data so that users can more easily locate resources that they need. This would work in a similar way to the myExperiment website discussed in Section 8.3.1.

The general architecture of the VPH-Share project is illustrated in Figure 8.10. This illustrates the project's aim of taking the best available components from the community and integrating them together to form a single infrastructure that allows users to develop their computational concepts and processes, share data and visualize results.

The de facto model for interacting with cloud-based resources is currently that of web services, Therefore, and since it is already popular in the research community, Taverna has been selected as the workflow engine of choice. We have already discussed some of the problems associated with this workflow engine as applied to computational biomedicine, and these issues are being addressed in modifications to the system. One common issue is the quantity of data that needs to managed or transferred through the workflow engine. The biomedical community is increasingly working with data volumes that are on the gigabyte scale or more. As we have already discussed, this is an issue for a workflow engine that is primarily data-driven. The VPH-Share project seeks to overcome some of these issues by providing a low-level storage service for the data that means that, by default, the workflow only ever passes data references. The lower-level computational infrastructure takes care of moving or 'staging' the data to where it needs to be. This feature allows projects to take advantage of the excellent workflow composition tools while dealing as little as possible with execution problems.

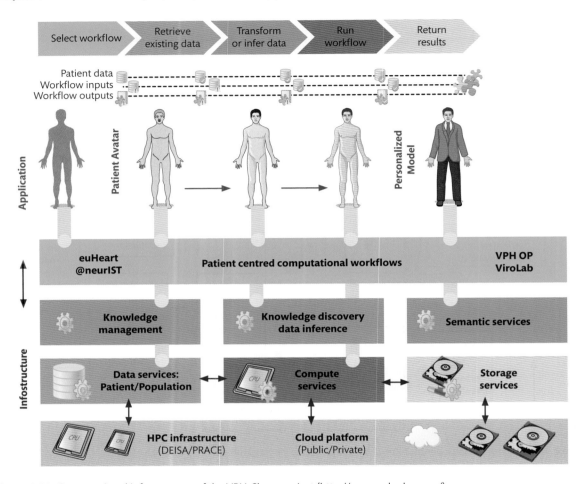

Figure 8.10 Computational infrastructure of the VPH-Share project (http://www.vph-share.eu/).

With cloud-based infrastructures it also becomes possible to address issues of algorithm distribution and sharing. In this paradigm, the people producing the tools do not simply distribute source code for compilation; they select an operating system, compile the code, and build a fully functioning server that is complete with access via web services if these are required. As most workflow runs involve only one or a small number of files, this 'virtual machine' may be easily transferred to any resource that is capable of running it and will run consistently anywhere it is installed.

The VPH-Share software has a large amount of control over the cloud infrastructure and therefore over how resources are allocated within a project. Since these project workflows often deal with large biomedical datasets, the system has been set up so that it is often faster as well as easier to move the compute service in the form of this virtual machine to the data than the other way around. The converse situation is the default for most workflow engines. This situation, however, allows complex and efficient resource plans to be generated easily, reducing overall execution time. It also offers a seamless transition plan into the production phase as the services can move from implementation on a research infrastructure onto one or more commercial platforms with little or no modification to the workflow or any of its components.

Perhaps the most interesting development in this project from a workflow perspective is that it aims to annotate all of the services that have been registered using semantics. This formal definition of, for example, the task performed by each tool and its input/output parameters presents some very interesting possibilities.

At its simplest level, this type of service definition will allow the workflow composer to suggest to users alternative components that have similar functions.

Beyond this, it should be possible to achieve what is termed 'goal-oriented service composition', in which the user simply specifies what they need to achieve and the system automatically searches for the appropriate collection of components, or even pre-existing workflows, to deliver their requirements. As the tools developed by the computational biomedicine community are necessarily very complex, this type of assistance is necessary if this domain is to be opened up to wider groups of non-expert users such as clinicians.

8.5.3 BrainSpan

In contrast to the other two projects discussed in this section, BrainSpan is a project involving bioinformatics rather than biomedical modelling and simulation. It is discussed in this section, however, because the project participants are aiming to develop standard scientific and computational processes to be shared with the wider research community, and because those participants have chosen a formal workflow representation to achieve this.

The BrainSpan atlas has been designed to relate specific patterns of gene activity to processes of normal human brain development and thus to facilitate translational research into human neurological disease. It therefore provides a valuable public resource

for researchers, educators, and clinicians. It consists of several main features:

- a survey of gene expression in specific brain regions measured using cutting-edge RNA sequencing, exon microarray, and DNA microarray techniques (see Chapter 2);
- cellular-resolution *in situ* hybridization image data pinpointing gene expression in the developing and adult human brain;
- histological reference atlases for different developmental stages to provide a neurodevelopmental and anatomical context for interpreting the gene-expression data;
- an integrated suite of web-based tools that allow researchers to view, search, and mine gene-expression patterns in space and time in the anatomical context of human brain development, as well as linking and integrating all data modalities.

One aspect of the first item in this list—RNA sequencing—is of particular interest from a workflow perspective. The BrainSpan project has developed an RNA-Seq analysis workflow for single-ended Illumina reads that has been termed RseqFlow (Figure 8.11). This workflow includes a

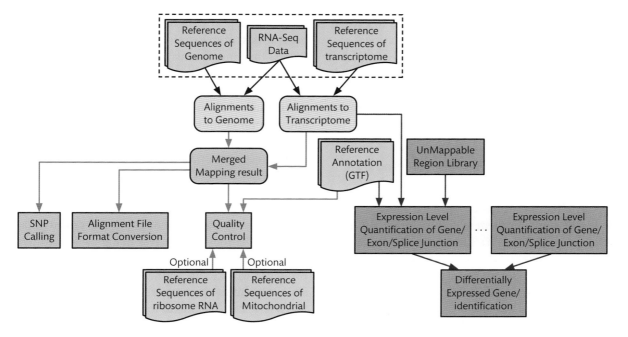

Figure 8.11 RseqFlow, an RNA-sequencing analysis tool in the BrainSpan project, (doi:10.1093/bioinformatics/btr441) is managed using the Pegasus Workflow Management System.

set of analytical functions such as quality control for sequencing data, signal tracks of mapped reads, calculation of expression levels, identification of differentially expressed genes, and single nucleotide polymorphism (SNP) calling from coding regions. This workflow is formalized and managed by the Pegasus Workflow Management System, which maps the analysis modules onto available computational resources, automatically executes the steps in the appropriate order, and supervises the whole running process.

The project partners have also opted to distribute a pre-built virtual machine on which all their software has been installed ready to use. This is becoming a common paradigm for sharing complex applications; it does, to some degree, help with the issue of encapsulating the whole computational process and ensuring that it can be repeated by different users. However this approach does not make it easy to integrate a single process into a wider workflow which is why a more abstract definition can often be preferable.

8.6 Some Key Considerations in Workflow Design

If—or when—you come to be responsible for selecting a framework or system in which to execute a process or set of processes in computational biomedicine via a workflow, you should find it helpful to consider the following questions. (Note that we only discuss the two major workflow paradigms here, and that it is possible that neither of these may meet your requirements entirely.)

8.6.1 Is your Process Best Represented by a Sequential Workflow or a State Machine?

This is probably the most challenging question to answer, since either system is capable of operating in the domain that is more suitable for the other. It is a key question, however, principally because selecting the less appropriate type can lead to faults in the design process and to workflows that take a lot of effort to edit and maintain. In general, we have found that processes involving a large number of human interactions, optional steps, or runs for very long

periods of time are best served by a state machine. Sequential workflows, however, are generally better able to cope with workflow designs that are primarily data-driven, and these are very common in scientific applications.

8.6.2 Is your Workflow Required to Pass Significant Volumes of Data Between Tasks?

A number of important issues can arise if the answer to this question is 'yes'. These can be purely technical constraints, including the fact that many web servers, and therefore web services, have a 2 GB limit on the size of input/output arguments that they can process. Working outside this limit is possible but requires special configurations that are not often implemented by systems administrators. In addition, encrypting large volumes of information is very resource-intensive, particularly if the level of security required is sufficiently high to need the use of the https:// protocol. Data-flow-oriented systems such as Taverna have still not implemented entirely satisfactory solutions to this problem, although much work is under way to address this problem.

In state machine workflow engines it is usually a mistake to try to embed anything other than simple metadata within the workflow engine itself. While this might appear to be a significant disadvantage, it becomes less of an issue in tightly controlled environments. Imagine, for instance, that you are working within a medical imaging department in a hospital. Then you would expect all your source, and possibly destination, data to be held on a picture archiving and communication system (PACS). In this case, the processing tasks themselves can be adapted so that passing a reference to the data such as the patient identifier or study ID will trigger the movement, or staging, of the images to a location where the appropriate stage of the workflow code can operate on them. This is a slightly more involved process that relies on the tools themselves being aware of their environment, although wrapper scripts can often be produced to take care of these staging tasks. If possible, it is almost always more efficient and reliable to use references (metadata) to pass data and to allow more appropriate technology to handle data staging. This principle becomes even more important when you are dealing with cloud-based systems. It is definitely worth considering whether your environment can, and whether it needs to, support these types of processes before deciding on a solution.

8.6.3 For How Long Will Your Process Need to Run?

Timing can be a serious issue when considering a workflow framework, as the longer a workflow takes to execute, the more chance there is that something will fail during this period. At the time of writing (August 2013), the Taverna website states that its longest-running workflow to date runs for 2 weeks. The general requirement is that the workflow state needs to persist for the whole time that the workflow is running. Without this facility, which is called persistence or check-pointing, significant data and processing can be lost, and this is likely to lead to an increase in costs. Workflow persistence can prevent or at least minimize these problems.

Another factor related to timing is simply that the longer a workflow runs the more resources it requires. This can cause significant problems, particularly if a large number of instances of a workflow (or of different workflows) need to run concurrently on the same system. If the current state of a workflow persists while the workflow is waiting for a response to be received or a task to complete, resources that could be used for other purposes—primarily system memory—will be supporting it. This can be a huge constraint, particularly in a production-level system.

8.6.4 Do You Need to be Able to Modify the Workflow While it is Being Executed?

This is not generally required in scientific workflows. However, it should be distinguished from a similar but more commonly used concept, which is termed *steering*. Steering a workflow involves manipulating the data inside it so that it will take a different path or decision tree, usually when it is in some form of iterative loop. Modifying the workflow means more than this: it means introducing another step or steps into the workflow while it is executing. This is technically very challenging but can sometimes be necessary, particularly in workflows that run for very long periods of time. One example of this involves the use of workflows to represent the processes involved in clinical trials. These can last for many years, and it is not uncommon for the trial protocol to be modified during the course of the trial. The researchers concerned will then be faced with the issue of what to do with trial subjects who went through the procedure before the protocol was changed and are therefore currently represented by an out-of-date process. None of the commonly used workflow engines support this kind of update/migration process natively and if it is a requirement it is likely that significant development work would be required by the user. In conclusion, it is always worth thinking about the potential need to modify the running workflow at the start of the design process, as these problems can be particularly difficult to solve at later stages.

8.6.5 What Level of Provenance and/or Audit Do You Need?

The primary factor in assessing the level of provenance and audit that will be required for a workflow is considering how far you are expecting to go down the path from a research to a production scenario. 'Production' in this case might mean the development of an actual product for use outside the research team, but it might also mean, for example, the use of the workflow in a clinical trial. It is important to understand that, although provenance information can be difficult either to manage properly or to interpret, it becomes more important the closer you get to an application that can be used in the real world. These issues are discussed at more length in Section 8.5.

There is, however, a clear need for provenance at all stages of the process of workflow design. The difference is that in the earliest stages of this process the provenance might simply act as an aide mémoire for the researcher, where at the end it might be a formal record that is submitted as part of a clinical trial.

One subtle issue that is becoming more of a problem for the developers of today's workflows is the use of third-party web services as workflow components. This is one of the capabilities that has made social networking approaches such as the myExperiment project possible, but it is also the source of potential problems. If the developers of a workflow are not also the owners of the code that that workflow will execute, they may not be informed if there are updates or changes to it. Provenance information such as when a service is called or the data that are used as its input or output may be seriously compromised if the code behind the web service is not the same every time it is run. Software version numbers can provide useful provenance information, but these may not always be available. This type of issue is likely to be less of a problem when a new workflow is being developed than when it is in regular use. It is, therefore, useful to consider constraints that will apply during routine use early in the development process in order to avoid costly re-engineering.

We hope that this chapter has given you some insight into the theory and application of computational workflow technologies and how they can be applied both in computational biomedicine research and in the clinical application of the tools developed within such a research programme. We have also highlighted some of the key issues that must be considered when selecting a workflow system, and provided a few typical case studies.

8.7 Conclusions

The concept of computational workflows is a very powerful one. It allows the details of research processes to be encapsulated in a different and complementary way from a research publication. It also provides researchers with the ability to impose the level of repeatability and rigour that are necessary for some aspects of biomedical research more easily and efficiently than with wholly human driven processes. We have explained, however, that the development of workflows can be a difficult process and that there are a number of key issues that must be considered in doing so. For example, implementing a wholly effective system for tracking the provenance of all the data used requires absolute control of the code that is executed in the workflow. This can be particularly challenging in a large distributed development project. We believe, however, that these tools can be extremely valuable in, for example, aiding collaboration between research groups and encapsulating the intellectual property of a project. We would therefore

encourage our readers to explore this option for all but the most straightforward computational modelling projects.

 ## List of Projects Referenced

@neurIST	http://www.aneurist.org/
BrainSpan	http://www.brainspan.org
myExperiment	http://www.myexperiment.org/
Pegasus Workflow Management System	http://pegasus.isi.edu
Taverna Workflow Management System	http://www.taverna.org.uk/
Triana	http://www.trianacode.org/
VPH-Share Project	http://www.vph-share.eu/

 ## Recommended Reading

Cheney, J., Stephen, C. et al. (2009) *Provenance: a Future History.* New York: ACM.

Fowler, M. and Parsons, R. (2010) *Domain-Specific Languages.* Harlow: Addison-Wesley.

Goble, C. (2002) Position statement: musings on provenance, workflow and (semantic web) annotations for bioinformatics. In *Proceedings of the Workshop on Data Derivation and Provenance.* Chicago, October 2002.

 ## References

1. Murata, T. (1989) Petri nets: properties, analysis and applications. *Proceedings of the IEEE* 77(4), 541–580.
2. Chapman, N. (1997) *Petri Net Models.* ISE-2 Surprise 97 Project 1997 27-05-1997. http://www.doc.ic.ac.uk/~nd/surprise_97/journal/vol2/njc1/.
3. Lo, P. (2008) *Identification of Differential Genes using the LIMMABioconductor Package Within R.* http://taverna.knowledgeblog.org/files/2010/12/casestudy_microarray_workflow.png.
4. Jordan, D., Evdemon, J. et al. (2007) Web services business process execution language version 2.0. *OASIS Standard*, 11.
5. Oinn, T., Greenwood, M. et al. (2006) Taverna: lessons in creating a workflow environment for the life

sciences. *Concurrency and Computation:Practice and Experience* 18(10), 1067–1100.

6. Wang, Y., Hu, C., and Huai, J. (2005) A new grid workflow description language. In *Services Computing, 2005 IEEE International Conference on Services Computing*, vol. 2, pp. 257–258. Los Alamitos: IEEE Computer Society Press.

7. Deelman, E., Blythe, J. et al. (2003) Mapping abstract complex workflows onto grid environments. *Journal of Grid Computing* 1 (1), 25–39.

8. Brooks, C., Lee, E.A. et al. (2008) *Heterogeneous Concurrent Modelling and Design in Java, vol. 1, Introduction to Ptolemy II.* Tech. Rep. UCB/EECS-2008-28, April. Berkeley, CA: EECS Department, University of California.

9. Moreau, L., Plale, B. et al. (2007) *The Open Provenance Model.* Southampton: University of Southampton.

10. Goble, C. (2002) Position statement: musings on provenance, workflow and (semantic web) annotations for bioinformatics. In *Proceedings of the Workshop on Data Derivation and Provenance.* Chicago, October 2002.

11. Simmhan, Y.L., Plale, B., and Gannon, D. (2005) A survey of data provenance in e-science. *SIGMOD Record* 34(3), 31–36.

12. Simmhan, Y.L., Plale, B., and Gannon, D. (2008) Karma2: provenance management for data-driven workflows. *International Journal of Web Services Research* 5(2), 1–22.

13. Cheney, J., Stephen, C. et al. (2009) *Provenance: a Future History.* New York: ACM.

14. Moreau, L. et al. (2006) *An Open Provenance Model for Scientific Workflows.* Presentation at the HPC'06 Workshop (High Performance Computing and Grids), July 2006. http://www.gridprovenance.org/talks/hpc06.ppt.

15. Moreau, L., Clifford, B. et al. (2010) The open provenance model core specification (v1.1). *Future Generation Computer Systems* 27(6), 743–756.

16. Heinis, T. and Alonso, G. (2008) Efficient lineage tracking for scientific workflows. In *Proceedings of the 2008 ACM SIGMOD International Conference on Management of Data.* New York: ACM, pp. 1007–1018.

17. Missier, P., Paton, N.W., and Belhajjame, K. (2010) Fine-grained and efficient lineage querying of collection-based workflow provenance. In *Proceedings of the 13th International Conference on Extending Database Technology.* New York: ACM, pp. 299–310.

Chapter written by
Daniel A. Silva Soto, University of Sheffield, Sheffield, UK;
Steven Wood, Sheffield Teaching Hospitals, Sheffield, UK;
Susheel Varma, University of Sheffield, Sheffield, UK;
Rodney Hose, University of Sheffield, Sheffield, UK.

Distributed Biomedical Computing

Learning Objectives

After reading this chapter, you should:

- be able to select the most appropriate computational resources to execute the different components of a proposed workflow;
- understand how to map different algorithms to relevant computational resources;
- be able to launch applications on high-performance computing e-infrastructure;
- be able to incorporate those resources and applications into complex workflow scenarios;
- understand the clinical relevance of distributed computing.

9.1 Introduction

The future of biomedical science and clinical medicine is bound to be heavily dominated by the pervasive presence of information technology and predictive computational methods, which can provide a personalized aspect to treatment [1]. As we have seen already, many of these physiological simulations are inherently multi-scale. The previous chapter showed us that simulations can often be carried out effectively using workflows. In these, models at different scales are connected together using a workflow engine that manages the execution of the different parts and components of the simulation, transferring data between each computational model. Workflows can be extended for other tasks: for example, retrieving datasets from storage resources and visualizing the results of simulations.

It is clear that simulating complex human physiology and disease progression requires many different types of modelling at many different scales. These require very different codes and each needs

computational resources of varying sizes and powers, ranging between the desktop workstations that are now found in almost every home and office in the developed world and the most powerful supercomputers. It can be a challenge for scientists developing workflows to select the most appropriate of these resources for each component of the workflow.

The resource types that we have mentioned here are accessed in different ways. It is usually possible to run a simulation code installed on a desktop workstation by simply double clicking on the icon representing the executable file. In contrast, running a simulation code on a supercomputer will probably require a much more complex series of tasks, creating a script of commands to set up and run the simulation in the background (in batch) and then submit that job to a batch queuing system. Many large computers have service-oriented interfaces that allow users to submit jobs and stage data in an automated fashion.

In this chapter we consider the variety of resources that are available to researchers who wish to execute biomedical workflows. We will explore how to choose resources that are appropriate for different tasks and how to port applications to resources of different scales. We will also examine some of the practical issues involved with running simulation codes on remote resources (that is, resources beyond a user's desktop) and incorporating these into simple simulation workflows.

9.2 Parallel Applications

We discussed the benefits of parallel computing briefly in the previous chapter on workflows. One important advantage of resources that are more powerful than simple desktop computers is that they often use multiple computer processors simultaneously, and so can exploit this kind of large-scale parallel algorithm. A multi-scale biomedical simulation processed using a workflow will typically be composed of a number of different application codes. When it is, each is likely to require a different level of parallelization. Some will run quite efficiently on single-processor workstations, others will require access to a cluster of processors, and still others will only run effectively on a large supercomputer that may be located in a different country from the developer or user. Even if an application only requires a single code, this code may be best executed in several different ways: it may be optimized to be serial, *data parallel*, or *task parallel*. It also may run optimally on a single workstation or a large cluster.

As a researcher in computational biomedicine, you will find that no one computer set-up or degree of parallelization will be appropriate for all application scenarios that you need to run. Indeed, as we have implied, you may find that the different components of a single application cover all possible types of parallelism and need a wide range of resources to run effectively.

The two parallelization schemes mentioned above are those that a researcher in biomedical simulation is most likely to make use of. In more detail, these involve the following.

- *Data parallelism*: this form of parallelism splits a dataset into small packets that are distributed across multiple processors (or

compute nodes), with the same algorithm run over each data packet. Typically there is little to no communication between the different data-processing nodes while the computation is taking place. The output from each of the separate computations is only assembled once all the computations have completed. The terms *embarrassingly parallel* and *delightfully parallel* are synonymous with data parallel.

- *Task parallelism*: task parallelism is achieved when the execution processes that operate on a single dataset are distributed across a set of compute nodes. This usually requires a significant amount of synchronization between the execution processes, so there must be a high-performance communication infrastructure between all compute nodes. Task parallelism leads to more tightly coupled applications in which the levels of communication between compute nodes varies as a simulation progresses. Task-parallel code may also be termed *inherently parallel*.

The form of parallelism that is most appropriate for a particular simulation depends on both the nature of the simulation algorithm and the data that it is to be operated on. The enormous range and scale of the computational resources that are available today implies that it can be possible to incorporate both data and task parallelism in the same scheme. In this case, multiple task-parallel codes will usually be executed simultaneously on different sections of a dataset.

A discussion on the parallelization of algorithms is beyond the scope of this book. If you need to know more about this the books by Peter Pacheco listed under Recommended Reading will be a good place to start. We assume that once you are ready to start planning to run a simulation using a range of resources, including parallel ones, you will have at least some familiarity with the codes that you are intending to use and the types of resource that you will need to run them on.

9.2.1 Achieving Parallelism

Task-parallel applications typically rely on specific libraries of algorithms and frameworks to implement parallelism. They run many applications at a time and these will often need to communicate with each other. The Message Passing Interface (MPI) [2] has been developed as an application programming interface (API) specification that allows these processes to communicate with each other through sending and receiving messages. It is typically used by parallel programs that run on clusters of processors and on supercomputers.

This interface is a platform-independent protocol for communication between processes on a parallel system. It sits at level 5 of the Open Systems Interconnection (OSI) network model, which was developed in 1977 as a means of standardizing computer networking. A high-performance computing (HPC) resource will typically include an implementation of MPI that has been configured to work efficiently with the technology that connects the nodes in that resource. It should therefore provide the best possible performance when running applications. MPI applications use multiple, tightly coupled processors to run tasks, and these tasks communicate with each other using this interface. Often the nodes of such a cluster will be connected by a fast network interface, such as InfiniBand[1] or Myrinet[2]. Researchers using a system like this will compile their applications using the system's in-built MPI libraries, and then launch those applications using a tool that is supplied with the MPI system. This tool will start instances of each application (see previous chapter) on the appropriate set of nodes, and will configure the environment to allow these processes to communicate with each other in an appropriate way.

The MPI interface provides a set of services that are independent of the computer language in which the application code is written in. These allow processes to be mapped to nodes or servers in a language-independent way; MPI library functions also provide facilities for other general operations including exchanging data between pairs of processes, choosing between a Cartesian or graph-like topography for logical processes, combining together the partial results of computations, synchronizing nodes, and obtaining network-related information about an application, such as the number of processors involved in a particular computing session and the identity of the processor on which each process is running. This is currently the de facto standard model for parallel programming on most HPC machines.

Task parallelization can also be achieved by using shared-memory multiprocessor programming environments such as OpenMP[3]. This environment implements thread-based parallel processing in which a master thread initiates multiple 'forked' threads to execute the parallel parts of a code. These will be joined back to the main master thread when the forked, parallel sections of the code have completed. Hybrid schemes that combine both OpenMP and MPI style techniques can achieve performance improvements by optimizing communication between and within nodes.

Data parallelism requires a scheme that can split input data into chunks that can be operated on independently; this is often application-specific. Once the data has been partitioned, the execution of the code on each node can be managed using a workflow engine or a dedicated job-submission manager such as Condor[4] (see later in this chapter). MapReduce is a framework that has been designed to automate the process of distributing data-parallel tasks between compute nodes. A MapReduce program is comprised of two parts. The first is a map() function, where a master node takes a set of inputs, divides it into smaller sub-problems, and then distributes them between compute nodes. Each of the compute nodes then processes its sub-problem and passes the results back to the master node.

The second part of a MapReduce program is the reduce() function, which assimilates the results from the computation of each of the sub-problems and produces an answer to the problems being computed. Since MapReduce applications typically do not have any interdependence between sub-problems, limited network bandwidth is required between compute nodes. This means that problems can be run across a single cluster, or even a set of separate clusters.

MapReduce manages fault tolerance within a distributed system; if one or more sub-problems fail, due to a fault on a particular cluster node for example, MapReduce can rerun the failed tasks on another compute node. MapReduce programs can also implement multiple levels of hierarchy, with

---

[1] http://www.infinibandta.org/
[2] http://www.myricom.com/

[3] http://openmp.org/wp/
[4] http://research.cs.wisc.edu/condor/

sub-problems that consist of their own `map()` and `reduce()` functions.

9.3 The Computational Ecosystem

The landscape of computer facilities that is available to researchers in biomedicine (and other scientific disciplines) is constantly shifting. The machines available on researchers' desktops are becoming more and more powerful. Desktop machines now often incorporate multicore central processing units (CPUs) or even multiple CPUs. Even these high-performance desktop machines, however, are not powerful enough to run many of the simulations that biomedical modellers wish to use in a sensible time frame. The length of elapsed (or so-called wall-clock) time that is needed for a simulation becomes even more important if the results of that simulation are to be used in the clinic. Almost all decisions about the optimum treatment of an individual patient have to be taken against the clock.

The obvious solution to this dilemma is to run the simulations on more powerful computational resources beyond the individual researcher's desktop. It is likely that a researcher in a well-funded university department, for example, will have a range of such resources available both locally and remotely. These will include networks of high-performance workstations (such as Condor pools), loosely coupled compute clusters (also called Beowulf clusters), and supercomputers in which compute nodes are very tightly coupled together, sometimes making use of shared memory. All these are loosely grouped together under the term *e-infrastructure*.

The boundaries between these different types of resource can be rather blurred. However, it is still useful to use these terms when you are thinking about which resources will be most appropriate for the different parts of a simulation workflow. A more detailed description of these different types of computational resource is given below.

- *Multicore computing*: a multicore processor is a processor that includes multiple execution units on the same chip. These processors differ from those that are known as superscalar and which can issue multiple instructions per cycle from a single instruction stream. In contrast, a multicore processor can issue multiple instructions per cycle from multiple instruction streams. A single workstation may contain one or more multicore chips.

- *Symmetric multiprocessing*: a symmetric multiprocessor (or SMP) is a computer system that includes multiple identical processors that share memory and connect via a *bus*. This term is used to refer to any subsystem that transfers data between components within a computer, or between computers.

- *General-purpose computing on graphical processing units (GPGPU)*: the graphical processing unit, or GPU, in a computer typically handles only the computation that is involved in generating computer graphics. General-purpose computing on GPUs is a technique in which the GPU is used to perform types of computation that are traditionally handled by the CPU. The GPU is a powerful unit and is most typically used to run data-parallel applications. Specialized libraries such as CUDA and OpenCL allow applications that exploit the parallelism of GPGPUs to be easily created, and OpenACC simplifies the development of parallel programs on heterogeneous CPU/GPU systems.

- *Heterogeneous multicore*: many newer processor architectures are adopting a heterogeneous multicore architecture, designed for parallel processing. Chips such as the Cell or Intel Phi combine traditional compute cores with specialized processing cores, such as GPUs or other coprocessors, to improve performance and power efficiency while maintaining the versatility of a traditional CPU.

- *Cluster computing*: a cluster is a group of loosely coupled computers that work so closely together that, in some respects, they can be regarded as a single computer. Clusters are composed of multiple standalone machines connected by a network.

- *Massively parallel processing*: a massively parallel processor (MPP) is a single computer that has many networked processors. MPPs have many of the same characteristics as clusters, but they use specialized interconnect networks to communicate between processors. In comparison, clusters typically use commodity hardware for networking.

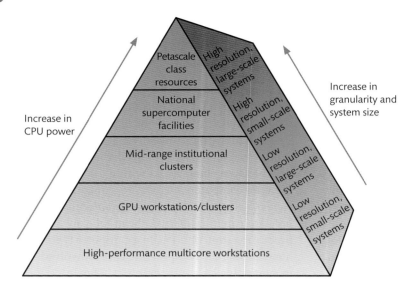

Figure 9.1 The Branscomb pyramid, showing the relative abundance of computational resources.

It should be clear that the cost of these resources will vary widely. Single-processor desktop workstations are the cheapest: the most powerful—the supercomputers—are the most expensive. Researchers will therefore find that they have more desktop workstations at their disposal than supercomputers. This fact was noted by Lewis Branscomb in a report back in 1993 [3]. This introduced the idea of a Branscomb pyramid (Figure 9.1), which shows the relative abundance of different classes of computational resources.

9.4 Computing Beyond the Desktop

In this chapter we use the term *high-performance computer* rather generically to refer to everything above the level of the desktop workstation. We have already seen that there are significant differences between different levels of high-performance systems, but there are also many similarities between them in terms of how they are accessed and used.

HPC resources are traditionally based on tightly integrated parallel computing systems; manufacturers of machines on this level include the well-known names of Cray, IBM, and Silicon Graphics. They increasingly consist of 'clusters' of commodity hardware linked together using a fast interconnect and running software such as ROCKS. The rise of the distributed-memory supercomputer has inevitably led

to the development of computational frameworks that can harness the power of this type of machine. MPI, which was discussed in the previous section, has now become the de facto standard for communicating between parallel processing on a distributed-memory cluster machine.

9.4.1 Allocating Resources to Workflow Components

It should be self-evident that researchers who are planning to execute a workflow will need to select the most appropriate resources for each section of that workflow. This means that they will need to understand the degree of parallelism that is used by each section of the workflow. This information will then be used to select resources of the most appropriate scale. It is important for researchers in this situation to consider the scalability of their code as well as the degree of parallelism. While some parallel codes may scale up to thousands of computer cores in a linear or near-linear fashion, other codes will only perform well on a small number of cores.

It is also important to consider the amount of time that is needed to run a particular simulation code. Many HPC resources impose time limits on jobs that are run on those machines. Many resources include a number of batch queues, and each queue will be have a time limit with it that means that jobs in those

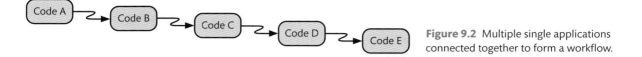

Figure 9.2 Multiple single applications connected together to form a workflow.

queues will only be allowed to run for that amount of time (this is given as wall-clock time rather than CPU time).

Let us consider an example. The simple workflow shown in Figure 9.2 is composed of five different simulation codes that need to be executed sequentially. The output of each code provides the input for the next. The codes used in this example are as follows:

- Code A is a scalar code that can run on a single workstation.
- Code B is a task-parallel code that scales up more or less linearly to 32 cores. Adding more compute cores after that provides no extra speed.
- Code C is a linearly scaling parallel code, and the size of the problem being tackled means that the simulation will complete in 12 h of wall-clock time (as is necessary) if 1024 cores are used.
- Code D is a data-parallel code that splits the output data from code C and processes them in an embarrassingly parallel fashion.
- Code E is a serial code that gathers the output from code D and performs some post processing to produce output for the application (e.g. graphs).

One logical and efficient way in which these five codes could be executed is as follows. Code A is executed on the user's local workstation and generates the initial input for code B. The user then stages these data to a compute cluster in his or her home institution that includes a total of 128 compute cores with InfiniBand interconnects between the nodes, and executes code B at this site. Once code B is completed, the user moves the output data from the local compute cluster to a national HPC facility with 10 000 compute cores and a high-performance interconnect between nodes, and uses this machine to execute code C. Once that simulation is complete, the data is moved to an institutional Condor pool where code D is executed. The results from code D are then returned to the user's workstation where code E can be executed.

9.5 Executing Simulations in a High-Performance Environment

In simple examples such as the one described above there is no need to use a workflow engine to manage the execution of the different components of the simulation. Instead, the user moves the data between the various resources and launches each of the simulation codes manually. This method is both laborious and error-prone, but it can often be appropriate, particularly when a researcher is designing and developing a workflow and testing the different parts of the simulation chain.

The commonest method of submitting jobs to computer clusters and supercomputers is still the batch-processing model. This model involves the use of a queuing system to ensure fair access to the resource for all users. Queuing systems such as PBS[5], SGE[6], and LSF[7] are configured by the administrators of each resource to define the priorities of the queuing system and the lengths and sizes of each queue. In most systems, all users are treated with equal priority and jobs are run on a first-come, first-served basis. Some use other systems, for example prioritizing so-called *capability* jobs that make full use of the whole resource.

Each HPC machine will have a number of different processor queues, and each of these will have an associated maximum wall-time limit; that is, a maximum time that jobs submitted to that queue will be allowed to run for. Users who submit jobs are also able to specify the maximum time that their job will run for; if they do not, the default run time will be allocated by the resource administrator.

All queuing systems use a scheduling algorithm, or a combination of such algorithms, to order jobs within a queue. Examples of commonly used algorithms include First Come, First Served and Back

[5] http://www.pbsworks.com/
[6] http://gridscheduler.sourceforge.net/
[7] http://www.platform.com/workload-management/

Fill. The first case is the simplest, in that jobs are run on resources in the order in which they are submitted. With Back Fill, the scheduler will attempt to fit holes in the queue with shorter jobs submitted later. This means that the jobs will not necessarily be run in the order in which they arrive.

Users who need to access HPC machines will typically use a Unix protocol such as a secure shell (ssh) to establish an interactive login session on the remote machine. Alternatively they might use some form of middleware to launch their jobs. This process will be described in the next section of this chapter.

Users who wish to run parallel scientific applications on HPC resources using a secure shell will typically need to perform a number of complex tasks in order to do so. To start with, if the required application is not already installed on the resource, the user will need to download the source code and build the application. This process itself might need the user to download and install extra software libraries, or to work with the administrators of the machines concerned to make sure that they know the locations of these libraries and have the required permissions to use them.

It is likely that each resource that a researcher needs to use will come with its own set of compilers and libraries. Users will need to familiarize themselves with these before building applications. This is often a difficult process, particularly for inexperienced users. Many, if not most, scientists working in computational biomedicine are not themselves experienced programmers, but they will often need to take extra steps to optimize code for the particular architecture of the machine(s) that they are working on. It should be obvious that this will require a clear understanding of that code and what it is designed to do.

Once a user is satisfied that the required applications have been installed and will run on the correct machines, then he or she will often need to generate a script in order to submit the job into a queue. The multitude of HPC resources available to researchers today use many different queuing systems and a scientist may have to create different scripts to run the same application on different machines. It should at least be possible to 'clone' a script for each subsequent run of that application on that machine, simply by changing parameters and the names of the input and output files. Before submitting a job the user should also make one last check that the correct libraries have been installed on the machine to be used and the correct scheduler options set.

Finally, the user must stage all the data to the resource being used and submit the jobs to the queuing system. Depending on the load and policies of the machine, it could take from minutes to several days for the job to run. Once the job is complete, the user will usually need to retrieve the output data onto their local file system in order to examine them, and for further data processing, perhaps involving visualization.

Practical Example 9.1 shows you how to launch an HPC job.

Practical Example 9.1 Launching an HPC Job

We have shown that the process of executing a job on an HPC resource is complex. Once a user has checked that the required application is installed on the resource and the required input data have been staged to the machine, the user must next create a job-submission script that describes the application to be executed and its various parameters, to be sent to the machine's queuing system.

Scripts generally use one of several Unix scripting languages, and an example is given below.

```
#!/bin/bash
#$ -N job_name
#$ -S /bin/bash
#$ -lh_rt=00:30:00 #30 min run
#$ -pempi 4
#$ -cwd
#$ -ojob_name.out
```

... continued

```
#$ -ejob_name.err
#$ -notify
/opt/mpich2/gnu/bin/mpirun -np 4 /usr/local/bin/cpmd.x 110.cpmd.inp
```

Here, the first line of the script tells the system which interpreter to load (in this case the bash shell, /bin/bash). The following lines, which start with the characters #$, pass parameters to the queuing system; for example, the command -pempi 4 tells the queuing system to use the MPI parallel environment with four CPUs. The final line tells the queuing system to execute the CPMD[8] code for *ab initio* molecular dynamics simulation, using the mpirun execution tool with four CPUs and taking input from the file 110.cpmd.inp.

The above example has been written for use with the SGE batch queuing system; other queuing systems use similar approaches and therefore similar scripts, but it is rarely possible simply to copy a script written for one system and use it in another. The exact commands to use vary from machine to machine and application to application.

Once the script has been created, the user must submit it to the queuing system: on most queuing systems this is done with the qsub command:

```
qsub myscript.sh
```

The job will then be assigned a reference number, which is communicated to the user in the following way:

```
Your job 19012 ("myscript.sh") has been submitted
```

Submitted jobs can be monitored using the qstat command. Once the job has completed, users will most often need to copy the output files that are generated to their desktop computers for further analysis.

---

[8] http://www.cpmd.org/

9.6 Case Study: Calculating Drug Binding Affinities

Since the mid-1990s the development of drugs targeted to several parts of the HIV life cycle has transformed the prospects for people suffering from HIV infection and AIDS. However, this can be compromised by the development of resistance to these antiviral drugs[9]. The enzyme HIV protease, which plays a crucial role in the maturation of the virus, is a particularly attractive target for anti-AIDS therapy and forms a part of several effective complex multidrug regimes [4]. There are nine inhibitors of the protease of HIV-1 (by far the most common form of the virus) in current clinical use, and drug resistance is a problem with them all.

The efficacy with which an inhibitor binds to its enzyme target is determined by its biomolecular

binding affinity (or free energy). Experimental methods for determining biomolecular binding affinity are well established and these are now being used to study the resistance conferred by particular viral mutations *in vitro*. This experimental data can provide information that is invaluable in clinical decision making, but its use is limited as the studies are generally performed on characteristic mutations only. It is now fast and easy to obtain viral sequence from any individual infected with HIV and thus to determine the exact mutations in his or her unique viral protease. However, an exhaustive experimental determination of drug-binding affinities making use of this individual sequence would be far too costly and time-consuming to perform in a clinically relevant way.

Computational methods for determining biomolecular binding affinities are much faster than experimental ones and are now precise and accurate enough to provide useful input into clinical decision making.

---

[9] http://www.chain-hiv.eu/

One recent study tested the binding energy of one protease inhibitor, saquinavir, against the wild-type HIV-1 protease and three observed drug-resistant strains using free-energy methods in molecular dynamics simulations [5]. The protocol implemented in this study gave accurate correlations to experimentally determined binding affinities of this drug.

This study made use of a tool called the Binding Affinity Calculator (BAC) [6,7]. This enabled the molecular dynamics simulations to be prepared, deployed, and implemented and the resulting data processed rapidly and automatically across a range of supercomputing resources. Using the BAC tool allows researchers to automate binding-affinity calculations for all the nine HIV-1 protease inhibitors currently in the clinic against a protease sequence with an arbitrary number of mutations. Although this method has so far only been validated with saquinavir and its ability to produce equally accurate results with any protease inhibitor still needs to be tested fully, it clearly has enormous potential as a clinical tool. Potentially, at least, it offers an automated *in silico* method for assessing the drug resistance for any given viral strain.

With optimal access to computational resources, calculations using BAC take approximately 4 days for each combination of drug and viral strain. This may seem time-consuming, but it is much less so than experimental determinations of drug efficacy and is more than adequate for the timescale that would be required to support effective clinical decision making. Increasing the computer power available so that these calculations can be performed on a routine basis will allow clinical decisions about the choice of HIV-1 protease inhibitors to be made in a truly patient-specific manner and on an appropriate timescale.

The BAC tool consists of a complex workflow that requires a number of computational codes that run on a range of resources at different scales. For a particular combination of drug and protease mutation(s), BAC initially runs a series of set-up scripts on a local cluster to create the necessary input files for the required set of simulations. This is achieved by taking a default model and customizing it to use the mutations and drugs under consideration. Next these data are staged to a supercomputer class resource, and the NAMD[10]

molecular dynamics code is used to run equilibration steps to prepare the model. This is followed by some nanoseconds of molecular dynamics simulation, which uses between 32 and 64 processes in each simulation but still takes several hours of wall-clock time. If the resources are available it can be possible to run several simulations in parallel, increasing the sampling effectiveness. Once the simulation chain is complete, the terabyte quantities of output data that have been generated are staged to a smaller-scale cluster for post processing, using the Amber molecular dynamics package to calculate a drug-binding affinity.

A typical clinical scenario will involve running this workflow nine times, testing each of the nine currently approved HIV-1 protease inhibitors against a protease model including an individual patient's set of mutations and then placing the drugs in a ranking order depending on their calculated binding affinity.

9.7 Computational Infrastructures

We have already seen how isolated computational resources can be of only limited use. Deploying computational workflows onto high-performance resources involves choosing the most appropriate resource available for each section of the workflow.

It is clearly possible to execute such a workflow manually, transferring data to each resource and running each program in turn. However, this is a very arduous process. It is therefore extremely useful to automate these processes. Automation is made possible through middleware interfaces that allow simulations to be executed automatically in turn. These interfaces also typically group resources together into resource grids.

9.7.1 Grid Computing

About 10 years ago, grid computing was the only established paradigm for accessing datasets and running jobs automatically on high-performance resources. Since then, cloud computing has taken over this role to some extent, but grid computing is still valuable and popular.

There are several different useful definitions of grid computing. It has been described as 'distributed computing that is conducted transparently by

---

[10] http://www.ks.uiuc.edu/Research/namd/

disparate organizations across multiple administrative domains' [8]. Its goal is to give the end user transparent, uniform access to computational resources that are owned and operated by disparate organizations. These resources may have different security and access policies at an institutional level. They should, however, be set up to give reasonable security assurances to the institutions participating in the grid.

Foster and his coworkers have defined the underlying problem that is associated with grid computing as follows [9]: 'Co-ordinated resource sharing and problem solving in dynamic, multi-institutional virtual organisations'. A *virtual organization* can be thought of as a group of institutions or companies that have come together to share their resources and that use some kind of software system to manage the resources and sharing policies. This quote neatly summarizes the key concept of grid computing, which is that resources are shared between different institutions with varying security and access policies. Grid providers may want to share a wide range of resources, not all of which, strictly speaking, are 'computers'. The resources employed on and shared by grids may also include powerful scientific instruments such as microscopes and radio telescopes. These resources are very often geographically dispersed and seamless access to them is provided using the public Internet or dedicated, high-speed point-to-point networks [10].

The term *grid* is used to describe this arrangement of inter-institutional resource sharing by analogy with the electrical power grid. Just as power grids provide dependable, consistent, pervasive, and (hopefully) inexpensive access to electrical power, a computational grid is designed to provide the infrastructure for dependable, consistent, pervasive, and inexpensive access to high-end computers and other scientific resources.

The different types of scientific resource that make up grids, including scientific instruments, data repositories, and computers that range in size from desktop machines to large-scale parallel supercomputers, are made available at national and international levels. The grid model is particularly appropriate for, and has been very widely implemented in, the sharing of HPC resources. The main reason for this is that the grid model is a good way to get the most from machines that are expensive to purchase and run.

In the previous sections we described a common scenario for launching applications on HPC machines. This is fairly simple if only a single resource is involved, but when, as often happens, a heterogeneous grid of computers is involved, the complexity of the procedure will grow with the number of nodes on the grid. However, the power of a grid also lies in its complexity: in the fact that the grid gives its users access to many resources that they can use in combination to solve challenging problems.

You might think that the perennial and exponential growth in computer power will be making grid computing obsolete. Certainly, the latest most powerful machines are each capable of executing over 10^{15} floating point operations per second (1 petaflop). However, computational power and speed are not the only attributes offered by grids. The power of a grid arises largely from its inhomogeneity, as it links visualization engines, data storage, and instruments as well as HPC machines.

The software used to tie a grid together is referred to as its middleware. There are many forms of middleware available, but the majority of production grids in existence today use one of only three middleware stacks. These are Globus[11], UNICORE[12], or gLite[13], which are all designed to provide the comprehensive set of services that are necessary to build a variety of different types of grid (all are discussed in Section 9.8.1). These are all defined as 'heavyweight' middleware stacks, which means that they are complex and require expert system administrator support to install and maintain. They are therefore difficult for end users to deploy and, in fact, they often contain far more functionality than any individual end user will require.

Grid computing is a fairly specialist discipline and requires rather different tools and techniques from conventional HPC. Simulation code that uses conventional MPI is not, in itself, able to exploit the power of the grid. This code can, however, be used as building blocks to create grid-based workflows. Such a workflow might be designed, for example, to

[11] http://www.globus.org/
[12] http://www.unicore.eu/
[13] http://glite.cern.ch/

perform a molecular dynamics equilibration protocol such as the one described in Section 9.6 by running several different molecular dynamics codes on different-scale grid resources. It is also possible to build 'grid-enabled' applications that take some intrinsic advantage of the grid infrastructure, for example by utilizing some form of cross-grid MPI or by including real-time visualization and computational steering.

These applications, however, involve a large number of different steps and therefore can be particularly difficult to manage and deploy. For example, building an MPI code to run across different sites requires the user to build a different version of the code for each different site on which that code is to be deployed. The data must be staged to each site in advance, each machine's operator must be contacted to check that the sites will all be available for running the application at the same time, and then the components of the application must be launched at the same time. Troubleshooting problems with an application like this can be very difficult and time-consuming, and this can often present significant challenges to researchers who need to carry out large-scale scientific studies.

Resource Brokers

In an ideal scenario, a computational grid would be set up so that it appears to its users to be as seamless to use and as transparent as the electrical power grid after which it was named [9]. A grid component termed the resource broker (or just 'broker') is essential if this is to be achieved. This is the component of the grid system that is responsible for distributing jobs efficiently between the resources on the grid. It automatically takes into account factors such as the cost of each resource and the load on each machine. The broker provides a point of contact between the user and the grid, placing application instances submitted by the user on resources that are both appropriate and available. The grid users, therefore, do not need to deal directly with each appropriate machine, logging on to examine the load of each in turn in order to select the one with the least load. It also enables expensive HPC resources to be used as efficiently as possible, ensuring that one machine does not lie idle while another equivalent one has a long queue of jobs.

The grid broker is not the only piece of software that is involved in scheduling jobs, however. Most HPC resources on a grid will still usually have their own local scheduling mechanisms that are responsible for allocating jobs between available processors. The broker will interact with these local scheduling mechanisms as it decides where to place jobs.

9.7.2 Cloud Computing

Cloud computing was developed after grid computing, and is growing in popularity as an alternative schema for running computational applications on a wide range of remote machines of different types. The unifying idea behind all the many forms of cloud computing is the use of a business model to provide access to distributed computer power and resources in return for monetary payment or an equivalent. These resources might include CPU, memory, and storage (in which case the clouds are known as Infrastructure as a Service or IaaS clouds) or applications (in which case the clouds are referred to as Software as a Service or SaaS clouds). For example, users of IaaS clouds can gain access to a virtualized server and have complete control over that server as if it were their own machine, even though it runs in an administratively distinct domain.

Global software and hardware companies such as Microsoft, Amazon, Google, and IBM have made major strategic investments in cloud computing over recent years. These are one of the two main types of cloud provider, selling access to industrial and academic users on a commercial basis; some academic institutions also provide access to cloud computing resources using a research funding model. The relatively low cost of data storage and sharing offered by both types of cloud can make this model a particularly attractive one, although there are still remaining questions over the security and legal issues that surround it.

Virtualization is a broad term used in many areas of computer science and information technology—not just cloud and grid computing—to describe the abstraction of resources. The key benefit of virtualization is that the details of an underlying hardware or software system can be hidden from the users. The benefits of this are manifold: developers can code to a single virtualized interface or system rather than for its implementation on a specific piece of hardware, multiple virtual instances of a system can often be run side by side on a single physical system (in machine virtualization, for example), and physical resources can be protected.

The growth of virtualization technologies, along with service-oriented architectures (SOAs), has powered the development of cloud computing. In an IaaS cloud, the specific details of a cloud's architecture are hidden from the cloud users through virtualized interfaces and systems. Several cloud computing models that take the form of virtualized servers running on hardware platforms managed by a cloud-hosting company have been developed. In these, each user is given access to and control of one or more virtual servers. This has the effect of separating users of the system from each other, since each has access to his or her own, effectively separate, virtualized system. This also provides a degree of *elasticity*, as the number of virtual machines in a cloud environment can be greater than the number of physical servers available to the hosting entities.

Examples of cloud software stacks include Eucalyptus, OpenStack, and OpenNebula. A user who wishes to employ an IaaS in a scientific workflow will need to install or upload their scientific code onto an instance of a virtual machine in the cloud along with the input data needed to run the application. The user will then launch the virtual machine instance on the target cloud platform, and then—just as with grid computing—retrieve the results onto their desktop machine once the execution of the application has completed.

The interfaces used to access cloud resources differ from platform to platform, and an extensive discussion of these is beyond the scope of this book.

9.7.3 Gaining Access to Resources

It is obvious that the first step in running a simulation on any of the types of machine discussed in this chapter—HPC, grid, or cloud—is simply to gain permission to access the resources required. The procedure for doing this will depend on the type of infrastructure involved. The biggest supercomputers usually require potential users to submit a formal proposal describing the science behind the simulation or other job to be run and estimate the amount of CPU and storage that is needed. This process can resemble the procedure used in applying for grants. Institutional-level and other lower-level resources are often free to use and may only require the submission of a simple web form. It is often possible to pay for time consumed on public cloud computing resources by using a credit card.

9.8 Distributed Applications

The e-infrastructure described in this chapter will typically consist of multiple compute clusters each made up of tightly coupled nodes containing single or multiple processor cores. Such a cluster will consist mainly of nodes that solely carry out computation ('worker nodes'), with fewer submit nodes where users log in, stage data, and prepare and submit jobs. The cluster will be managed by a local resource manager (LRM) such as PBS from Altair Computing, LSF from Platform Computing, or Load Leveler from IBM. The LRM will automatically allocate jobs submitted to the cluster among the worker nodes and queue newly arrived jobs until appropriate resources are available. Users will typically submit those jobs via a job-submission interface such as the Globus Resource Allocation Manager (GRAM). Alternatively, if the resource provider allows, users may simply log in to a job-submission node and submit their jobs directly to the required queue.

9.8.1 Middleware Systems

The middleware—that is, the software that sits between the provider institutions and the users, allowing resources to be shared between users in a seamless and uniform way—is an essential part of services for sharing computer resources between institutions.

The Open Grid Forum has proposed the development of an Open Grid Services Architecture (OGSA) [11]. OGSA aims to build on the practical experience of early users of grid systems to address some of the perceived deficiencies of earlier middleware systems. It has been designed as an architecture for producing interoperable, service-oriented middlewares using open standards and industry-standard technologies such as web services. This means that emphasis is put on the services offered by the resources being shared rather than the physical resources themselves. All resources—for storage, computation, and data sharing—are represented as services, and all the entities in the grid are described in terms of their interface and behaviour.

The most widely deployed middleware systems today include Globus, gLite, and UNICORE. We will discuss each of these very briefly in the next sections.

Globus

The open-source Globus middleware has been deployed widely by many international grid projects. From version 3 of the Globus toolkit, Globus has followed a service-oriented approach, tracking the development of the OGSA. The most widely deployed version of the toolkit is currently version 4 (GT4), which comprises a set of web services and a web services container and has a set of non-web-services-related tools. These services are used to manage tasks such as transferring data and launching jobs. Recently, Globus has moved away from web services architecture, with the latest version of the toolkit, GT5, reverting to the client–server system that has not been used since GT2. The client tools supplied with all versions of Globus are command-line-based.

All versions of Globus include a security infrastructure that is based on X.509 public key cryptography. Users must first convert their digital certificate into a format known as Program Editor macro (PEM) and place it, with the correct file permissions, into a specified subdirectory of their home directory. Prior to issuing a Globus command, users must create a short-lived proxy version of their full certificate that allows middleware tools to perform actions on their behalf. The client tools invoke a mutual authentication protocol to establish the user's identity, via the proxy certificate, with the Globus job manager running on the remote resource.

UNICORE

UNICORE (for Uniform Interface to Computing Resources) is a middleware system that has been adopted for use by several international e-infrastructure initiatives. It has been used in, for example, the EU-funded PRACE platform[14]. It implements a three-tier client-gateway server architecture. Jobs are then passed around the infrastructure as serialized Java objects known as abstract job objects. The first tier of the architecture consists of a client that the researcher uses to prepare and submit jobs and that also receives output back from the jobs. The middle tier consists of a gateway controlling authentication on the target resource. This tier also contains the Network Job Supervisor (NJS), UNICORE User Database (UUDB), and the Incarnation Database (IDB). The NJS manages jobs and authorizes users on their target resources.

The UUDB checks logins onto the resource against user certificates, and the IDB translates the Java objects into platform-specific commands that the resource understands. Authenticated jobs are then submitted to the server tier (Target System Interface or TSI) which takes care of running the job on the target resource. With the exception of the TSI, the UNICORE system is implemented as a Java API and a set of related programs. Version 6 of the UNICORE middleware presents a web services interface based on the OGSA framework.

As with Globus, security in the UNICORE system is maintained through the use of X.509 certificates. Prior to using the system users must configure their client to use their digital certificates, imported into the Java keystore format. Users must also import the certificate authority root certificates for any of the resources that they want to access. When the client is launched, each user is prompted for a password to unlock the keystore. When submitting a job to the UNICORE system the user's client uses the certificate to digitally sign the abstract job object before it is transmitted to the NJS. From here, the signature is verified using a copy of the public part of the user's certificate maintained at the NJS site, thus establishing the identity of the submitting user.

gLite

gLite is the product of the Enabling Grids for E-Science (EGEE[15]) project, and uses components from a number of different sources to produce a middleware stack with a wide range of basic grid services. It also is widely used in important international projects. For example, it is the middleware underlying the EGI grid that was designed to process

---

[14] http://www.prace-project.eu/

[15] http://www.eu-egee.org/

the predicted 15 petabytes of data produced annually by the large hadron collider based at CERN in Switzerland.

Condor

Condor is a software system for the management of computing environments that offer large amounts of processing capacity delivered over long periods of time and that are made up of collections of distributed computing resources. These can be described as high-throughput computational environments. Condor can be used either to manage workload on a dedicated cluster of computers or to distribute work to pools of idle desktop computers (in a process known as cycle scavenging), and it is often suitable for running embarrassingly parallel-type jobs. Where Condor systems are made up of pools of desktop machines the users of those machines are given full control over their use. This allows these users to, for example, kill Condor processes on their machines when they return to them and wish to resume interactive work. This ensures that use of the machines by programs delivered through Condor does not interfere with the machines' use by their main owners. Originally Condor pools consisted of computational resources contained within a single administrative domain.

9.8.2 Distributed Application Requirements

A growing number of researchers working on many projects that involve exploitation of HPC resources are finding that the batch-processing model is not sufficient to support their investigations. They may, for example, require simultaneous, interactive access to resources for both executing an application such as a simulation and for visualizing its results. Other cases where batch processing is inadequate involve the investigators changing the parameters of a simulation while it is running. In these cases, the scientists concerned will need to access the resources interactively, while the programs are running. In a system of batch queues, jobs will often start in the middle of the night.

The MPI specification is primarily concerned with process communication between linked HPC machines. The growing prevalence of grid middleware tools such as Globus on these resources has led to the development of versions of this specification that use this middleware to create secure communication channels between resources on different systems that are part of the same grid. Several versions of MPI that can be implemented over grid middleware have been developed, including MPI-g (formerly MPICH-G2), which was developed at Argonne National Laboratory.

Developing versions of MPI that run on a grid provides two main benefits. Firstly, it allows problems that require more memory or CPU to run on a single HPC resource to run instead on a 'meta-computer' that is made up of multiple actual resources. Secondly, it allows large problems that require long CPU times to run, and that therefore typically sit for long times at the top of job queues waiting for sufficient resources to become available, to 'harvest' processors from other machines. This enables sufficient resources to be made available more quickly, enabling these long jobs to run faster.

9.8.3 On-Demand Access Mechanisms

Middleware distributed between different systems is also used in cross-site applications in which a single simulation is performed using multiple resources. MPI-g applications are parallel codes that are distributed across a number of grid resources, and they are often used because no single machine has enough CPU cores or memory to run the required application on its own. A computational grid with advanced reservation policies is needed to support computationally intensive projects that require simultaneous access to multiple resources. Reserving time on a single specified machine in advance is often technically possible using the machine's own queuing system (such as PBSPro) but this facility is quite rarely offered to 'ordinary' users rather than system administrators. This implies that average end users will need to jump through administrative hoops to reserve advance time on these machines.

QCG-Computing[16] is a job-submission and advanced reservation interface that has been designed to allow users to reserve blocks of processing cores or nodes on a cluster. Reservations are made for a defined period of time, starting at a time specified by the user. Users are able to co-reserve access to numerous

[16] http://qoscosgrid/trac/qcg-computing

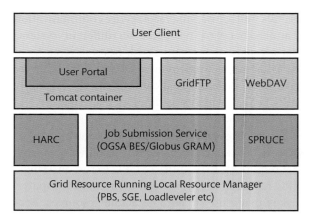

Figure 9.3 A typical multitiered set of grid services, with higher-level services building on the capabilities of those lower in the stack.

machines to run a workflow, just by making multiple reservations at different sites. Potential users should note, however, that the technical ability to make cross-site advanced reservations does not imply that they will always be given immediate interactive access to the resources they need to carry out their work.

The concept of job priority and pre-emption is related in its aims to advance resource reservation. It is based on the idea that urgent jobs—for example, the simulation of a hurricane that is actually approaching—should take priority over regular jobs. In order to set up job pre-emption, resource providers will need to put in place mechanisms to flag jobs as urgent and run them immediately. The TeraGrid project in the USA, for example, deployed the Special Priority and Urgent Computing Environment[17] (SPRUCE) for this. Using SPRUCE, researchers with requirements that the grid administrators recognize as urgent can gain immediate interactive access to a machine without having to make an advanced reservation.

The services that have been described in this section all combine and build on the functionality of the lower-level services to provide what has been described as a rich service ecosystem. Figure 9.3 shows an example of such a system.

9.8.4 Grid Security

Users who need to access grid resources will generally need to secure this access through personal authentication and authorization based on X.509 certificates (which have already been mentioned briefly in Section 9.8.1). Anyone needing to access the resources on a particular grid will need a certificate that is recognized by that particular grid. These certificates are generally issued on a national basis, by a national research certificate authority (or CA). The International Grid Trust Federation[18] ensures that there is mutual trust between national certificate-issuing bodies, so certificates that have been issued in one country will be accepted by grid resources in another.

Once a potential grid user has obtained a security credential, this will need to be converted into an appropriate format to use with the required grid middleware. For example, the Globus middleware uses certificates in PEM format, whereas the UNICORE middleware requires certificates that have been converted into a Java keystore. Details of how these conversions are made will always be provided by the grid that you intend to use.

In this chapter we have shown how you can become authenticated to use a grid or cloud system, gain access to one of these systems, and execute an application. We have not, however, discussed the equally important issue of how to protect data stored on such a public system. This is a particularly pressing concern if the data to be stored are personal, clinical data. A full explanation of how to deal with the security and privacy issues that arise from using medical data in simulation, including when those simulations are run on a cloud or grid system, is available in Chapter 10.

9.8.5 Infrastructure Platforms

We have already seen how expert users or system administrators will be primarily involved in installing applications on HPC or supercomputer systems, tuning and testing them and making them available to researchers. This scenario is common to many different research disciplines that need to use parallel applications on high-performance computers, not only computational biomedicine. Clearly, researchers who wish to use a particular application will use an e-infrastructure on which that application has

---

[17] http://spruce.teragrid.org/

[18] http://www.igtf.net

already been made available, or, if that is not possible, one on which it can readily be installed.

As an example, imagine a user who wishes to launch one particular application that is frequently used in biomedical simulations—the molecular dynamics program NAMD—on a remote resource. This user will first select computational resources on which the program has been pre-installed and which have enough available power for the simulation she wishes to perform. She may log in to each of a set of resources to check the load on each machine and select the one she believes will deliver the fastest turnaround time. She will then upload the completed NAMD input files from her local machine to the target resource, using, for example, the GridFTP[19] protocol. She will write an appropriate job-submission script to run the NAMD application with her chosen input files and then submit it to the scheduler on the target resource. Job submission can either be direct or via a middleware interface such as Globus GRAM. Once the application has been launched the user will monitor it until it completes and then either download the NAMD output files back to her local machine for analysis (using, for example, a molecular visualization tool) or leave them on a remote file system for further computation.

The primary concerns of the end users of an e-infrastructure system are to run their chosen applications in a timely fashion and to obtain results that—with any luck—further the objectives of their research. All the services and facilities offered by those who provide and administer this infrastructure must be subservient to this end. If a user obtains adequate results that are delivered within a useful time frame—whether this is the months (or sometimes years) required to write and publish a scientific paper, or the much shorter time required for a simulation that will aid clinical decision making—it will not matter in the end which machine or machines have been used.

Practical Example 9.2 outlines how to execute simulations using Globus middleware.

Practical Example 9.2 Executing Simulations with the Globus Middleware

All versions of the Globus toolkit provide command line clients that allow users to interact with resources that run the Globus middleware. Version 5 (GT5) is the latest version of this toolkit, but many if not most systems that use Globus are still running GT4 (see Section 9.8.1).

Before you use any of the Globus toolkit components you will need to create a proxy certificate. This is a short-lived security credential that will show that you are authorized to use the e-infrastructure that you require. In order to create a proxy you will need to use the command:

```
grid-proxy-init
```

You will then be prompted for your certificate password.

Once you have gained access to the machine but prior to launching a job you will need to stage any input files required by the job to the remote machine you are going to use. This requires use of a GridFTP service (which is based on standard FTP) and this uses the `globus-url-copy` command. This command is used as follows:

```
globus-url-copy file:///tmp/foo
gsiftp://node1.mygrid.com/tmp/bar
```

Like standard FTP, this command takes two arguments: the path to the source file and the path to the destination it is to be copied to. Some systems allow the use of environment variables in place of full path names.

You will then need to submit your job to the system. The simplest command for job submission in Globus is `globus-job-submit`. The minimum parameters used by this command are the name of the

... continued

[19] http://toolkit.globus.org/toolkit/docs/latest-stable/gridftp/

machine on which you want to run the job and the application to be run. In pseudocode, therefore, the command is

```
globus-job-submit hostname application parameters
```

Here `hostname` is the name of the machine you want to use, `application` is replaced by the full path to the application, and `parameters` are the parameters to be passed to that application. For example, the command

```
globus-job-submit node1.mygrid.com /bin/hostname -f
```

will run the application `/bin/hostname` on the machine `node1` on `mygrid.com` with the single parameter `-f`.

The `globus-job-submit` command will return a reference handle to the job that allows it to be monitored. This resembles a standard URL and looks something like this:

```
https://node1.mygrid.com:64001/1415/1110129853/
```

Jobs are monitored with the `globus-job-status` command followed by the reference handle of the job to monitor, e.g.:

```
globus-job-status https://node1.mygrid.com:64001/1415/1110129853/
```

The command will report the status of the job. When the job is complete, the status will be reported as DONE.

Once a job is complete, the output files generated can again be retrieved using the `globus-url-copy` command, with the local and remote components reversed, e.g.:

```
globus-url-copygsiftp://node1.mygrid.com/home/grid-bestgrid/output.txt file:///home/yhal003/
output.txt
```

This command will copy the file `output.txt` from the directory `/home/grid-bestgrid/` on the machine `node1.mygrid.com` into the user's home directory. Clearly, not all output files will be text files, although many will be.

9.9 Orchestrating Workflows from Distributed Applications

So far in this chapter we have seen how to select appropriate resources on which to run the different components of a workflow. We have looked at how each component of that workflow can be executed individually, either directly on the chosen machine or via a grid or cloud-based system. However, executing workflows individually is not scalable in a context where many instances of each workflow need to be run. Many scientific applications involve the routine, production use of workflows. The HIV protease example discussed in the first case study in this chapter is a small, simple example of this in which the optimum protease inhibitor for treating a particular patient with HIV is selected using nine instances of the same workflow. Automation would help greatly even in this case: in larger examples, automation can be said to be required.

The simplest way to automate the process of running a workflow is to create yet another script, this time to automate calls to a grid client and commands to chain together multiple runs of a workflow and manage data staging between them. This would allow a run of workflows to be executed as a single application from a workstation by using a single command.

The only disadvantage of this method is that it is not easily used for sharing workflows between collaborating scientists; that always requires a workflow engine. An illustration of a scripted workflow is given in Practical Example 9.3.

9.10 Case Study: Computational Investigations of Cranial Haemodynamics

Cardiovascular disease causes more deaths in the developed world than any other disease [12]. An enormous amount of effort has therefore been invested in understanding the function of the heart and circulatory system in both health and disease and in developing new therapies. Several applications of modelling and simulation to the function of the heart and circulatory system have already been described in this book and we look at another here.

Understanding the behaviour of the cerebral blood flow plays a crucial role in the diagnosis and treatment of vascular disease. Anomalous blood flow around the brain in the neighbourhood of bifurcations and aneurysms is known to lead to strokes, although the mechanism is not understood in detail [13]. Experimental studies of this in humans are often impractical and may be unethical, although it is now possible to acquire static and dynamic imaging data non-invasively using computed tomography (CT) and three-dimensional rotational angiography (3DRA) respectively.

Despite—or perhaps because of—these advances in measurement and data-acquisition methods, modelling and simulation have a crucial role to play in studying haemodynamics. Simulation can usefully complement experimental methods that are necessarily limited. It offers the clinician the possibility to perform non-invasive, patient-specific, virtual experiments to plan courses of surgical treatments and study their effects. Simulation is of no danger to the patient and may provide support for the diagnosis, therapy, and planning of vascular treatment [14]. These techniques also offer the prospect of modelling blood-flow patterns, which are poorly understood both in the normal brain and in neurovascular pathologies including aneurysms and arteriovenous malformations.

The data that are used as an input to these simulations are provided by CT and 3DRA scans, and from patient-specific models of the vascular structure of the brain that are built using this data. The HemeLB [15] application is used to create these models, which are used in advising the surgeon before surgery; it is likely that these models will eventually be used also during surgery. HemeLB also includes an in-built ray-tracing engine which provides real-time visualization capabilities, with the visualization frames rendered on the same processor cores as the simulation. A simple desktop display and client software connects to the application at run time to both display the visualization and allow the user to interact with it.

This is an example of a scenario that cannot be run by simply using the traditional 'batch' model of submitting jobs to computational resources that is supported by the majority of HPC providers and that has been discussed in detail elsewhere in this chapter. Simulations that are used to support clinical procedures require large numbers of processor cores but must be run whenever a clinician requires. Surgical procedures that may be life-saving simply cannot be postponed until a simulation job reaches the top of a batch queue. Instead, emergency medical simulations must either pre-empt all other jobs running on a machine or be able to be scheduled into the clinical workflow. Clinicians are not concerned with where simulations are running, nor with the details of reservations; application launching, data marshalling, and all other aspects of managing the submission of routine and emergency jobs must all be done behind the scenes.

This is also an application in which the timescales involved in the simulation are of great concern. Clinical practice generally allows for only 15–20 min to elapse between the acquisition of a three-dimensional dataset of blood flow (which may be 2–4 Gb in size) and the start of a surgical procedure to treat an arteriovenous malformation or an aneurysm. If computational modelling is to play any clinically relevant part in this procedure it needs to fit into this timescale. Clinicians, therefore, require grid and cloud providers to allow them to access computational resources as and when required, so that modelling can be incorporated into their day-to-day activities. They therefore need to make use of interactive methods of accessing HPC resources, such as the advanced reservation and 'emergency' computing discussed

Practical Example 9.3 Scripting a Simple Workflow

This chapter (and, indeed, this book) is not long enough to describe all the practicalities involved in using each of the workflows and workflow engines described above. However, as noted earlier, simple workflows may be developed quite easily by writing scripts to call different commands on the middleware available for staging data to and executing and monitoring jobs on different machines. This is particularly useful when developing and testing a new workflow.

The script below is an example of a very simple workflow application. It simply executes one application using the Globus middleware client, waits for it to complete, and then executes a second application. The lines beginning with # (apart from the first one, which invokes the bash shell) are comments that explain what each of the commands in the script involve.

```
#!/bin/bash
# First copy the input file from the local machine to the remote site
globus-url-copy file:///tmp/foo gsiftp://node1.mygrid.com/tmp/bar
# Launch the first application using the globus-job-submit command
JOBID=`globus-job-submit node1.mygrid.com /bin/app -f`
# loop and perform monitoring command until status of first job is complete
STATUS=""
while ["$STATUS" == ""]
do
echo "Polling $JOBNAME"
# wait for 1 min before polling
sleep 60s
#execute the monitor command, and store the output in the variable jobStatus
jobStatus=`globus-job-status  $JOBID`
STATUS=`echo $jobStatus | grep DONE`
done
echo "Job $JOBID complete"
# Once the first application is complete, move the output data to a second grid resource,
node2.mygrid.com
globus-url-copy gsiftp://node1.mygrid.com/tmp/bar gsiftp://node2.mygrid.com/work/bar
# Launch the second application using the globus-job-submit command
JOBID=`globus-job-submit node2.mygrid.com /bin/app2 -f`
# loop and perform monitoring command until status of second job is complete
STATUS=""
while ["$STATUS" == ""]
do
echo "Polling $JOBNAME"
sleep 60s
jobStatus=`globus-job-status  $JOBID`
STATUS=`echo $jobStatus | grep DONE`
done
echo "Job $JOBID complete"
```

This script initially stages an input data file from the user's local machine to one chosen computational resource and then executes the first of the simulation codes in the workflow. It next polls the status of the job every minute until the job is complete, and when it is the script stages the output data from the first application to a second machine that will run the second application. The second application is then launched in the same way on the second machine, and again the script polls the job status every minute until the workflow is complete.

above, that allow them to be used in a more interactive manner.

The HemeLB bloodflow simulator is capable of running a single simulation that is distributed across multiple resources. Tying together multiple resources in this fashion can decrease the turnaround time substantially but still requires CPU time on all the disparate resources that are to be used to be reserved in advance. Since the resources provided by a single e-infrastructure may not always be sufficiently powerful or appropriate for running large-scale distributed models, an application like this may need resources that are provided by multiple e-infrastructures and that must be federated to conduct a particular investigation. This might, for example, involve using resources that are physically based in several countries.

An application of this type therefore needs to allow extremely flexible access to high-performance computational resources, and these resources must typically be connected using dedicated links so that simulations can be performed on multiple machines and so that visualizations can be performed rapidly, if possible in real time. In addition to these hardware requirements, the clinicians will require suitable middleware tools that can hide all the components of the e-infrastructure that are not strictly necessary for simple job submission from them.

The software environment that is deployed to support these clinical simulations can bring to the forefront details of research-level processes that clinicians need to be aware of. These may include, for instance:

1. the process of image segmentation to obtain a three-dimensional neurovascular model;
2. the specification of pressure and velocity boundary conditions;
3. 'real-time' rendered images of the vasculature.

Although clinicians need not understand the way that resource reservation and computational steering works (in fact, they will almost certainly prefer not to) they will need to use these processes to run, monitor, and interact with simulations in real time.

A client tool that executes the entire workflow will be required to fully automate the process from the perspective of a clinical user. This client, run by the clinician on a desktop workstation, would automate the whole process of retrieving imaging data from a data repository, running a segmentation tool to create a HemeLB model, staging the model to a Grid-FTP server, launching the application on the remote resource, and, finally, allowing the clinician to steer and visualize the resulting simulation as it runs.

It is clear that security will be a prime concern when dealing with data from individual patients. All developed countries require datasets used by clinicians and researchers to be anonymized. In the UK, for example, this involves adhering to a protocol that is specified by clinical partners in the UK National Health Service (NHS) before data are allowed to leave the confines of the secure NHS network. This protocol ensures that the datasets used to run simulations on remote resources contain no data that could be used outside the NHS to identify the individual patient concerned.

9.11 Conclusions

We have seen that multi-scale simulation scenarios of the types that are most often used by biomedical and clinical researchers can require multiple computational resources of different sizes and powers. Exploiting these disparate resources using a workflow as described in Chapter 8 requires each different component of that workflow to be mapped to a resource of an appropriate scale. Each of these may use a different mechanism for allowing user access and may be deployed in different administrative domains. While it is possible to execute such a workflow manually, copying data to the resources and launching simulations by hand, this cannot be scalable. It also prevents workflows from being executed in a routine manner, which can be essential.

Many resource providers federate their high-performance machines into grids, using common middleware software to allow users to interact with the resources remotely. These service-oriented interfaces allow resources to be used in a more automated fashion to orchestrate workflows using higher-level tools, which are known as *workflow engines*.

After reading this chapter you should have a basic understanding of how algorithms are parallelized, how different types of application are mapped to resources of appropriate scales, and how the complete range of resources in the computational infrastructure can be orchestrated to execute workflows and perform powerful simulations. You should also have a basic knowledge of how some of this software can be used in practice, and understand a few examples of how such tools can be applied in clinically relevant scenarios.

Recommended Reading

Carr, N.G. (2008) *The Big Switch: Rewiring the World, from Edison to Google*. New York: WW Norton & Company.

Foster, I. and Kesselman, C. (eds) (2003) *The Grid 2: Blueprint for a New Computing Infrastructure*. San Francisco: Morgan Kaufmann.

Hey, A.J.G., Tansley, S., and Tolle, K.M. (eds) (2009) *The Fourth Paradigm: Data-Intensive Scientific Discovery*. Redmond, WA: Microsoft Research.

Pacheco, P.S. (1997) *Parallel Programming with MPI*. San Francisco: Morgan Kaufmann.

Pacheco, P.S. (2011) *An Introduction to Parallel Programming*. Amsterdam: Elsevier.

Silva, V. (2006) *Grid Computing for Developers*. Hingham, MA: Charles River Media.

Sotomayor, B. and Childers, L. (2006) *Globus Toolkit 4: Programming Java Services*. San Francisco: Morgan Kaufmann.

References

1. Sadiq, S.K., Mazzeo, M.D. et al. (2008) Patient-specific simulation as a basis for clinical decision-making. *Philosophical Transactions of the Royal Society A: Mathematical, Physical and Engineering Sciences* 366(1878), 3199–3219.

2. Gropp, W., Lusk, E. et al. (1996) A high-performance, portable implementation of the MPI message passing interface standard. *Parallel Computing* 22(6), 789–828.

3. Branscomb, L., Belytschko, T. et al. (1993) From desktop to teraflop: exploiting the U.S. lead in high-performance computing. *Final Report of the National Science Foundation Blue Ribbon Panel on High-Performance Computing*, National Science Foundation, Arlington, VA. http://www.nsf.gov/pubs/stisl993/nsb93205/nsb93205.txt.

4. Alfano, M. and Poli, G. (2004) The HIV life cycle: multiple targets for antiretroviral agents. *Drug Design Reviews-Online* 1(1), 83–92.

5. Stoica, I., Sadiq, S.K., and Coveney, P.V. (2008) Rapid and accurate prediction of binding free energies for saquinavir-bound HIV-1 proteases. *Journal of the American Chemical Society* 130(8), 2639–2648.

6. Sadiq, S.K., Wright, D. et al. (2008) Automated molecular simulation based binding affinity calculator for ligand-bound HIV-1 proteases. *Journal of Chemical Information and Modeling* 48(9), 1909–1919.

7. Wright, D., Hall, B., Kenway, O., Jha, S., and Coveney, P.V. (2014) Computing Clinically Relevant Binding Free Energies of HIV-1 Protease Inhibitors, *Journal of Chemical Theory and Computation* 10(3), 1228–1241.

8. Coveney, P.V. (2005) Scientific grid computing. *Philosophical Transactions of the Royal Society A: Mathematical, Physical and Engineering Sciences* 363(1833), 1707–1713.

9. Foster, I., Kesselman, C., and Tuecke, S. (2001) The anatomy of the grid: enabling scalable virtual organizations. *International Journal of Supercomputer Applications* 15(3), 200–222.

10. Coveney, P.V., Giupponi, G.. et al. (2010) Large scale computational science on federated international grids: the role of switched optical networks. *Future Generation Computer Systems* 26(1), 99–110.

11. Foster, I., Kishimoto, H. et al. (2003) *The Open Grid Services Architecture*. Open Grid Forum. http://www.ogf.org/documents/GFD.80.pdf.

12. Guilbert, J. (2003) The world health report 2002: reducing risks, promoting healthy life. *Education for Health (Abingdon)* 16(2), 230.

13. Thubrikar, M. and Robicsek, F. (1995) Pressure-induced arterial wall stress and atherosclerosis. *Annals of Thoracic Surgery* 59(6), 1594–1603.

14. Taylor, A.C., Draney, M.T. et al. (1999) Predictive medicine: computational techniques in therapeutic decision-making. *Computer Aided Surgery* 4(5), 231–247.

15. Mazzeo, M.D. and Coveney, P.V. (2008) HemeLB: a high performance parallel lattice-Boltzmann code for large scale fluid flow in complex geometries. *Computer Physics Communications* 178(12), 894–914.

Chapter written by Stefan J. Zasada and Peter V. Coveney, Centre for Computational Science, University College London, London, UK.

10 Security and Privacy in Sharing Patient Data

Learning Objectives

After reading this chapter, you should:

- be able to articulate the security and legal issues that are specific to e-health systems;
- understand the privacy issues that arise when processing patient data;
- know and be able to compare how the IT architectures of e-health systems impact security and legal frameworks;
- understand why e-health security is a problem that cannot be solved by technology alone and that the legal system has to define the boundaries of acceptable use of data;
- understand how data-protection frameworks must be designed according to the specific needs of different data-sharing scenarios.

10.1 Introduction

One key characteristic of computational biomedicine as it has been described in this book is its close relationship to the clinic. In particular, many projects in computational biomedicine rely on data collected from individual patients. These data—which include biological, imaging, clinical, and genomic data of types described in detail in Chapters 2–4—are unique to each individual. This necessarily means that legal and ethical issues to do with privacy, security, and data protection must be considered whenever this data is shared. The sharing of patient data among experts from different fields that is a necessary component of computational biomedicine initiatives will help in developing targeted, personalized therapies that are linked to the specific genetic and other characteristics of each individual patient. Thus 'personalized medicine' can

be expected to improve outcomes for the patients involved.

The importance of data sharing has been recognized by the British government, among others; late in 2011 the UK government announced that every patient in the country should be 'a research patient' with their medical details shared with private research and healthcare firms [1]. This priority is also present in the European Commission, which has invested many millions of euros in IT infrastructure for medicine. Projects funded by the European Union (EU) include the Human Brain Project (HBP, €1 billion over 10 years), Personalized Medicine (P-Medicine, €13.3 million) [2], VPH-Share (€10.7 million) [3], and the European Data Infrastructure Project (EUDAT, €10.7 million) [4] among others.

The main challenge faced by those who design and develop data-sharing platforms within the general discipline of 'e-health' (which is wider than computational biomedicine as such) is how to provide a legal and security framework that enables seamless and secure access to shared patient data. Unfortunately, high-profile security breaches and data losses are frequently headline news. The need to store and access patient data from a wide range of platforms both within and between hospitals bears a high risk of that data being misused with considerable negative effects for all the stakeholders [5]. These may include data providers such as hospitals and care centres, universities, research institutes, and their industrial partners who wish to access and use the data in their research; and, of course, patients, who need to give their consent.

One well-known example of a breach of patient privacy was reported in Finland in 2008 [6]. There, a nurse's medical history was released in an unauthorized way by a colleague in the hospital where she worked. The nurse's medical history showed that she had been treated for HIV. This revelation resulted in her contract not being renewed by the hospital and her colleagues at work knowing about her disease, which clearly shows the potential personal impact of such breaches of privacy. The hospital was ordered to pay the nurse €14 000 in damages and €20 000 in costs. An unauthorized disclosure of data of this type may not only lead to legal liabilities and fines, but may also provoke negative publicity and a loss of trust in the credibility of the medical profession. Incidents of data loss reported by the Information Commissioner's Office in the UK [7] include loss of over 100 patient records on a compact disc left at a bus stop. This type of incident may also cause a decline in the willingness of patients to participate in clinical trials and share their data more generally.

Health data and in particular genetic data records are highly sensitive. Not only do they contain information about the health status of the patient concerned, but they may also reveal information about the identity of the patient and the genetic constitution of his or her close relatives. Individuals may not want their data to be shared beyond the clinicians involved in their care, especially if there is a possibility that they may be identified personally. Much effort has therefore gone into arrangements whereby patient data can be *anonymized* to hide patients' identities for research purposes. Data can also be

pseudonymized, in which case it becomes possible to re-identify patients, if this comes to be necessary, through the use of a key. Issues of data anonymization and pseudonymization have been discussed at more length earlier in this book.

Another problem with the release of health data that can be linked to individuals is that it may lead to possible *discrimination* (e.g. if insurers require people who are vulnerable to a particular disease to pay higher premiums). Furthermore, the integrity and correctness of medical data must be assured, as unauthorized modification of patient data may lead to incorrect patient treatment. Identity theft (stealing a patient's identity to cover for a crime, fake death or claim free treatment using an insured patient's data) is another concern.

For all these reasons, it is essential that a high level of data protection and data security is established for all platforms that might involve the sharing of patient data, to protect the patient's privacy rights and to ensure that the data are used for the intended purpose to which the patient consented. Patient consent is a key issue in this, but, unfortunately, patients are not in control of their medical records and so it is not easy for them to decide whether and with whom to share their data. As a result, many legal constraints have been set up to control when and how data can be exchanged. These affect the way that the computational architectures for data storage and sharing can be set up.

Currently, almost every time an individual goes to the doctor he or she has to enter data on forms. Insurance companies and pharmacies often require medical information as well. It is not easy for patients to remember all the details that are needed, particularly when these refer to family members. If patients were able to access their own and possibly also their families' medical records, medical care might be improved in a variety of ways. This would make patients feel more involved, better informed, and in closer control of their own care. It would also help to keep the medical profession informed, when, for example, a patient switches physicians or visits a new doctor for the first time. Patients who were in control of their own medical records would feel more empowered to share their data with third-party organizations and research institutes, and this might make them keener to participate in research such as clinical trials. This would also ease some of the legal constraints on the design of experimental and computational platforms that are used to generate new medical knowledge. This might impact, for example, discovery science, the development of novel mechanisms for the design and conduct of randomized controlled trials, evaluation of quality and patient outcomes, and decision support for policy making in health and social services. Although few, if any, individuals have access to their own medical records, many patients have willingly provided their personal medical data to Facebook-like sites such as patientslikeme.com [8] and 23andme.com [9]. These sites support active communities of engaged patients with particular conditions who can take part in medical research.

In the next sections we present a brief overview of the legal background to data security in order to justify why security controls are required. This is followed by a brief overview of information security goals, and common legal and security requirements for a large class of data-sharing environments. We discuss the data-sharing life cycle, describe several possible architectures for data warehousing (which may be centralized, distributed, or cloud-based) and present their legal implications.

10.2 The Legal Background

We aim here to provide an overview of the data-protection rules that have to be taken into account when dealing with data in e-health data-sharing platforms. Furthermore we will define technical security terms and some of the terms that we have already used, such as personal, pseudonymous, and anonymous data, on a legal basis. Understanding

these rules should make it clear to you why we need security controls to ensure compliance with legislation.

The emphasis in this section is on the legal framework in the EU. Similar laws exist in the USA and all other developed economies and regions.

10.2.1 A Brief Overview of Relevant Laws

In the EU, the use of data is governed by EU Directive 95/46/EC (see Box 10.1) on the protection of individuals with regard to the processing of personal data and on the free movement of such data [10]. This states the general rules for processing personal data anywhere in the EU and must therefore always be taken into account when processing data in the EU. However, this Data Protection Directive is not directly applicable to individuals or even countries, but rather directs the member states that they need to harmonize their national laws in order to comply with the Directive. It therefore sets standards for the processing of personal data that need to be transposed into national law by all member states of the EU [11]. Even though this national transposition can vary from country to country, all the national laws on data protection in the EU reflect the basic rules set by this Directive [12].

The EU Data Protection Directive is not the only piece of European legislation in the field of data protection that is relevant to clinical trials. There are several other Directives that contain specific rules on data protection in specific fields. The Clinical Trials Directive [13] (Directive 2001/20/EC) seeks to simplify and harmonize the administrative provisions governing clinical trials by establishing a clear, transparent procedure for them. It is also necessary

Box 10.1 Regulations and Directives

European (EU) law distinguishes between directives and regulations.

Regulations are legislative acts of the European Parliament that are directly applicable in all member states. They lead to a high level of harmonization, but do not leave space to maintain particular national laws of the member states.

Directives, in contrast to regulations, are not directly applicable but require each EU member state to achieve a

particular result, for example a certain level of data protection. Each member state has the right to choose how to achieve these results by means of national legislation that transposes these requirements into national law. This allows each state to adapt the directive to fit with its own particular requirements.

to consider Directive 2001/83/EC, which sets out the community code relating to medicinal products for human use.

We describe below the most important legal concepts that are given in the Data Protection Directive, including the different categories of data and the requirements for fair and lawful data processing. You should remember, however, that this directive only sets the framework and that this framework is interpreted in different ways by different countries; and, furthermore, that many countries outside the EU have by now established a similar legal framework for data and its processing.

10.2.2 Categories of Data

The Data Protection Directive defines several different categories of data. Since this directive is only applicable to the processing of 'personal data', the most important distinction that is made is between personal data and anonymous data [14]. 'Pseudonymous data' and 'sensitive data' define special categories of personal data and so are also covered by the directive.

Personal Data

Personal data, which is sometimes also named *Personal Identifiable Data* (or PID), is defined as 'any [data] relating to an identified or identifiable natural person ("data subject")' [15]. An identifiable person in the sense of this provision is one who 'can be identified, directly or indirectly, in particular by reference to an identification number or to one or more factors specific to his physical, physiological, mental, economic, cultural or social identity' [16].

The main criterion appearing in these definitions, therefore, is that of identification: whether it is possible for the data or information concerned to identify an individual. This can be achieved through certain 'identifiers', the most common of which is the person's name. However, identification can also be achieved using a combination of other information, such as:

- address, full postcode or zip code, date of birth, ethnic category;
- local or national hospital numbers [e.g. the National Health Service (NHS) number of an individual in the UK];

- photographs, videos, audiotapes, or other images of service users;
- rare diseases, drug treatments, or statistical analyses that have very small numbers within a small population and that may therefore allow individuals to be identified;
- anything else that may be used to identify a patient or health service user directly or indirectly.

An example of this is seen in the USA where the gender, date of birth, and the five-digit zip code of an individual alone are enough to identify 83% of the population [17].

Anonymous Data

According to the wording of the Data Protection Directive data are to be considered as anonymous if and only if the data subjects are in no way identifiable from the data. This is the case if there is never a link that refers to the data subject within the dataset, or if an existing link has been irrecoverably erased. The question of whether data have to be regarded as identifiable can pose severe practical problems. In answering the question of whether data can be considered anonymous, the directive states that 'account should be taken of all the means that are reasonably likely to be used either by the controller or by any other person to identify said person' [18]. Thus, if there is no or negligible possibility of 'reasonably likely' steps being taken to identify an individual, the person is not considered identifiable and the information concerned is not considered to be personal data.

This clearly implies that anonymization does not require that the person to whom the data belong must be identifiable by no means whatsoever. Rather, it is generally acknowledged that data are to be regarded as anonymous if they can be attributed to an identified or identifiable individual only after a disproportionate amount of time or expense (Figure 10.1).

In each case, the scientists involved in constructing a model using personal data will need to use their judgement in assessing whether it is possible to identify the individuals involved through 'reasonable means'. The factors they will use for this include the cost of such an identification, the intended purpose, the way the data processing is structured, the

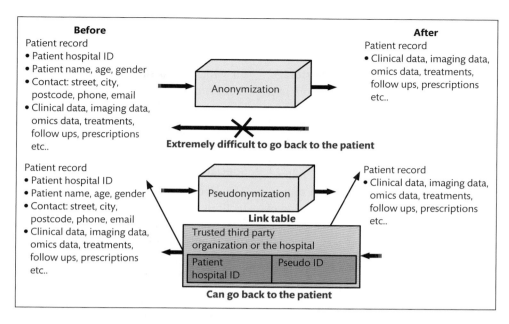

Figure 10.1 Anonymization and pseudonymization of data. With anonymization, it is difficult to re-identify a patient whereas with pseudonymization it is possible to re-identify a patient using the link between the patient identifier in the hospital and the pseudo ID number. This link is usually stored in the hospital or with a trusted third-party organization.

expected advantage for the controller in making an identification, and the interests that are at stake for the individuals concerned as well as the risk of organizational dysfunctions (e.g. breaches of confidentiality) and technical failures [19].

Pseudonymous Data

Data pseudonymization represents a kind of 'halfway house' between personal and anonymous data, although pseudonymous data are treated legally as a special case of personal data. To pseudonymize a personal dataset the identifying characteristics of each individual (such as a name) are replaced with a label to preclude identification of the data subject or to render such identification substantially difficult (as shown in Figure 10.1).

The use of pseudonymous rather than anonymous data can have particular benefit for the patients concerned, because it becomes possible to re-identify patients who (for example) are thought most likely to benefit from a particular newly developed treatment. It is possible to protect a patient's privacy rights by guaranteeing that medical data will only be transferred to the researchers under the pseudonym, with the table that matches each patient's identity and pseudonym

remaining stored by a trusted third party (TTP). This may be, and often is, an institution that is independent of the hospital concerned and of the management of the research project that will be using the data.

Sensitive Data

Another sub-category of personal data is data that can be regarded as particularly sensitive. This includes, for example, data that could reveal information about an individual's racial or ethnic origin, political opinions, religious or philosophical beliefs, trade union membership, sexuality, or certain aspects of their medical history. Data-protection laws typically state that the processing of this type of data is only allowed in exceptional circumstances. In the EU this is stated in the Data Protection Directive and the exceptional cases are set out in laws that are enacted in each member state [20].

10.2.3 Data-Processing Rules

Processing

The term 'processing of personal data' (or just 'processing') as used in the EU Directive is extraordinarily broad. It is designed and set out to cover 'any operation or set of operations which is performed upon personal data, whether or not by automatic means, such as collection, recording, organization, storage, adaptation or alteration, retrieval, consultation, use, disclosure by transmission, dissemination or otherwise making available, alignment or combination, blocking, erasure or destruction'. This definition therefore includes virtually any operation that might be performed on personal data: interestingly, following this definition, even the anonymization of personal data can be regarded as a type of processing of personal data.

Principles for the Lawful Processing of Personal Data

The Data Protection Directive establishes several principles concerning the fair and lawful processing of personal data [21]. It sets out criteria for the legitimate processing of these data [22] and makes provision for technical and organizational measures that ensure the security and safety of the data [23]. Data processing may only be conducted within a legal framework and in compliance with all relevant legal requirements. In addition, the rights to data protection have to be balanced against the interests that others will have in processing the data. Furthermore, the purpose for which personal data is to be collected and processed must be explicitly specified, and it must, of course, be legitimate. These data, once collected, must not be further processed for other purposes and any further processing and use of data will only be allowed if it is closely connected to the original purpose. As a consequence of these restrictions the purpose for which the data is to be collected must always be clear from the start [24].

This strict limitation of purpose might pose a problem for research projects in which data are made accessible to project partners who may wish to use the data for other purposes. This would, for instance, be the case if data collected for a specific research project in cancer are to be used for research in a different field of medical research. The Directive does state, however, that further processing of data for historical, statistical, or scientific purposes shall not be considered as incompatible as long as appropriate safeguards are provided. In the EU this is generally covered by national data-protection laws.

Last but not least the Data Protection Directive states that personal data must in general not be stored for longer than is strictly necessary for the purposes for which the data were collected or for which further processing has been agreed [25]. Here again the Directive offers the possibility to store data for longer periods if it is required for historical, statistical, or scientific use.

Special Protection for Sensitive Data

The Data Protection Directive grants special protections for the processing of the types of data that were defined as 'sensitive' in the previous section. You should have understood from this list that data taken from patients will very often fall into one of these categories, so these rules and principles must be considered carefully in every project that involves patient data.

As a general rule sensitive data (including health data) may not be processed except under certain clearly defined circumstances [26]. Sensitive data will usually be allowed to be processed if the patient or individual concerned (the data subject) has given his or her consent to this processing. Patients in particular will often benefit considerably from the collection and processing of their sensitive data as it is often needed to plan the best treatment for those patients, and special arrangements are often made in cases like these. Data processing that is designed to support medical research and optimize the treatment of future patients (such as clinical trials) is not in the vital interests of the patients who provide the data, and these exemptions rarely apply to cases like these. Computational biomedicine projects generally fit into this latter category in which the use of sensitive patient data is very tightly controlled.

In the EU further exemptions for the processing of sensitive data for medical research may be introduced by the national laws of the member states. It is even possible for supervisory authorities at a less than national level to decide to allow the processing of sensitive data, as long as they can provide specific safeguards to protect the fundamental rights and the privacy of the individuals concerned [27]. As these rules vary from country to country and are often subject to changes, it is not considered good practice to base research projects that are designed to have clinical applications (that is, translational research projects) on them. These exemptions, however, can serve as a fallback scenario for unforeseen problems, such as when the legal data-protection framework that has been set up for such a project fails.

10.2.4 Patient Consent

When a patient gives permission for some or all of his or her health record to be shared with third-party organizations for a clinical trial or other research project, that patient is said to have given consent. All legal data-protection frameworks state that a patient's consent is a necessary precondition for that patient's data to be used in research. Furthermore, consent has to be given one project or procedure at a time. Any medical intervention, whether preventive, diagnostic, or therapeutic, and any medical research project involving patients (or, indeed, healthy volunteers) is only acceptable if the individuals concerned have given their consent and if that consent is based on adequate information. It is important to realize that consent should be expressed explicitly rather than assumed, and that it may be withdrawn by the person concerned at any time and for any reason without disadvantage or prejudice.

Patient consent, therefore, should be informed. In order to obtain patients' consent the researchers involved in the study will need to provide them with all the information that is necessary for them to understand what will be done with their data. This requirement ensures that the use of their data is transparent to the patients [28]. It is meant not only to prevent legal uncertainty but also to generate trust between patients and researchers.

Researchers must understand, however, that a patient's ability to consent to the processing of his or her data is not unlimited or under no control. The Data Protection Directive already establishes requirements that have to be met by the patient's consent. Firstly, this consent has to be freely given and indisputable, without any doubt, and secondly, the patient should know and understand as far as possible exactly what he or she is consenting to.

This presupposes that a patient whose consent is requested must be capable of making decisions and must therefore be capable of understanding the information that he or she is given in order to do so. Patients should be able to appreciate the value of the

work and how, if at all, taking a particular decision will benefit their medical condition. In order to do this fully, patients must be able to reason, understand cause and effect, and, often, probabilities and percentages. They must also be able to make decisions and communicate them clearly to the research teams involved [29]. It is clear that children, for example, will therefore never be able to give informed consent. The law presumes that capacity for giving informed consent is achieved at the age of majority, which in many countries is fixed at the age of 18. It is generally assumed that people who have reached the age of majority will, if they are patients, be empowered to exercise their rights of consenting (or withdrawing their consent) for the use of their sensitive personal data for medical research and similar purposes.

More and more often, however, minors—that is, children and teenagers—are becoming involved in decisions concerning their health and are being granted the right to take decisions by themselves. This is clearly a very risky business, and children's rights in this area must be interpreted in a very restrictive way. Furthermore, minors are not the only category of people for whom questions are raised regarding capability to consent. These are also raised in cases involving mentally ill and geriatric patients, for example; in the latter case, certain patients may have the capacity to consent, but be losing it. Most countries have laws that authorize a legal representative to give consent on behalf of a minor or incapacitated patient.

Patient consent is required for ethical as well as legal reasons. It is only ethical that data collected in the course of medical treatment should be used for research purposes only if the patient has previously agreed. It is a basic precondition that data shall not be processed against the expressed wish or without the knowledge of the patient, even if it can be assumed to fall within the definition of 'anonymous data' given in Section 10.2.2.

Informed consent has to meet specific requirements in order to be valid. The patient, obviously enough, has to be provided with enough information to enable him or her to make the decision. This information should be comprehensive and understandable and should include at least the main intentions of the research project, the range of possible uses of the data (for example, whether there is any likelihood of its being sharing with third parties), measures taken to protect patients' personal rights, the possible risks and benefits of the research, and any further implications of participation. Furthermore, consent can only be given voluntarily. This means that the patient must explicitly sign a paper or tick a box on a web form allowing the data use, he or she must be aware of the purpose for which the data will be used, and he or she must be able to give informed consent.

10.3   A Brief Overview of Information Security Concepts

Once you have understood the rules for data protection that we have described in the previous section, you will have become aware of the need for mechanisms to enforce those rules. In any research project involving patient (or any personal or sensitive) data, the controller or processor of that data needs to implement appropriate technical and organizational measures to protect the data against accidental or unlawful destruction or accidental loss, alteration, unauthorized disclosure, or access. This is particularly important where the data processing will involve the transmission of data over a network, be it either the global Internet or a local intranet. Some key concepts underlying information security are summarized in Box 10.2.

Technical measures deal with the practical use of the methods that are implemented to secure the data that is being processed (including methods for preventing physical access to the hardware by unauthorized people, including secure premises and access control [30]). They also include the use of encryption, secure connections, firewalls, and identification of people who are authorized to access the data using biometric or similar methods. Organizational measures refer to rules that enable data security by regulating authorization and authentication procedures; that is, access policies and identity management for the IT system that processes the data [31].

Several different types of measure are involved in data security. These involve, for example, taking appropriate measures to prevent the misuse of patient data (protection), identifying possible damage to data, such as alteration or deletion of patients' records (detection), and recovering data after such damage has been done (reaction). Concepts, techniques, mechanisms, tools, and standards for several crucial aspects of security such as confidentiality, integrity, availability, authenticity, and accountability have evolved over the years.

The security measures that have to be provided in a specific case depend on several factors including the 'state of the art' and the category of data that is being protected. The Data Protection Directive does not regulate what precisely state-of-the-art technology is, for various reasons, the most important being the fact that the technical discipline of data security is to a large extent evolving very fast. Data belonging to each of the categories discussed in Section 10.2.2 have different legal requirements with different degrees of strictness. For example, anonymous data is the most flexible format and this type of data can usually be readily moved between countries, whereas personally identifiable and genetic data and genetic

---

Box 10.2   Information Security Overview

Some fundamental aspects of information security are defined and described below.

- *Confidentiality*: only authorized users are allowed to *view* patient data. For example, only the doctor treating a patient can usually view a patient's history and previous treatments. Patients are able to give consent for other data users to view their data.
- *Integrity*: only authorized users are allowed to *modify* patient data. For example, only the doctor treating a named patient can write a new prescription for that patient.

- *Availability*: authorized users should be able to access patient data upon demand.
- *Authenticity*: this is the process of verifying a claimed identity by, for example, using a username and password pair or a security token such as smart card and a pin.
- *Authorization*: this is the process of verifying whether a named person is allowed to perform a particular task on a patient data record.
- *Accountability*: the ability to trace users who were responsible for performing a particular task on the data, for example, who accessed or modified a named record at a given point of time.

data types have strict requirements due to the risks posed by theft of such data, including identity theft.

Data that have been pseudonymized should still be used within a secure environment and accessed only on a need-to-know basis. If a patient or hospital identifier is required or if there is a need to trace data back to a specific patient (which is common in translational research) the reasons for and usage of the data should be fully documented and approval obtained from the appropriate data owner. This auditable trail of access to a patient's records is required so that patients can be informed of who has accessed or seen their data. This audit should provide accurate data in the event of a breach of data security or similar incident. The key items to be documented include who has access to each database containing identifiable data, the particular data accessed, the time of access,

and reason(s) for and output(s) from the access. This audit should be kept within a separate structured database that is accessible for query and audit. The log of accesses must be regularly audited via sampling users or types of data accessed to check for unusual patterns of access. Unusual patterns of access should be reported via the Data Governance Department.

10.4 The Data-Sharing Life Cycle

Researchers who seek to share data, particularly patient-specific data, in a project are required to set up one or more computational environments in ways that are secure and legal from a data-protection perspective. The data-sharing life cycle shown in Figure 10.2 illustrates the process of setting up such

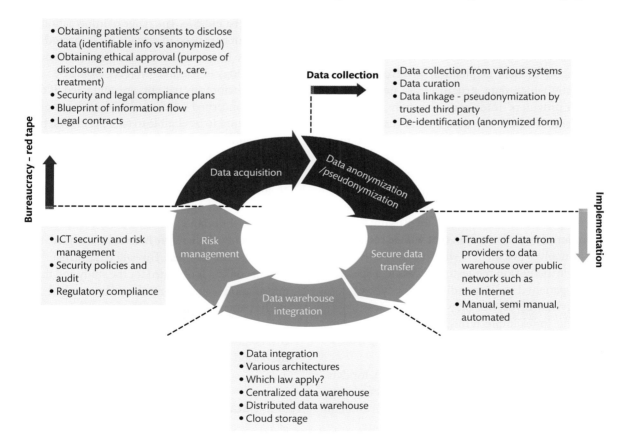

Figure 10.2 The data-sharing life cycle. The starting point is obtaining patient consent, then applying for ethical approval for the research project or clinical trial and preparing a legal contract between the entities sharing the data. The process also involves identifying the data structure to be shared, ano/pseudonymization, transfer, storage, and integration of data, and secure access to the shared data by deploying security controls and setting up a risk-management process to ensure that the data are stored in compliance with relevant legislation at all times.

a data-sharing platform that is secure enough to hold personal and even sensitive data.

The process starts with *data acquisition*. This is not a simple task; it involves identifying the data to be shared, obtaining consent from the patients involved, and obtaining ethical approval from local and national agencies. The procedures involved in this depend on whether the researchers involved in the project are all located in one country or whether the project is expected to span several countries.

The second step involves *collecting the data* from the data providers' IT environments and de-identifying the data. The final step, *implementation*, involves secure transfer of patient data to the data-sharing environment, providing a uniform interface to the data, and, finally, ensuring that the security controls that have been put in place to protect the shared data are adequate.

10.4.1 Data Acquisition

The process of data acquisition involves identifying exactly which data structures are needed from the data providers' information systems; providing a blueprint for how the data will be used, stored, and protected; and obtaining consent and ethical approval from the appropriate bodies. The next few paragraphs explain these processes in more detail.

- *Data scoping*: the first step in setting up a data-sharing environment involves identifying the content of the information that will be requested. This involves researchers within a consortium or a single institution requesting access to a specific set of patient data. Depending on the type of research involved, this may be more or less specific; for example, it might focus on data from all cancer patients or from those suffering from a particular type of tumour.
- *Data flow*: while developing a framework for data protection and data security it is first necessary to identify the data flows in the shared environment that is envisaged. This is to highlight where during the research process personal data might be exchanged, which safeguards need to be taken and where. For instance, the data flow will show where the data will be transferred to and stored, for example within the hospital network, outside

the hospital private network, or with a TTP in the same or a different country. Each of these scenarios has different and serious implications for the legal and security framework that is most appropriate for the project.

- *Consent*: the next step is to obtain patient consent for the disclosure of their information to third parties involved in the research project. The patients should know whether the data will be held in a format that may identify them (i.e. pseudonymized data) or in a format that would make identification extremely difficult (i.e. anonymized data). We have already seen that in order to obtain patient consent for using their data in (for example) a clinical trial the patients must be provided with all the information that is required for them to decide whether to take part, and that this information must be comprehensive and understandable. The range of information that researchers need to provide for informed consent to be given indicates that the form and methodology of research projects should be concrete before any patients are contacted. If the project concerned is a clinical trial, each patient who is found to be eligible to take part in the trial will be contacted separately by his or her physician.
- *Ethical approval*: before a project that involves sharing patient data between organizations can go ahead, it is necessary that the objectives of the data sharing are approved and endorsed from both a scientific and an ethical perspective. The experiment will therefore have to be analysed by an ethics committee. This will require that patient consent is obtained for the study and that information security controls will be installed on the patient data-sharing platform. The organizations enter into a legal agreement that covers the data shared, the purpose of the study, and the period of time for which the data will be kept.

10.4.2 Data Collection

The data-collection step self-evidently involves collecting the patient data that is needed for the study from existing information systems within the hospital involved. The collection of patient data can be a very challenging task from a technical point of view,

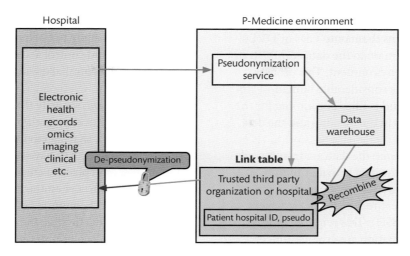

Figure 10.3 Data being pseudonymized during transmission from the hospital to the data warehouse and the link between the patient identifier and the pseudo number stored on a TTP organization site specializing in this type of application. This enables scientists to re-identify a patient in case of a new finding. This mode is used in the P-Medicine project which is described in detail in Section 10.5.4.

given that the data required may be stored on different information systems that have been developed and installed in the hospital by different contractors. The main challenge is therefore to establish the technical interoperability of the different data sources with the system to be set up for the project that is envisaged. The collection process may involve de-identification of the data if the data are to be used for research purposes. The data will then be transferred from the data provider site to the shared platform as discussed below.

Pseudonymization/Anonymization

After the data have been acquired it is almost inevitable that they will be either anonymized or pseudonymized. It may not be technically possible to fully anonymize the data (for example, if it is genomic data that contain the whole DNA sequence of each patient; see Chapter 2) or it may not be desirable, for example if it might be beneficial to contact individual patients following the results of a clinical trial. In these cases it is most likely that the data will be used in a pseudonymized form (Figure 10.3).

10.4.3 Implementation

The implementation of a data-sharing environment includes several steps: unlocking the data from the

hospital IT systems and transferring them to the data-sharing environment where they will be hosted. This section describes some of the approaches to data transfer and data storage that are currently in use.

- *Data transfer*: several methods are available for transferring data from a provider to a shared data environment (and potentially also back again, particularly in translational research) but they are not equally secure. Data transferred without encryption on DVDs and USB data sticks clearly are at risk from those media being lost or stolen. Pseudonymization or anonymization generally takes place in the course of the transfer of the patient data to the platform. If the patient data are merely pseudonymized, the keys that provide the links between the data and the patients are usually stored at a TTP that is independent from the hospital transferring the data, the data warehouse, and the partner organizations that will be working with the data. The involvement of a TTP is intended to ensure that the patients' privacy rights are respected; they are therefore regulated in detail by contracts. Two data transfer solutions are commonly used, as follows.

  1. *Manual transfer*: the data manager in the hospital trims and anonymizes or

pseudonymizes the patient data that are requested, burns them onto CDs or DVDs, and passes them on to the data administrator of the shared environment. This process clearly requires considerable human intervention and is time-consuming for the data managers in the hospital and the data environment.

2. *Electronic transfer*: the data are transferred directly from the data provider to the data-sharing environment over a secure connection (i.e. the Secure Socket Layer protocol SSL/SSH). This requires approval from an information governance officer because it involves opening firewall ports to transmit the data which may increase the risk of compromising their own resources.

- *Data integration*: once the data have been transferred to the data warehouse, the latter should be set up to accommodate a heterogeneous set of data types. These may include imaging data, text files (e.g. for gene sequence data) and Microsoft Excel spreadsheets, among other formats. Imaging data, whether derived from magnetic resonance imaging (MRI), X-ray computed tomography (CT), positron emission tomography (PET), or another technique will generally be held in the standard Digital Imaging and Communications in Medicine (DICOM) format that is described in more detail in Chapter 4. There are several technologies that enable this range of heterogeneous data types to be stored, such as relational database management systems and file servers. Tools for handling, storing, printing, and transmitting DICOM medical imaging data in a standard format are available; open-source examples include OSRIX [32] and dcm4che [33].

- *Data storage*: it is important to decide exactly where data will be stored. Questions that should be asked include whether it should be hosted behind or outside the firewall of a hospital, on the premises of a TTP, or on another partner's premises. If the data are to be hosted in the cloud it may be necessary to know whether the machine hosting the data is located in the same country or in a different country to the patients whose data are to be used and/or the researchers who will be

using it. These choices have serious implications for which laws apply to the data-sharing environment and control access to it. We will consider both centralized and distributed data storage here.

- *Risk management*: security threats are evolving at a fast pace. Since research partners access the patient data-sharing environment over the Internet, it is critical to have security controls and processes in place to ensure that the system is resilient against cyberattacks. In January 2012 the World Economic Forum described these as their fourth most important global risk [34]. This is not generally considered a high priority for research projects, but clinicians who develop hospital-based data-sharing platforms that are accessible over the Internet will need to ensure that their systems are resilient against worst-case scenarios, cyberattacks, and novel threats.

10.5 Data-Sharing Platform Architectures

In this section we describe three different architectures for data warehousing platforms. These all enable patient data to be shared across a mixed set of organizations that have different security domains. Very many of the data warehouses that are in current use employ one of these architectures.

The data-sharing platforms that we will discuss are:

- traditional, centralized data warehouses;
- distributed data warehouses with components located in one or more countries;
- TTP storage: generally security-certified storage facilities in trusted organizations;
- the cloud storage that is provided by large software vendors such as Amazon may be considered as a special case of both a centralized data warehouse and TTP storage.

These architectures use different types of access control mechanisms, which can involve username/password pairs, Kerberos, Shibboleth, and certificate-based public key infrastructures to prevent unauthorized access and therefore protect the data.

For each of these different data storage architectures, we will look at:

- data-storage options;
- authentication and authorization options;
- legal implications.

10.5.1 Centralized Data Warehouses

The most common type of architecture for storing patient data is a centralized warehouse. In this, all data are transferred from the providers to a centralized environment for hosting the data. One variation of this model is the cloud, which provides a centralized data-hosting environment that is managed by a generic software vendor. Companies involved in providing cloud storage include such household names as Google, IBM, Microsoft (Azure), and Amazon (EC2).

The physical infrastructure of this model involves a dedicated set of resources such as database servers, file servers, and DICOM servers for managing the different types of data that are to be stored. This might include, for example, medical records, omics data (including DNA sequences), imaging data, and

the corresponding metadata. The capabilities and credentials of the expected users of the data that allow access to the environment are also stored centrally. Each partner organization involved in the project connects to the data warehouse through a public network, most likely the public Internet as shown in Figure 10.4.

There are several options for hosting a central data warehouse.

- *Option 1:* the data warehouse is stored on a third-party site that is independent of all partners. In this case, the warehouse will be maintained and operated by personnel who are not members of any of the partner organizations in the project who are sharing the data. It might be difficult for some hospital organizations in particular to accept this because patients tend to trust hospitals rather than third parties with their data. This issue is more one of social acceptance than a technical one. One way to address this issue is through giving patients assurance that their data will be protected adequately, perhaps by showing that the TTP has a security certification from an international or national organization. ISO

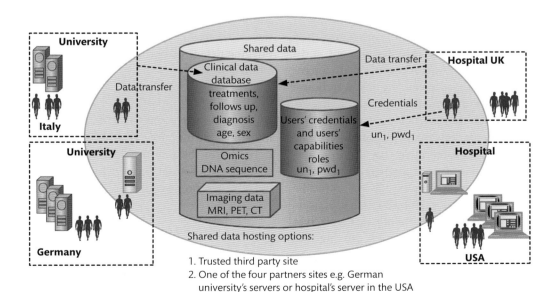

Figure 10.4 A centralized data warehouse architecture with various data-hosting options. Authorized users with valid credentials (e.g. un_1, pwd_1 = username/password) can access the data warehouse to perform scientific tasks.

270001/2 is an internationally accepted security standard that certifies that a facility has state-of-the-art security controls.

- *Option 2*: the data warehouse is stored on a server belonging to one of the partners that is involved in the data sharing; for instance, a local university data storage facility or hospital network. In this case, the data warehouse will be maintained by members of staff who belong to one of the partners. This model is more common than the above one, especially in translational research projects. Sharing data on a hospital local network is not a common practice; it introduces more challenges into data-access controls such as authentication, as we explain below.
- *Option 3*: the data warehouse is stored on the servers of a cloud provider. This option is discussed in more detail in Section 10.5.5.

Whichever option is selected, this generic architecture is considered to be the easiest to set up in terms of authentication and authorization because the whole system is hosted by one entity under one security policy.

Authentication

Every time a user requires access to the data-sharing environment that includes patient data their access will need to be authorized. All users from the consortium involved in the research project need first to register with the data-sharing environment, and a database entry will be set up for each user. Once a request for registration is received, the data warehouse administrator will contact a representative of the organization to which the user belongs (most usually, the human resources division or the intended user's head of department) to confirm that the information that the user has given in registration is valid. Once users have been accepted, they will either be given a dedicated set of credentials (such as a username and password pair) to authenticate themselves to the warehouse or use credentials provided by their local organizations. Shibboleth, for example, enables users to invoke credentials from their local organizations to access remote resources. This, however, requires that the data-sharing environment must have the capability to verify the credentials that are provided by each user (see Figure 10.4). If

the user is from a different country the data-sharing environment should incorporate a server that can contact a server at the user's home institution in order to verify the identity of the user. This clearly requires that the data-sharing environment should be trusted by the identity provider. This approach is becoming popular because it is usable and standardized. Its main disadvantage is that authentication has to rely on the identity servers being up and running all the time.

Authenticating users becomes particularly difficult if the data warehouse is hosted behind the firewall of a hospital because it will require giving users belonging to other organizations access to the hospital's security domain. This is not acceptable to many hospitals because it weakens their security defences. This scenario is most often accepted in cases where a university has one or more teaching hospitals and researchers need to share data between them.

Authorization

User authorization relies on the data-sharing environment and depends on the authentication mechanism. The most common approach for determining which users are permitted to perform which tasks is based on a role-based access control mechanism. Users are given usernames and assigned to roles, and each role is assigned some tasks. For instance, a user can hold the role of *physician* and be able to query the data warehouse, to run simulations, and to view imaging data (but, perhaps, not to modify the simulation code).

Auditing

Auditing sensitive tasks in a data-sharing environment must be monitored centrally. This requirement is similar to the way in which mobile phone users are billed by their service providers for using phones outside their home country.

10.5.2 Example of a Centralized Data Warehouse

An example that adopts this model with the type of hosting discussed in option 2 above occurs within the ContraCancrum (Clinically Oriented Translational Cancer Multilevel Modelling) project that was funded by the EU under the FP7 programme. This project

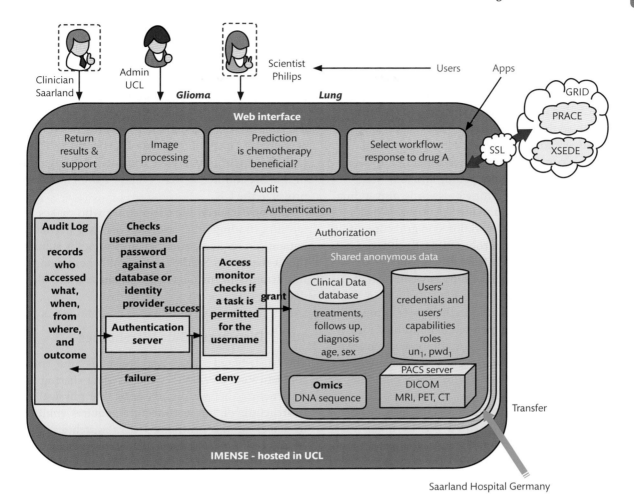

Figure 10.5 The ContraCancrum data-sharing environment (IMENSE). Any request to perform a task involving patient data within the IMENSE system, e.g. run a simulation on a grid resource, has to pass through various security controls – auditing, successful authentication, and authorization – otherwise the request fails.

uses a heterogeneous set of patient data, in particular data from patients suffering from lung cancer and glioma. This data was provided by the University of Saarland Hospital in Germany and hosted on servers in University College London (UCL) in the UK. The data-sharing platform, which is known as IMENSE [35] (Figure 10.5) allows clinicians and other end users to bring together and query patient data, edit the data, upload and download imaging data, and to use patient data as input to workflows and simulations on a grid infrastructure (see Chapter 9 for more details of this set-up).

In this example, medical data from the participating patients are stored in a centralized environment as shown in Figure 10.5. The stored data types include:

- image scans (i.e. MRI, PET, CT) stored in DICOM format;
- patient records (therapy details, follow-up diagnosis, treatments);
- histopathology data;
- DNA profiles stored in different formats (database, file servers, raw files, XML).

The Audited Credential Delegation [31] access control system was used to secure the IMENSE environment. This is designed as a wrapper around the data warehouse to control access to data.

- *Authentication*: users from the consortium partners were required to register in order to get an account on the IMENSE environment. Account creation only occurred after face-to-face interviews with all end users. Each user was provided with a username/password pair that enabled them access to the environment.

- *Authorization*: this was based on role-based access control. Two roles were identified: the Administrator role for adding and removing users' accounts, resetting passwords, and adding new functionality, and the Scientist role that allows users to run workflows on grid computing infrastructures in Europe (PRACE) and the USA (XSEDE) seamlessly from the web interface using the application hosting environment that is described in Chapter 9.

- *Auditing*: sensitive tasks in the data-sharing environment were required to be audited for accountability purposes. Audit logs, which are a common way for auditing all events on a system, were used for this.

Let us illustrate this with an example. Consider a user named John Smith who wishes to run an image segmentation of a patient using a high-performance computing (HPC) resource. The request to perform this task is first intercepted by the authentication wrapper which checks the credentials of the user against an authentication service (in this case ACD). The outcome of the authentication process is recorded in the audit log. After John has been authenticated, his role as Scientist is determined from the authorization wrapper, and the system therefore determines that he is permitted to perform the image segmentation task. The result of the access control check is also recorded in the audit log. Once access has been granted the task is performed on the HPC resource. Once the segmentation finishes, John is notified so that he can download the result. This ensures that only authorized personnel can run tasks in the patient data-sharing environment and that each user can only access the result of their tasks. More details of this process are available in [35].

Legal Implications

The data warehouse for ContraCancrum is held at UCL in the UK and data is transferred from the University of Saarland in Germany. In order to transfer the data to the UK, researchers and administrators in Saarland had to go through all the steps described in Section 10.4, including data scoping, ethical approval, data collection, and eventually data transfer using DVDs. Patient data were pseudonymized and the table linking data to patient identifiers was kept at Saarland Hospital. The data transferred to UCL's IMENSE environment were considered anonymous in the UK and hence were exempt from ethical approval in the UK.

Any physician in the consortium who makes interesting findings that are relevant to a particular patient can request patient identification by submitting a request to Saarland Hospital, where data administrators in turn assess the request and make a decision.

10.5.3 Distributed Data Warehouses

The distributed data warehouse is the most sophisticated of all the types of data-sharing infrastructure that we discuss here. The physical infrastructure of this architecture requires multiple instances of the centralized data warehouse that is described above, hosted in various locations (Figure 10.6). There is additionally a need for a central secure searchable hub that allows scientists and clinicians to seamlessly query and search patient data irrespective of where the data are hosted.

This data-sharing environment requires a separate administration authority and a dedicated database server for storing user capabilities for each instance of the warehouse, allowing access to the environment, as shown Figure 10.7. The administration is also responsible for maintaining a persistent identifier for each component (see Section 10.5.2) that allows patient datasets from different data warehouses to be reliably identified.

The identifiers that we are discussing here can be understood by an analogy with the widely used International Standard Book Number (ISBN) system for identifying books. Each book is assigned an ISBN when it is published, and this can be used as a permanent and citable reference. Different versions of a text (for example, hardback and paperback versions of the same book) are assigned different ISBNs. Similarly, here, data from the same patient, or even copies of the

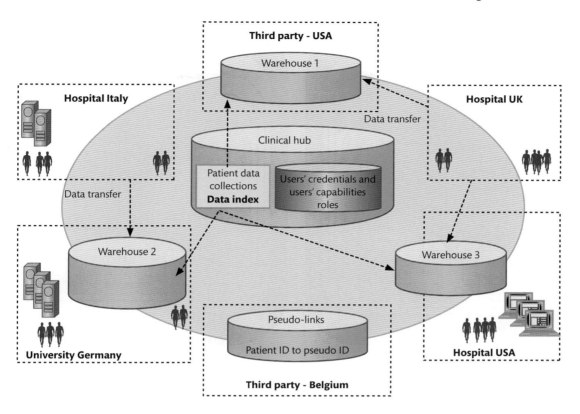

Figure 10.6 A distributed data warehouse architecture where several data warehouses are hosted in different countries and managed by different institutions (commercial, educational). The dispersed data warehouses are joined by a hub that controls access to each data warehouse and maintains an up-to-date index to all data including the DOI numbers of published papers.

same data stored in different formats or on different platforms, may be assigned different identifiers.

This type of architecture will often consist of several data warehouses. Each warehouse may be hosted via any of the options discussed in Section 10.5.1; that is, within a hospital or a partner site, at the site of a TTP, or in the cloud. These warehouses will often be in different countries; it is possible to have one data warehouse stored on a cloud in Germany, another hosted by a TTP organization in the UK, and a third stored on a local university system in France. A unified interface to the entire data warehouses can be provided to the end user, who does not necessarily need to know where each dataset is stored.

- *Authentication*: in this type of architecture, users from the research consortium partners will use a dedicated credential to access the data warehouse. Federated identity solutions such as Shibboleth

can allow users to access the shared environment; this system is particularly widely used in academia. Each instance of the data warehouse must trust security tokens that are issued by the central hub in order to identify legitimate requests to query or update individual data warehouses.

- *Authorization*: decisions on whether to grant permission for a user to perform a task will be decided separately for each resource or data warehouse. After successful authentication, the user will be issued a token such as a SAML token that confirms that the user has been authenticated. This authentication may involve a username/password pair, a digital certificate, or any other valid mechanism. The token will include the entity that confirmed the user's identity and the role of the user in the consortium, which is allocated when the user first registered. A token for John Smith in the previous example might include

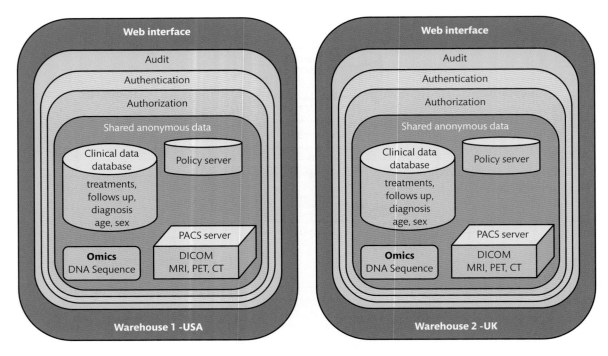

Figure 10.7 Two instances of a data warehouse with security controls. Instances can be hosted in geographically distributed locations, outsourced to a commercial company specializing in data storage or a cloud provider.

the data 'John Smith', UCL, clinician, the serial number of the token, its expiry date (lifetime), and the token issuer signature. Each instance of a data warehouse, such as those shown in Figure 10.7, will have its policy server configured at the time when it is designed by the administrator of the consortium. This policy will include the range of acceptable roles and their corresponding tasks, and which token issuer to trust. Therefore, when a warehouse receives a request by a specific user to access data and perform a specific task, the token and the request are evaluated by the policy server according to the consortium rules. If the request is issued by an authenticated user holding a relevant role, and if the task requested is one that is permissible for holders of that role, the request will be granted; otherwise it will be denied.

- *Auditing*: in this architecture, each data warehouse keeps its own audit log. If there is a dispute, the administrator can request the audit logs from each individual data warehouse in order to investigate the incident. This is similar to the 'roaming' system when using a mobile phone abroad in which the bill is sent to the foreign operator that

is linked to the user's home operator in order to pay it. This can become more of a management problem if data warehouses are hosted by third parties or within hospitals since the release of information will be subject to separate rules for information governance in each organization that hosts an instance of the data warehouse.

10.5.4 Example of a Distributed Data Warehouse

The European P-Medicine project is based on a distributed data warehouse concept (Figure 10.8). It is in some ways a larger-scale version of the ContraCancrum project involving a different range of cancer types including the Wilms' tumour (a type of kidney cancer that occurs most commonly in children), breast cancer, and acute lymphoblastic leukaemia. The aim of this project is to develop an open-source platform that allows consortium members to access patient data stored in various data warehouses distributed across the EU in a seamless way. The data are accessed from a single entry point via a web portal without the need to move the data

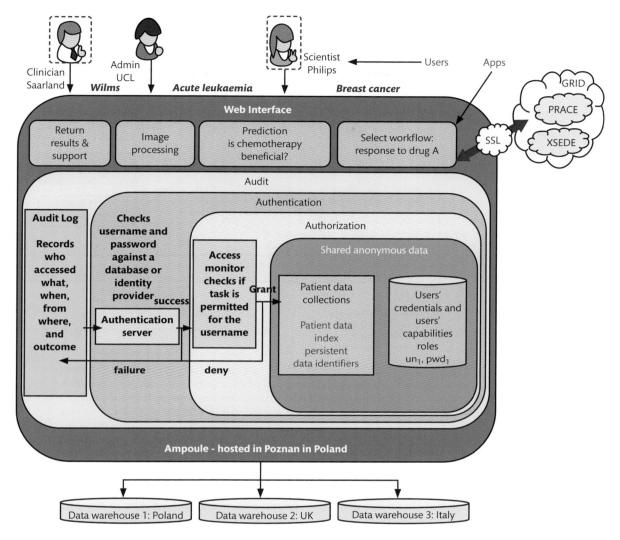

Figure 10.8 The architecture of the P-Medicine data-sharing environment, which consists of several data warehouses, hosted in different countries, and a unified web interface that enables clinicians and other end users to search, and to execute applications on the various data warehouses without knowing where each dataset is coming from.

from one warehouse to another. Each warehouse may use a different interface with different features, enabling users to access tools for data mining, clinical trials, imaging, workflow analysis, and different types of simulations and to access and search the distributed data warehouses using a standard ontology framework. (Ontologies are described in more detail in Chapter 2.)

Legal Implications

As the P-Medicine research project aims not only to build up an infrastructure for personalized medicine but also to enable researchers to contact patients directly if new findings have implications for their treatment, complete data anonymization cannot be an option for this project. To provide a solution that ensures the highest possible level of data protection while retaining the possibility of contacting patients in this way, the framework is based on the use of de facto anonymous data.

De facto anonymization is achieved by pseudonymizing the patient data as they are transferred to the data warehouse, and by ensuring that the researchers who have access to the data warehouse and

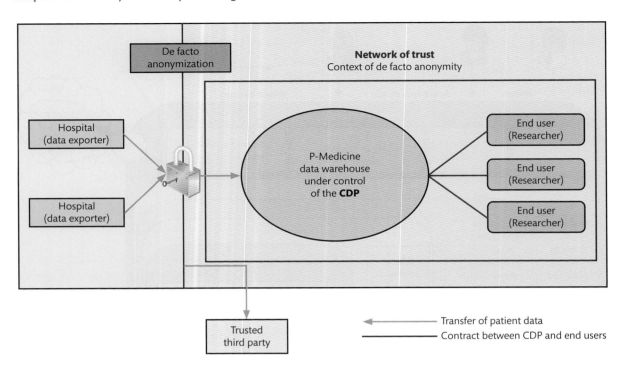

Figure 10.9 The P-Medicine project as an example of a network of trust. Every participant in this network is contractually obliged to respect patients' privacy rights and to comply with the necessary data-protection and data-security rules. In particular it is assured that the partners and users in the network will not undertake any measures to reveal patient identities by matching patient data transferred to the P-Medicine data warehouse or received from it with other personal datasets.

controller of the data warehouse itself have neither access to the links between the data and the patients nor the possibility of procuring these links by reasonable means. Furthermore, no other person who is allowed to access or process the data should be able to re-identify the patient. All this has to be safeguarded within the data-protection and data-security framework set up for P-Medicine. Therefore, it is essential that the data transferred to the P-Medicine infrastructure can only be accessed by researchers who are contractually bound by the data-protection rules that have been set up for the P-Medicine project (and thus who are in a network of trust).

In order to build up this type of data-protection network it is necessary for a legal entity to be established that is able to conclude binding contracts with the project participants and that is empowered to inflict penalties for infringement of this code. This entity necessarily has to be of a legal nature. In addition this legal body needs to be empowered by the consortium. As it acts on behalf of the consortium and as central data controller for the framework, it

needs to be independent in its decision-making. The legal entity that has this function for the P-Medicine project is the Center for Data Protection (CDP; **www. privacypeople.org**), which acts as the central data-protection authority for the project.

To summarize, building up a legally binding net work of trust for data sharing within P-Medicine— such as that summarized in Figure 10.9—mainly requires three components: (1) a legal entity that is able to conclude contracts on behalf of the project; (2) a TTP that stores the pseudonymization key; and (3) contracts that must be concluded in order to ensure that all users abide by confidentiality, data security, and compliance with data-protection legislation. In this example, three contracts are needed: the first concerns the transfer of patient data to the P-Medicine infrastructure (a Data Transfer Agreement); the second concerns data processing within the P-Medicine framework (a contract on data protection and data security within P-Medicine); and the third is an agreement regarding the duties and role of the TTP that holds the pseudonymization key.

Since patient data in the environment controlled by the CDP can be regarded as anonymous data in the sense of the Data Protection Directive, the data-processing restrictions of the Directive do not apply provided that the network of trust's rules are obeyed. The main advantage of this from a research perspective is that the data can be used freely within the scope of the research purposes of the project by (and only by) all registered users of the project infrastructure. This ensures that the P-Medicine infrastructure as a whole has to be highly flexible in terms of the physical location of its databases. It should therefore be possible to shift data from one database to another under the control of the CDP without any restrictions, provided that both databases are located in EU member states.

Fallback Scenarios

A legal basis has been provided for the unlikely event that this network of trust fails and data legally defined as 'personal' are processed within P-Medicine. All patients are asked to consent to the processing of their data within the project, and if this consent is given it will allow analysis of personal data. Seeking patient consent not only allows for such a fallback scenario but also fits best ethical practice. Furthermore, if the informed consent given by a particular patient should be invalid for an unforeseen reason the national exemptions for data processing in medical research discussed in Section 10.1 can be used as a second fallback position.

Patient Identity Management System

As different hospitals use different tools for data pseudonymization, different pseudonyms are likely to be allocated to the same patient by different hospitals. This is in general unlikely to cause serious problems, as it is rarely necessary to combine pseudonymized information from different hospitals about the same patient in a common database. However, the attribution of different pseudonyms to the same patient may affect the quality of the data within collaborations, as would the possibility that the same pseudonym is allocated to different patients. This may be a particular problem in large-scale collaborations such as P-Medicine in which it is necessary for geographically dispersed data to be shared. To provide a valid dataset on, for example, the long-term evaluation of data concerning the effects of a specific treatment, a secure and lawful system for managing identities and pseudonyms will always be needed.

This task is fulfilled by using a patient identity management system or PIMS (Figure 10.10). The main function of this type of system is to guarantee that a single pseudonym is allocated to each patient when his or her data are transferred to the framework. This is done by checking whether that specific patient has already been registered in the common PIMS database. A new pseudonym will only be created if no such patient has been registered so far; if the patient is already registered the existing pseudonym will be copied across. PIMS therefore avoids the creation of different pseudonyms for the same patient (or synonyms) as well as the creation of the same pseudonyms for different patients (or homonyms).

10.5.5 Cloud-Based Data Warehouses

In a cloud-based environment, the service and data maintenance are provided by a third-party vendor, which is often a software giant such as Amazon. This potentially leaves the client ignorant of where each process is running or even where the data reside. There are many security issues in cloud computing that have yet to be resolved; these are concerned with data storage, compliance of the cloud system with relevant legislation such as the Health Insurance Portability and Accountability Act (HIPPA) in the USA, and information assurance [36,37].

An instance of a data warehouse as described in Section 10.5.3 can be hosted on the cloud. However, in this scenario the location of the instance can become the source of problems. Knowing data-storage locations is very important as this is necessary to identify the laws and regulations governing the data [38]. Only recently have Amazon and Microsoft started offering data-storage options guaranteeing that the data will be held in data centres located in Europe to address legal compliance issues (Figure 10.11). Users of cloud services therefore often have to trust their chosen providers as to where and how the data are protected and whether the security controls are adequate and in place. Most third-party vendors cannot guarantee in their service level agreements that the data will always remain stored in a specific location for business continuity reasons.

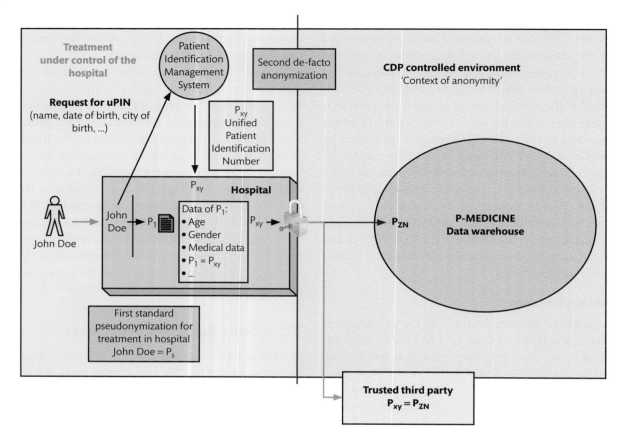

Figure 10.10 The structure of a patient identity management system (PIMS). Each patient participating in a clinical trial will be assigned a unified patient identification number (P_{xy}) in addition to the pseudonym created within the hospital where the patient is treated. Only the PIMS pseudonym will then be used to transfer patient data to the data warehouse so that all data belonging to a particular patient will be stored under the same pseudonym.

10.6 Conclusions

There are many legitimate reasons why research in pharmaceutical and health organizations and academic researchers will need to store and use data from individual patients. These data—which can take a variety of forms, including clinical, omics, and imaging data—are usually protected by law. Researchers who need to maintain and access databases of patient-specific data need to do so in a way that is compliant with all relevant legal and regulatory constraints, and that is ethical. It is possible and can even be straightforward to process personal data in a legal and ethical way provided that these constraints are considered when the project is designed.

The most important aspects of a system for storing and analysing data from individual patients are given below.

- *Patient consent*: data should be obtained by lawful and fair means and with the informed consent of the data subject.
- *Usage*: patient data should be relevant to the purposes of the study for which they are to be used and should be accurate (this is the concept of data integrity).
- *Security safeguards*: patient data should be protected by reasonable security controls, even when that data have been anonymized or pseudonymized.
- *Accountability*: the organization hosting the data should be held accountable for complying with all legal measures that give effect to the principles stated above.

The legal and security framework that is necessary for computational data-sharing environments should be

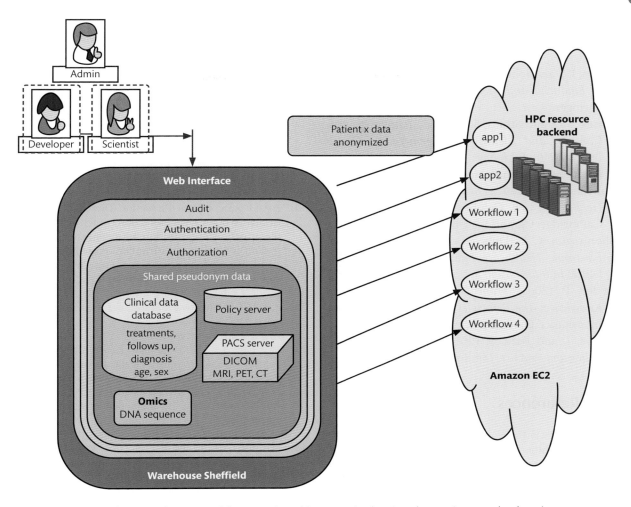

Figure 10.11 An architecture showing workflows running with anonymized patient data on Amazon cloud services.

seen as an enabler rather than an obstacle. The objective of such a framework is to protect patients' privacy and organizations' interests while developing new services and conducting research to improve health. Several appropriate models for data-sharing architectures have evolved in recent years, and researchers will choose the most appropriate one for each project based on the nature of that project. The slowest steps in setting up such an infrastructure are likely to be in obtaining ethical approval for the research and in obtaining patients' informed consent for the use of their data in that research. In many countries the ethical committees that have been set up to discuss issues of this type and give or withhold agreement only meet on average once every 3 months. Obtaining patient consent remains a challenge. Hospitals and family doctors (secondary and primary care, respectively) are useful sources of patient data although it is not always easy to unlock and interpret the data that are stored there. Another interesting source of patient data that is particularly useful in keeping patients involved and engaged with their own care is found in the medical social networks that are becoming popular. These include patientslikeme.com and 23andme.com [8,9] (both based in North America) and the POW Health network in the UK [39] in which patients share their medical data with other patients. These platforms can be combined with data stored in hospitals and used in research and analytics and, interestingly, there is no further need for patient consent as the patients give consent when signing up to the network.

Recommended Reading

Camp, L.J. and Johnson, M.E. (2012) *The Economics of Financial and Medical Identity Theft*. Berlin: Springer.

Hosek, S.D. and Straus, S.G. (2013) *Patient Privacy, Consent, and Identity Management in Health Information Exchange: Issues for the Military Health System*. Santa Monica: Rand Corporation.

Lowrance, W.W. (2012) *Privacy, Confidentiality, and Health Research*. Cambridge Bioethics and Law. Cambridge: Cambridge University Press.

Shoniregun, C.A., Dube, K., and Mtenzi, F. (2010) *Electronic Healthcare Information Security (Advances in Information Security)*. Berlin: Springer.

Taylor, P. (2006) *From Patient Data to Medical Knowledge: The Principles and Practice of Health Informatics*. Oxford: Blackwell Publishing.

Tingle, J. and Bark, P. (2011) *Patient Safety, Law Policy and Practice*. London: Routledge.

References

1. BBC (2011) Everyone 'to be research patient'. 5 December, http://www.bbc.co.uk/news/uk-16026827.
2. P-medicine. For further information please see http://www.p-medicine.eu.
3. More information available at http://www.vph-share.org.
4. The European Data Infrastructure, http://www.eudat.eu
5. Health Research Institute, PWC (2011) Old data learns new tricks: managing patient security and privacy on a new data-sharing playground. http://www.pwc.com/us/en/healthindustries/publications/old-data-learns-new-tricks.jhtml.
6. I v Finland [2008] ECHR 20511/03 (17 July 2008); EHealthInsider. 2008 European Court fines Finland for data breach. See http://www.e-health-insider.com/news.
7. The Information Commissioner's Office, http://www.icogov.uk/upload/documents/pressreleases/2010/rwh_nhs_trust_undertaking_240810.pdf
8. www.patientslikeme.com
9. Personal genetic information, www.23andme.com
10. Published in *Official Journal* L 281, 23/11 /1995, p. 31.
11. See Art. 32 Directive 95/46/EC.
12. In the respective landmark ruling of the European Court of Justice, C-101/01, Lindquist, 06/11/2003, European Court of Justice, C-101/01, Lindquist, 06/11/2003, ECR 2003, I-1297, the ECJ clarified that Directive 95/46/EC envisages complete harmonization of the data-protection regime within its scope. Member states are, however, allowed to set higher standards under specific circumstances. As a result the data-protection law is not completely harmonized within the EU.
13. Directive 2001/20/EC on the approximation of the laws, regulations, and administrative provisions of the member states relating to the implementation of good clinical practice in the conduct of clinical trials on medicinal products for human use, published in *Official Journal* L 121, 01/05/2001, p. 34.
14. Art. 3 para. 1 Data Protection Directive. See also Recital 26 of the Data Protection Directive: 'Whereas the principles of protection must apply to any information concerning an identified or identifiable person; …whereas the principles of protection shall not apply to data rendered anonymous in such a way that the data subject is no longer identifiable;…'.
15. Art. 2 lit. a) Directive 95/45/EC defines the term 'personal data'.
16. Art. 2 lit. a) Directive 95/46/EC.
17. Sweeney, L. (2000) *Simple Demographics Often Identify People Uniquely*. Data Privacy Working Paper 3. Pittsburgh, PA: Carnegie Mellon University. A paper by P. Golle gives the figure as 63%; see Golle, P. (2009) Revisiting the uniqueness of simple demographics in the US population. http://www.truststc.org/wise/articles2009/articleM3.pdf.
18. See also Art. 29 Data Protection Working Party, Opinion 4 /2007 on the concept of personal data, p. 15, http://ec.europa.eu/justice/policies/privacy/docs/wpdocs/2007/wp136_en.pdf. The Working Party has been established by Art. 29 of Directive 95/46/EC. It is the independent EU Advisory Body on Data Protection and Privacy. For further information see http://ec.europa.eu/justice/policies/privacy/workinggroup/index_en.htm.
19. Terstegge, Directive 95/46/EC, Art. 17 no.1, in Büllesbach/Poullet/Prins, Concise European IT Law Alphenaan den Rijn, 2006.
20. See Art. 8 of the Data Protection Directive.
21. See Art. 6 of the Data Protection Directive.
22. See Arts 7 and 8 of the Data Protection Directive.
23. See Art. 17 of the Data Protection Directive.
24. Kuner, European Data Protection Law, rec. 2.90.
25. Art. 6 para. 1 lit. e) Data Protection Directive.
26. Art. 8 paras 2-5 Data Protection Directive.
27. See Art. 8 para. 4 of the Data Protection Directive that has to be read in combination with Recital 34 of the Directive.
28. See outcomes of the FP7 project CONTRACT, http://www.contract-fp7.eu.
29. Appelbaum, P.S. (2007) Assessment of patients' competence to consent to treatment. *New England Journal of Medicine* 357(18): 1836–1837.

30. Golmann, D. (2011) *Computer Security*, 3rd edn. Chichester: Wiley.

31. Haidar, A., Zasada, S. et al. (2011) Audited credential delegation: a usable security solution for the virtual physiological human toolkit. *Interface Focus* 1: 462–473.

32. DICOM viewer, http://www.osirix-viewer.com/.

33. Open Source Clinical Image and Object Management, http://www.dcm4che.org.

34. World Economic Forum, Global Risks 2012 Seventh Edition, p.11, http://www.weforum.org/reports/global-risks-2012-seventh-edition.

35. Zasada, S.J., Wang, T. et al. (2011) IMENSE: an e-infrastructure environment for patient specific multiscale modelling and treatment. *Journal of Computational Science* DOI:10.1016/j.jocs.2011.07.001

36. Jensen, M., Schwenk, J. et al. (2009) On technical security issues in cloud computing. *IEEE International Conference on Cloud Computing*, September 2009, pp. 109–116. Washington, DC: IEEE Computer Society.

37. Tsai, C.L., Lin, U.C. et al. (2010) Information security issue of enterprises adopting the application of cloud computing. *Sixth International Conference on Networked Computing and Advanced Information Management*, August 2010, pp. 645–649. Washington, DC: IEEE Computer Society.

38. Arning, M., Forgo, N., and Krügel, T.A. (2009) Data protection in grid-based multicentric clinical trials: killjoy or confidence-building measure? *Philosophical Transactions of the Royal Society A* 367: 2729–2739.

39. http://www.powhealth.com

Chapter written by
Ali Nasrat Haidar, Centre for Computational Science, University College London, 20 Gordon Street, London, WC1H 0AJ, UK;
Nikolaus Forgó, Institut für Rechtsinformatik/Institute for Legal Informatics, Leibniz Universität, Hannover, Germany;
Hartwig Gerhartinger, Institut für Rechtsinformatik/Institute for Legal Informatics, Leibniz Universität, Hannover, Germany.

11 Toward Clinical Deployment: Verification and Validation of Models

Learning Objectives

After reading this chapter, you should:

- understand how computational biomedicine applications can be assessed within the healthcare system;
- be able to give an overview of how models are verified and validated;
- be able to explain the clinical accuracy of the models;
- understand the methodology used for the efficacy, risk–benefit, and cost–benefit analyses of new technologies that enter a healthcare system;
- appreciate how the impact of computational biomedicine on clinical care is of utmost importance for the sustainability of healthcare systems and to personalized medicine.

In this final chapter we look at some of the implications of involving computational models—such as those that are described earlier in this book—in clinical practice. As you have seen, almost all these models are still quite a long way from routine clinical use, although a few are approaching clinical trials. The focus of this chapter is on how trials of models, rather than of (for example) new drugs, are conducted, how the accuracy and efficacy of models can be assessed, and the implications of using such models in the clinic. Most of the examples we use here are of models and tools developed within the European Commission's Virtual Physiological Human (VPH) programme, and we have therefore chosen to emphasize the assessment procedures of the European Union (EU) and of the European Medicines Agency (EMA). We do, however, discuss procedures of the Food and Drug Administration (FDA) in the USA more briefly. Other developed countries and regions will use similar procedures.

11.1 Introduction: Health Technology Assessment

Health technology assessment (HTA) is a multi-disciplinary activity that involves systematically examining the medical and clinical implications of technologies in healthcare. This term, clinical implications, includes the efficacy, effectiveness, and safety of the technology concerned; economic, social, and ethical considerations of the technology; and the consequences of applying it (see Taylor and Taylor 2009, listed in Recommended Reading). The term health technology includes drugs, medical devices, and clinical or surgical procedures. Therefore, an HTA focuses on the value that a health technology adds to accepted clinical practice [1].

You will have come across the term Virtual Physiological Human in several earlier chapters. To recap, this has been defined as 'a methodological and technological framework that will enable collaborative investigations of the human body as a single complex system' [2]. Since 2007 the EU has funded a range of projects under this umbrella and has set up a 'research infostructure' (information infrastructure) that biomedical researchers use to describe biological processes from a systems point of view. This was designed to provide a scalable, robust, and reusable framework for researchers to develop computational models and tools for supporting both basic biomedical research and, eventually, clinical practice. Similar projects have been funded in other countries and jurisdictions, mostly by governments. Before any of the tools developed within these initiatives can be used in the clinic, however, they must be licensed and certified. The implications of this procedure form a major focus for this chapter.

The process of HTA forms an important part of the licensing and certification of any such technology. Put simply, the HTA process involves assessing the most appropriate health indications for a particular technology, from the point of view of policy makers [3]. In this book we have discussed some of the many computational tools that are under development for clinical (and biomedical) applications. However, none of these are yet in routine clinical use, and very few of them have been validated or certified.

For a biomedical technology, the *certification process* involves complying with all the relevant national or international regulations and standards. It is essential that any technology, including tools that have been developed through computational modelling, is certified and validated before it enters clinical use. Table 11.1 lists some of the standards and directives

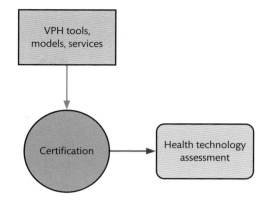

Figure 11.1 Schematic showing the relationship between the certification of a medical device or tool and HTA.

that clinical tools and devices need to meet before they can be used in clinical practice in the EU. These directives are part of the legal framework of the EU and are frequently revised; the last technical revision to be brought out was Directive 2007/47 EC [4]. After a tool or device has been certified, HTA will be applied to look at the overall benefits and possible drawbacks of the tool, in respect to its effectiveness, safety, and costs (among others). Figure 11.1 shows this relationship between certification and HTA schematically.

The definition of a medical device, as given in EU Directive 2007/47/EU, is shown in Box 11.1. You can see that this definition includes software; therefore, the computational tools and models that are designed to help in diagnosis or in clinical decision making and that have been discussed in detail in this book fit easily into this category.

The US FDA takes a rather different approach to the problem of defining a medical device, but they also recognize software as a category of such devices. The FDA branch responsible for this sector is the Center for Devices and Radiological Health (CDRH), which provides training material[1] on their regulations; more details of the US certification process for medical devices can be found there.

The most important bottleneck in holding up the future usage of computational tools in daily clinical

Standards	Description
ISO13485:2003 [5,6,7]	Foundation for all regulatory requirements for a quality-management system
Requirements	
Directive 90/385/EEC	Regarding active implantable medical devices
Directive 93/42/EEC	Regarding medical devices
Directive 98/79/EC	Regarding *in vitro* diagnostic medical devices

Table 11.1 Standards and regulations for certification of tools in Europe

---

[1] http://www.fda.gov/Training/CDRHLearn/ucm209135.htm

> **Box 11.1** EU Definition of a Medical Device
>
> Within the EU, Directive 2007/47/EC defines a medical device as 'any instrument, apparatus, appliance, software, material or other article, whether used alone or in combination, including the software intended by its manufacturer to be used specifically for diagnostic and/or therapeutic purposes and necessary for its proper application, intended by the manufacturer to be used for human beings. Devices are to be used for the purpose of:
>
> - diagnosis, prevention, monitoring, treatment or alleviation of disease;
>
> - diagnosis, monitoring, treatment, alleviation of or compensation for an injury or handicap;
> - investigation, replacement or modification of the anatomy or of a physiological process;
> - control of conception.'
>
> The equivalent definition from the US FDA can be found at http://tinyurl.com/FDAdefinition.

practice will be the certification process, whether in the EU or elsewhere. In all countries, computational biomedicine is largely government funded, with most of the research taking place in academic and government institutions rather than industry, and there is no dedicated funding for certifying the resulting tools. New methods of quality assessment for such tools need to be developed and applied in the EU, as in other jurisdictions. A possible solution to this dilemma is explained in Section 11.6.

11.2 Code and Model Verification

Most tools and technologies in computational biomedicine have at their heart a mathematical model that needs to be solved to derive information about an individual patient, such as a diagnosis or a recommended treatment. These models can be analysed and their precision or accuracy estimated using techniques that have been applied much more generally to engineering models. We refer here to what is known as the *inherent accuracy* of a technology, as distinct to its *clinical accuracy* (which will be discussed later). The inherent accuracy of a computational biomedicine technology is the accuracy with which it can make clinical predictions when used in the best possible conditions. It is, therefore, recognized that these 'best possible conditions' are not necessarily often found in routine clinical practice.

The determination of the inherent accuracy of a predictive model involves three steps:

1. *verification* (of both the code and the model);
2. *sensitivity analysis*;
3. *validation*.

A full discussion of this process is beyond the scope of this book; more details than we give here can be found in [7].

The predictive accuracy of any model is affected by a number of factors. Tools are developed within computational biomedicine by applying recognized physical, chemical, and biological theories to physiological phenomena and then building idealized mathematical models. The parameters and specifications for a numerical (computational) model are derived from this initial mathematical model, and a computer program is written to solve the numerical model according to these specifications. Each of these steps can be a potential source of error.

It is important to separate the contributions of these different types of error because it is possible for one error to compensate for another. In these cases, the overall accuracy of the model may seem excellent in one condition in which it is tested, but be very bad in others.

In practice the three steps listed above are involved in determining the accuracy of a computational model in the following ways.

- The error in solving a mathematical model numerically is assessed through the process known as *verification*.
- The appropriateness of the model specifications can be tested with *sensitivity analysis*.

- The error committed in assuming that the process to be modelled can be described by that particular combination of theories is determined through *validation*.

The verification of a model has two aspects: code verification and model verification. Code verification aims to set out numerically the accuracy with which a given class of mathematical problems can be solved by using a certain computer program. It is usually carried out by comparing the results from running the program to the exact solutions for a set of known problems that are called benchmark problems. There are some more sophisticated approaches; these involve advanced mathematical methods such as the method of manufactured solutions [8]. The degree of error that can be accepted in a model that is to be verified varies according to the application concerned, but it is almost always well below 0.1%.

But knowing that a program gives accurate solutions to a particular mathematical model over a range of biologically (or clinically) appropriate parameters is not enough. Once the computer program has been verified, the mathematical model itself must be analysed. The numerical accuracy and predictive power of a model are defined not only by the validity of the computer code, but by the way that the model itself has been constructed.

We can illustrate this using a trivial example. Imagine that you need to compute how the concentration of a substance, C, varies over the spatial coordinate x. This can be expressed mathematically by the simple derivative of C over x, and if this derivative is unknown it can be approximated using a finite-differences scheme:

$$\frac{dC}{dx} \approx \frac{\Delta C}{\Delta x} = \frac{C_2 - C_1}{x_2 - x_1} \rightarrow \lim_{\Delta x \to 0}\left(\frac{\Delta C}{\Delta x}\right) = \frac{dC}{dx}$$

You should see intuitively from this equation that the smaller the interval between x_2 and x_1, the smaller the error incurred by approximating the derivative in this way will be. This error, called the *discretization error*, is one of many types of error that are related to the model itself.

The process of model verification is a complex one. It requires a specific description of the type and mathematical form of the model being verified. There

is no space here to discuss even a small proportion of the known mathematical models of biomedical processes in detail. However, all these verification methods can be roughly divided into two broad categories, defined as *pre-hoc* and *post-hoc* verification. Pre-hoc verification methods provide information about the models that can verify them to some extent without actually solving them. This is particularly relevant when the models are very large and complex, and when solving even a single run of a model requires a lot of computer time. At their best, pre-hoc methods can provide upper boundary estimates of the numerical accuracy of the model. Post-hoc methods, which are very different from each other but which all involve solving the model with at least one set of input data, are required for the full validation of models.

In many cases there will be a trade-off between the numerical accuracy of a model and some other determinants of the model. These might, for example, involve the amount of computational resources that the model needs, and/or the time it takes to run. As most researchers who need to run these models have neither infinite computer resources nor infinite time, many will deliberately accept suboptimal accuracy of a model as a compromise. In these cases, however, it is particularly important to quantify the accuracy of a model. Where trade-offs between accuracy and resources are not particularly stringent it is reasonable to accept that the numerical accuracy of a model (incorporating the accuracy of the code) will be less than about 1%.

11.3 Sensitivity Analysis

Sensitivity analysis is the study of how the variation in the predictions made by a model can be attributed to variations in the data that are used as the input to the same model. In the context of assessing the accuracy of models, this form of analysis is used to verify that the predictions made by a model are not overwhelmingly sensitive to any particular input variable. Sensitivity analysis is not generally part of the formal assessment of the accuracy of a model.

There are two aspects of sensitivity analysis, and these tend to get tangled together in a way that is difficult to separate. The first is the causal relationship between the inputs and outputs of a particular model, and the second is the effect of uncertainties in the inputs to the model on its predictions.

Sensitivity analysis can be used, then, to verify that the cause/effect relationship between the model's inputs and its outputs is consistent with the biological or physical theories that underlie that model. Imagine, for example, a biomechanical model of a blood vessel. In this model the stress on the wall is expected to increase with its thickness in a linear fashion. If the output obtained by the model does not reflect this expectation, it is clear that the model is not behaving according to the theory that was used to encode it and that something else is going on. One possible reason for this is simply that a mistake was made in encoding the model. Another is that the model does not precisely reflect the biological situation and that other theories need to be incorporated into it. A third, and more subtle, reason is that the input data and parameters used in the model have reached the limits of their validity. Every model is only a mathematical idealization of a real biological situation, so even under the best conditions it will be valid only within a limited range of input values. In this context, sensitivity analysis and verification can be used together to explore these limits of validity.

Another point that must be considered in determining the accuracy of models is the precision of the input data used. Many (if not most) models in computational biomedicine rely on input data that have been measured experimentally. No experimental measurement can be obtained to infinite precision, so these input data values will be accompanied by a smaller or greater uncertainty. In many cases, this uncertainty will be small, such that changing the input values within this range will only affect the predictions of the model a very little. This is not always true, however, and the results should be treated with caution in these latter cases.

In general, altering the input data values within the range given by the known precision of the measurements used to generate those values should not change the values predicted (that is, the output data) by more than the known inherent accuracy of the model. For linear models, the coefficient of variation produced by uncertainty in input data values should not exceed 10%. The sensitivity of a model to uncertainties in the input data will also be affected by errors arising from the code and from the model.

Note that sensitivity analysis can also be used while models are being designed, to decide which types of input data affect the results obtained to the greatest extent and, perhaps, to devise strategies to simplify the model so that it can produce results that are as precise and accurate as practicable more quickly and using fewer resources.

11.4 Model Validation

Once researchers have checked the numerical accuracy of their modelling programs and made sure that the predictions are not too sensitive to small variations in the values of the input data, they will generally try to determine the inherent accuracy of the model by comparing its results with measurements obtained from controlled experiments. To do this accurately it is important to replicate in the experiment the exact process that is being modelled. This is particularly challenging because there can be a conflict between the need to control the experimental conditions carefully and the need to validate the model under conditions that are as close as possible to 'reality'. In most cases there will be a conflict between the needs for a control and for realism, and so designing an experiment to validate the model becomes a delicate exercise in seeking the best possible compromise.

One result of this compromise is that it will very often not be possible to completely validate the model. In these cases it will only be possible to obtain experimental data that are limited to certain predicted values or to a more limited range of possible input values than is really the case. Another potential problem that arises particularly often with biomedical models is that it is often not possible to validate the model on the 'real' organism for which it was designed; that is, a human being with a particular clinical condition. Instead, preclinical validation of models designed to be applied in the clinic will often involve comparing their predictions with measurements made on an animal model of that disease. Any model validated in this way will have a predictive accuracy that is systematically lower than if the model is used in clinical practice. It is safe to say that any model that performs poorly in these controlled conditions is not yet suitable for clinical testing.

The first step in designing a preclinical validation experiment for a model that has been designed to be used in clinical practice is, therefore, the choice of that experimental model system. There are usually four options:

1. *in vitro* models, using synthetic analogues and cultured cell lines;
2. *ex vivo* models, or human organs and tissues taken from cadavers or living donors;
3. *in vivo* models using animals;
4. *in vivo* models using human patients or volunteers.

In general, as we move down this list we get closer to replicating the reality of the clinical condition to be modelled, but it becomes harder to control the experiment properly.

The first option of *in vitro* validation should be considered only in special cases, for example where the anatomical variability of the tissues to be used is not important. Results from *in vitro* validation are generally only able to highlight truly bad models.

Ex vivo models are excellent only for organs and tissues that do not change their properties after death (or removal from the body in the case of donated tissue); unfortunately this can only be assumed in a few cases, primarily with connective tissues.

In vivo animal models can be very powerful, and it is often possible to use animals for experiments that would be illegal or unethical if carried out using human subjects. However, animal experimentation is becoming more restricted in many legislatures, and the differences between experiments that are possible using human subjects and those that are possible using animal models are gradually reducing. It is possible that animal experimentation will one day be banned in some countries, even for medical research. Another obvious problem with animal models is the differences between physical, chemical, physiological, and biochemical processes in humans and those in the animal species used in models. Our knowledge of these differences and how they should be allowed for in testing models is still quite limited.

Validation of a model using data obtained from animals, therefore, may not be predictive for how the model will perform on a human system. However, full validation of models *in vivo* in humans is generally not possible as the experimenters will not be able to control the conditions completely. There are some exceptions to this general rule, for example where it is possible to access an organ or a tissue during surgery and control the experimental conditions *in situ*. One example might be imposing an electrical stimulus of controlled intensity and duration on muscle tissue exposed during an operation.

In some cases it will be possible to use only one experimental system for validating a particular computational model; in other cases the researchers will have a choice of several appropriate ones. Whatever type of model is chosen, however, it is rare that the computer model will predict a quantity that can be easily measured. In most cases some post-processing of experimental data will be necessary before this can be compared directly with the results of the computational model (or vice versa). Special care should be taken with this step, as too frequently it is this post-processing, rather than the predictive accuracy of the model, that forms the major source of error.

Many models will give a prediction for the time course of a particular quantity; that is, values for that quantity at different points in time. In these cases it is vitally important that both experimentally measured and modelled values are obtained at precise points in both three-dimensional space and time. In simple words, this means that if a quantity in one location of an organ is measured, the same locale must be located with high precision on the computer model of the organ and the equivalent quantity modelled at that point. This means, of course, that the detailed geometry of the organ must be modelled as accurately and precisely as possible. However, it also implies that the organ must be positioned correctly with respect to the boundary conditions imposed during the experiment in both space and time, as these must also be reproduced accurately in the computer model. Therefore, if a model predicts the strain at each point of a bone that has been loaded by a given mechanical force, large errors in the model may arise because the force is applied to one point of the bone in the computer model and to another in the validation experiment. These errors will apply even if both the bone geometry and the force applied have been quantified very accurately.

The last point to consider is how the results of a model validation should be reported. It is unfortunate that in many cases the predictive accuracy of the model is expressed using a linear correlation coefficient between the quantities that have been measured and those that have been predicted in the same conditions. This, however, is wrong. It is important to look at how close the predictions actually are to the measured quantities, not just at whether they vary in the same way (which is what correlation coefficients measure). A good way of reporting this for many validation experiments in which the same quantity is measured repeatedly under different conditions is to calculate the root mean square error (RMSE) between the measured and predicted quantities over all conditions sampled. This gives the absolute value of an average error that can be normalized for the largest set of values measured. In one recent study, RMSE values were used to evaluate the differences between calculated and measured values of the strain in femurs under different load conditions [9]. An overall RMSE value of less than 10% showed that the model used could provide a good estimate of strain, but considerably larger errors in predicting strain in the lateral neck of the femurs suggested that this part of the model might need modification.

The RMSE alone, however, is not enough to quantify the inherent accuracy of a predictive model. It is also important to look at the worst case scenario, which provides an upper boundary for how wrong the prediction can be. Normally this should be cited as a peak error—that is, the largest difference between predicted and measured values normalized for the largest measured value. In a few cases, where the prediction is a global value but the validation is based on many local experimental measurements, the error can be treated as a statistical variable and reported in terms of a distribution (i.e. the 95th percentile). An example of a case in which this type of error reporting is used is a model that predicts the geometry of a growing tumour. It is validated by determining the geometry of a real tumour at many points over its surface and comparing these points with the predicted geometry of the whole tumour.

11.5 Validation of Integrative Models

So far in this chapter we have discussed how we can determine the inherent accuracy of a single predicted model that has a given set of inputs and outputs.

However, most of the models that are likely to be of practical use in the clinic—such as those that are now under development through the VPH programme—are actually *integrative* models. This means that they are composite models made out of various components, each of which defines the behaviour of one part of the physiological process that is being modelled. There will also be separate relational models that define how each component of the system interacts with the others. A computational biomedicine model can therefore be said to be a 'model of models'. It is not immediately clear how the procedure that we have outlined above can be used to validate such a composite model. In fact, the problem of validating complex predictive models is currently considered an 'open problem' [10]. The 'best guess' approach to this type of validation is currently to validate every single component of the model, and then, if possible, to try to validate the integrative model as a whole. This is inevitably a complex and time-consuming procedure.

When it is not possible to validate an integrative model—because no controlled experiment can be designed to validate it against, for example—one possible alternative is to use the model in a population validation study. Essentially, this means that the model is used with personal data to predict the outcome for many different individuals. These may be real people (patients or volunteers) or may be simulated by randomly sampling the input parameters across the known distribution of each quantity over a given population. The model is then run within a probabilistic framework such as a Monte Carlo method (see Box 3.3 for a brief description of this). The result of this exercise is a prediction for a population that can be compared with observed values that have been captured in epidemiological studies.

The last step in the procedure of validating a computational model is known as confirmation. Once the model has been tested with known experimental data it is used to predict something similar where there are no experimental data available. If experimental work subsequently reveals that the model has made a good prediction, that model can be said to be fully validated, at least for one set of closely related problems. This step, however, is not always possible, and the results may not be conclusive.

At the moment, therefore, it is not always possible to completely validate a full integrative model. In these cases, a combination of the validation of each component of the model with confirmation using, for example, a population simulation is generally seen as an acceptable alternative.

11.6 Clinical Accuracy

11.6.1 General Considerations

You should appreciate that preclinical validation of models, although essential, can tell only part of the story. Whenever a predictive computational model is used in clinical practice there are many factors that can degrade its predictive accuracy below the level predicted during preclinical validation using controlled experiments. Examples of the features that contribute to the degradation of predictive models in clinical practice are outlined below.

- The signal-to-noise ratio of data obtained from patients in a real clinical scenario can be much lower than that of data obtained in idealized test conditions, and these data can even be incorrect.
- The predictions produced by models must be transformed into clinically important information that is capable of improving a doctor's ability to diagnose, predict prognosis, plan or execute a treatment, monitor, or rehabilitate. This transformation can produce further inaccuracies.
- The physiological process that is predicted by the model does not happen in isolation but is interfered with by many other physiological processes in the same patient.
- Very often, compromises must be made to make a predictive technology practically usable in a hospital setting (for example, in the quality of the input data or the nature of the predictions made).

A model that has been validated using preclinical models and that has been proved to be inherently accurate enough for a specific clinical application must therefore be verified under real-world clinical conditions before it can be used in routine clinical practice. This type of accuracy is termed *clinical accuracy*.

Before computational biomedicine tools and models can be used in routine practice they must go through a process of quality assurance. Ideally, a tool that has completed this process will be authorized to be distributed throughout the world. However, different countries, and groups of countries such as the EU, have different laws and apply different standards to authorizing medicines and devices. For example, the requirements of the directives and ISO standards described in Section 11.1 must be guaranteed before a computational tool can be distributed and used throughout the EU.

Computational models in biomedicine are relatively new, and as such they belong to a broad family of health technologies for which a fully formalized process for clinical assessment has not yet been defined. The situation is completely different, however, for pharmaceutical products (drugs). The process of assessing candidate drugs before they are registered for clinical use is very tightly defined. We describe this in Section 11.6.2, making adjustments wherever possible between technology products and drugs.

11.6.2 Clinical Trials

Before a drug can be marketed it must be registered, and before it is registered it must undergo preclinical testing followed by three phases of clinical trials. A fourth phase of clinical trials often takes place immediately following registration, with the drug tested in large numbers of patients in more or less routine clinical use. We give a short summary of the clinical trials process here; more detail can be found in [11].

Translating procedures that have been developed for testing medicines in the clinic to make them relevant for computational biomedicine would involve assuming that each tool, model, or service developed using that approach is equivalent to a pharmaceutical product, and then developing procedures that are equivalent to preclinical and clinical drug testing. In drug development the starting point is always a clinical problem that needs to be solved. Clinicians are involved in every stage of the established procedure for the clinical trials of drugs, and they should be involved as early as possible in the process of developing and testing models and tools.

The preclinical and clinical assessment of models can be described by analogy with the procedures

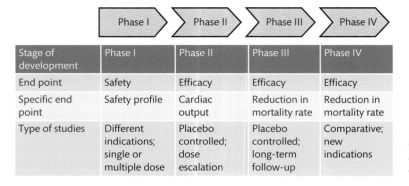

Stage of development	Phase I	Phase II	Phase III	Phase IV
End point	Safety	Efficacy	Efficacy	Efficacy
Specific end point	Safety profile	Cardiac output	Reduction in mortality rate	Reduction in mortality rate
Type of studies	Different indications; single or multiple dose	Placebo controlled; dose escalation	Placebo controlled; long-term follow-up	Comparative; new indications

Figure 11.2 Phases of clinical trials. The drug or therapy is registered, and enters clinical use, if it passes phase III clinical trials.

established for preclinical and clinical drug testing as described below. You should note, however, that this process is still experimental, and very few such models have entered—let alone completed—clinical testing as yet.

The process of assessing a drug or technology through the four stages of clinical trials is summarized in Figure 11.2. Note that this figure takes a cardiac therapy as an example; there are, of course, equivalent tests that are applicable to therapies for other indications.

Preclinical Assessment

We have already described the process of preclinical assessment of a model or tool that has been developed for use in the clinic (see Sections 11.2, 11.3, 11.4, 11.5). Before a tool enters the first stage of clinical assessment, the following points will be extracted from that assessment and assessed in turn by physicians. A tool will only begin clinical testing if the physicians who would be involved in testing it (and very likely also using it) have been assured that:

- the systems biology behind the model is as accurate as possible;
- the tool is delivering scientifically correct results;
- the mathematics behind the model have been calculated correctly;
- the software program is bug-free;
- the scientific thinking behind the model is up to date and based on state-of-the-art knowledge of the physiological process concerned;
- the model should be easily adaptable as the results of new research that might modify it become known.

Clinical Testing

Once a model or tool enters clinical testing it will pass through phases that are equivalent to the phases of drug trials. Only models that succeed in one phase of trials will be allowed to continue to the next.

In this section, the word technology is used to mean the computational biomedicine tool or model that is undergoing clinical tests.

Phase I: the first stage of clinical testing is simply a safety and usability assessment. Here, the technology is tested on a very small group of patients (fewer than 100) to answer the following questions.

- Can the technology be used on humans with the necessary levels of safety?
- Is the initial clinical question valid and appropriate, and has it been correctly addressed?
- Can the tool, model, or service be used to solve the clinical problem?
- What organizational impact would this technology have if it were deployed on a large scale; for example, will clinicians or other medical professionals find it easy to use, and would they require further training to make use of it?

Phase II: this is the first phase that is designed to test the predictive accuracy of a technology under full clinical conditions. Characteristics of phase II trials include the following.

- They are primarily prospective clinical trials that are typically conducted on small cohorts of patients. The patients are observed until the condition predicted by the technology under evaluation can be quantified.

- When possible, small phase II prospective studies may be preceded by, but cannot be replaced by, larger retrospective assessments. In these, the predictive accuracy of the technology is assessed over large databases of patients for which the clinical outcome of the condition predicted is already known. These studies are very effective in showing whether the technology exhibits—or fails to exhibit—sufficient clinical predictive accuracy.

- Phase II studies are also useful in assessing whether the technology is compatible with the organizational constraints that are imposed by the clinical pathway in which it will be applied. These are typically in terms of the time taken for the technology to produce its results, which (as we have discussed earlier in this book) may not be compatible with the timescale needed for clinical decision making.

A particularly delicate point in phase II studies is to decide whether or not the predictive accuracy of the technology under full clinical conditions is considered acceptable. This decision will ultimately affect the health of patients, and therefore it should always be made by the physicians responsible for the clinical study (Figure 11.3).

This decision, however, should be based on solid evidence; it cannot be arbitrary. In general, it is only

possible to compare the results of technology trials with the current standard of care. If the technology under assessment involves costs and risks for the patient that are comparable to the current standard of care, then the decision whether to continue developing that technology for that indication will only involve verifying whether the technology provides an improvement over this standard of care. If the introduction of a computational biomedicine technology or model will involve an increase in cost to the health provider or risk to the patient, the decision is more complex and a full cost–benefit and risk–benefit analysis will be required (see Section 11.7).

Phase III: as in drug development, phase III trials are randomized controlled clinical trials that typically involve cohorts of several hundred patients. The primary purpose of a phase III trial of a computational biomedicine technology is to confirm its superiority to the current standard of care under strict statistical and procedural conditions.

Again by analogy with drug trials, a phase III trial of a computational biomedicine technology would involve randomizing the standard treatment against the treatment that is predicted by the technology under test. At the end of the trial both treatment arms would be compared to see whether patients treated according to the model's predictions are doing better than those treated using the standard conservative approach. In this phase the model could also be used in the patients treated according to standard practices to see what treatment the model would predict in these patients. This allows for further validation of the model.

An example of this approach is given in Figure 11.4. This shows the experimental arm of a phase III trial of a model for predicting the change in tumour volume after preoperative chemotherapy in nephroblastoma.

In this trial the main question is whether preoperative chemotherapy leads to tumour volume shrinkage or not. Data from each patient will be fed into the model being tested—the Oncosimulator—and the results it gives validated using the real patient data. Patients will be randomized in two treatment arms. Arm A is the standard treatment arm in which all patients will receive the current preoperative chemotherapy. Arm B is the experimental treatment arm, in which treatment is

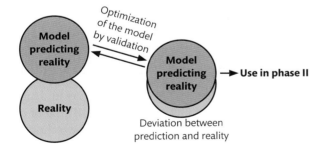

Figure 11.3 Schematic showing deviation between the prediction of a model and reality. The deviation between the prediction and the reality needs to be as small as possible before the model can be used in a phase II trial. This optimization is done in a feedback loop validating the model. A physician must always define the maximum acceptable deviation between the prediction and reality for the trial to go ahead.

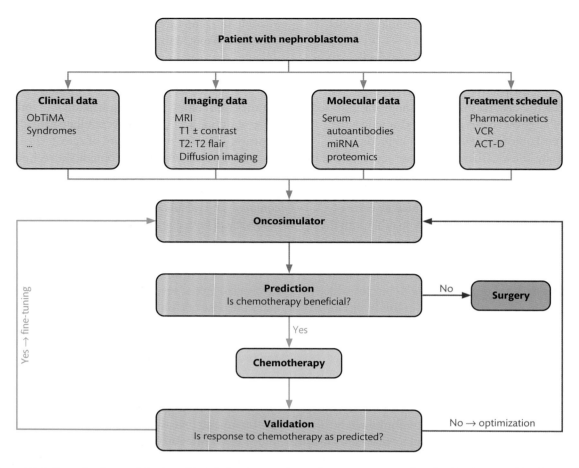

Figure 11.4 Example of a possible phase III trial with the experimental treatment arm as described in the text. This schema also includes the optimization and fine-tuning of the model according to the results of a phase II trial. This trial was under discussion (2013) in the SIOP Renal Tumour Study Group[2] (SIOP: International Society of Paediatric Oncology).

given according to the prediction of the model. If the model predicts that the volume of a particular patient's tumour will shrink, that patient will receive preoperative chemotherapy, but if the model predicts that there will be no shrinkage the patient will go forward to surgery without first receiving chemotherapy.

The model will also be used to predict the outcome of chemotherapy in patients treated according to the standard arm (Arm A: always preoperative chemotherapy). However, the results of the model in these patients will not be used for decision making but to compare the prediction of the model with what really happens. This will help to validate the

[2] http://web.visu.uni-saarland.de/rtsg/

tool in the standard treatment arm using prospective data.

Comparing the outcomes for the patients in both arms will answer the question of whether deciding treatment using the predictive model is more beneficial than the standard treatment. Just as in drug trials, successful completion of a phase III trial of a technology should be sufficient to allow the technology to be certified and to enter the market.

Phase IV: phase IV trials are also known as *post-marketing surveillance*. This means that they are carried out after the technology has been fully certified and authorized for commercialization. They involve monitoring the outcomes of patients treated using the technology to ensure that unexpected problems do not emerge later on. Post-marketing surveillance is required primarily because of the changes that are periodically made to the manufacturing process, which might introduce totally unexpected effects in some cases. It is also possible for the side effects generated by a particular drug or technology to be so rare and unexpected that they were not picked up during even phase III clinical trials. A recent example of this in the pharmaceutical industry was the withdrawal of the anti-inflammatory drug Vioxx™ (rofecoxib) due to rare but serious cardiac side effects that were not picked up during pre-registration trials. A third relevant reason for the use of phase IV trials to monitor technologies in computational biomedicine is the lack of appropriate training for medical professionals involved with the use of the technologies.

There is currently no computational biomedicine technology that has reached phase IV clinical trials. There is therefore no consolidated model in the regulatory industry for how to perform post-marketing surveillance of predictive technologies. However, one *possible* model for this involves anonymizing all data that have been processed with a computational biomedicine technology and storing it in a public database together with the quantified outcome predicted by the model. This can therefore be compared with data from real outcomes when that becomes available (and which might perhaps be obtained automatically from public health services). This could allow for continuous online analytical processing and comparison of predictive outcomes with actual ones, and therefore monitoring the predictive accuracy of the model. A more conventional outcome register similar to those that are already used for the post-marketing surveillance of other health technologies might be adequate for the reporting of complications and other unexpected outcomes.

It is apparent from what we have discussed so far in this section that a computational biomedicine model, tool, or service must be adopted into routine clinical practice before it can enter phase IV testing. Clearly, before this can happen the use of the tool must be accepted by the clinicians involved. Clinicians will not adopt tools that they cannot understand or find difficult to use, so it will be important to ensure that they get adequate training. Furthermore, the principles of computational biomedicine should be introduced into medical school curricula so that

issues of information technology for health, and more particularly computational modelling in biomedicine, are addressed early in medical education.

Today, however, it is not enough that a model or tool has been validated for clinical use in a particular indication and it is not even enough that the clinicians are given appropriate training and it is easy to use. With the budgets that are available for healthcare now being severely limited in many developed countries, the widespread adoption of any healthcare technology will only be possible if it is cost-effective and if practitioners can be reimbursed for its use. And the first information that is required before authorities can decide whether to reimburse for the use of a technology is its efficacy.

11.7 Efficacy, Risk, and Cost–Benefit Analysis

The term efficacy is used to refer to the therapeutic effect of a given medical intervention. This can, of course, be a drug but it might also be a medical device, surgical procedure, public health intervention or, as we discuss here, a computational model or tool. If a new intervention is to gain acceptance in clinical care it will need to be as good as or better than those interventions that are already in use. We have shown that comparisons of efficacy are based on the results of clinical trials, in particular phase III clinical trials. Efficacy must be distinguished from effectiveness, as it is possible to prove that a technology is effective without evaluating it in comparison with existing interventions.

Let us now consider two examples, which should indicate how the efficacy of computational models can be assessed. They both involve models that have been developed through the VPH initiative.

1. A computational model is being developed that will predict with extremely high clinical accuracy how the bone mass of a patient with osteoporosis will reduce over time. The questions to answer in assessing the efficacy of this model are: how will this additional information improve our ability to cure this patient, or at least to reduce her symptoms? The final goal is not to understand how a patient's disease will progress but to avoid osteoporotic bone fractures. We know that bone mass reduction is the best single predictor of fracture risk in patients with osteoporosis, but reduced bone mass is only the cause of around 60–65% of fractures that are observed. Known bone mass loss, therefore, is only one reason why a particular patient will experience a fracture. Therefore, an improved ability to predict how bone mass will change over time might help clinicians recognize which patients are more at risk of fracture. If this can be shown to be the case, the model, which was already known to be clinically accurate, would be said to have sufficient clinical efficacy to justify its use.

2.   Nephroblastoma is the most common kidney tumour in childhood. In most children with this disease, preoperative chemotherapy results in tumour shrinkage and down-staging, thus enabling less treatment after surgery. However, about 10% of patients fail to respond to preoperative chemotherapy, and their tumours continue to grow. It would be clinically beneficial for such patients to go immediately to surgery without receiving ineffective and potentially toxic treatment. If a computational model is able to predict the volume change in a tumour with chemotherapy at diagnosis, that model could be used to stratify the patients based on the characteristics of their tumours and decide which will be given chemotherapy (Figure 11.4). This model would have a high clinical impact.

Although the evidence that is used to evaluate efficacy is obtained from clinical trials, the perspective of efficacy assessment is more that of public health than that of clinical research. When researchers compare different interventions they need to take the risk that each intervention involves for the patient into account. Imagine, for example, a new intervention that has a higher efficacy than the current standard of care but that involves ionizing radiation. Even if this intervention has a higher efficacy than the current standard (which does not involve ionizing radiation) the current standard might be preferred because its *risk–benefit ratio* might still be preferable.

Additionally, it is impossible to assess the efficacy of a clinical intervention without considering its cost. All new interventions will also be compared to the current standard of care in terms of their *cost–benefit*

ratio, and they will only be adopted if this, as well as the risk–benefit ratio, is preferred.

An exhaustive discussion of either the assessment of efficacy or cost- or risk–benefit analysis is beyond the scope of this book. We must, however, stress here that the technologists who develop models need to keep in mind that a new technology will not be adopted unless it has a clear advantage over the current standard of care. Furthermore, this advantage must be defined not only in terms of the pure efficacy of the technology, but also in relation to the relative risks and costs involved.

The next two sections discuss first the impact and then the sustainability of computational biomedicine in the clinic, taking the VPH and its constituent projects as examples.

11.8   Impact

All new technologies, including health technologies, will eventually be assessed by policy makers in terms of their impact over the whole spectrum of the life of a society. The types of question that will be addressed include whether the new technology is likely to improve the length and quality of the lives of the citizens of that society, and the collective wealth of the country.

The next section, which has a European focus, describes how the impact of the VPH initiative is being assessed in terms of various priorities accepted by the European Commission.

The VPH Network of Excellence (VPH NoE [12]) is a project that has been funded within the overall VPH initiative to foster, harmonize, and integrate pan-European research in the field of patient-specific computer models for personalized and predictive healthcare: that is, computational biomedicine models. VPH projects are generally large and ambitious, involving many research groups from different countries and working in different research disciplines. This promotes the development of integrative models that range through a number of the levels of biological complexity that have been described earlier in this book. Furthermore, they directly involve many hospitals throughout Europe that together provide a large number of patient cases per year for the collaborative optimization and validation of such advanced, reusable, multi-scale models, tools, and services. The VPH initiative has proved successful in enrolling clinicians who interact with systems biology modellers to make sure that their research and development is focused on accepted pressing needs in healthcare. The successful development of these models should have significant and far-reaching implications, not only for European healthcare but in the wider European society. So, what is the potential impact of the VPH initiative on European society?

11.8.1 More Predictive, Individualized, Effective, and Safer Healthcare

One of the most important final goals of the VPH initiative is to provide clinicians with tools to support them in providing personalized treatment and therefore improving patient outcomes. Improving the ways in which biological information and knowledge can be shared between researchers and stakeholders will lay the groundwork for this more personalized approach to medicine and for a more predictive, effective, and safer healthcare system.

The initiative includes a number of projects that are exploring the legal and ethical rules for sharing data and biological material. Furthermore, the aims of VPH projects such as P-Medicine [13] (Table 11.2)

include *patient empowerment*. Better-informed patients are able to take better decisions concerning their health, and patients who are empowered to make these decisions will be able to take a more active role in their treatment, with positive effects on the outcome.

Finally, introducing standardized solutions for data mining will help to improve the quality of clinical research, which should also lead to better solutions for predictive healthcare.

11.8.2 Accelerated Developments of Medical Knowledge Discovery and Management, Development of Devices, and Procedures using *In Silico* Environments

Computational biomedicine models, such as those produced through the VPH initiative, aim to support clinicians in optimizing the treatment for individual patients by conducting experiments *in silico*. This can be expected to impact clinical research in many different ways.

- Firstly, *in silico* studies of disease mechanisms will help biomedical researchers to gain more sophisticated insights into complex disease phenomena, and thus to suggest treatment

Table 11.2 Summary of the intended impact of a computational biomedicine project on different levels of medical knowledge discovery and disease management

	Impact
On the patient level	Improved life expectancy and quality of life through personalized treatment planning
On the basic science level	Extending analytical and quantitative thinking to more biomedical applications, and particularly to predicting ones Developing multi-scale models that are relevant to physiology and medicine, including dynamic ones
On the technological level	Integrating a vast amount of heterogeneous data from different sources together through the development of novel technological platforms that are compatible with disparate data types, easy to use, and compatible with all relevant ethical and legal frameworks
On the educational level	Educating clinicians, researchers, and interested patients through multidimensional and visual renderings of disease dynamics and response to treatment
On the societal level	Rationalizing medical facilities to make them more efficient and effective for all citizens
On the financial level	Lowering the treatment cost per patient by using multi-scale models that incorporate individualized patient data to select the most effective treatment for each patient

strategies. This will help them in particular to design and interpret the results of clinical trials that rely on genomic data.

- Future *in silico* research will be accelerated by the development and provision of a set of open access, reusable, compatible modules that can be combined in different ways. In addition, the provision of secure infrastructure and architecture and data integration tools will help researchers to couple these modules together seamlessly and to use them with a wide range of sources of patient data. This will close the gap between clinical data acquisition and *in silico* research, thereby allowing computational researchers to concentrate on the development of new models instead of tiresome issues of manual data processing.
- The clinical value of *in silico* models as treatment support systems will enable disease phenomena to be visualized and modelled to enhance the training of doctors, researchers, and, if possible, also interested patients.

The introduction of VPH-like computational models can be expected to increase significantly the quality of *knowledge discovery* in biomedical research. Their impacts are likely to involve:

- the quality and practical relevance of medical knowledge discovery, as it will become easier to define patterns that standardize the evaluation of new tools and approaches;
- the efficiency of medical research as it will be possible to discover and reuse existing approaches, avoiding known 'dead ends' that stem from (for example) the inappropriate use of existing tools;
- the easier formation of collaborations to make better use of existing solutions, sharing and publishing not only computational tools but knowledge and 'best practice' concerning their use;
- adoption of *service-oriented business models* in which 'electronic services are seen as utilities that can be used but that are not owned by users' [14]. This paradigm can be applied to a wide range of biomedical research fields within the overall context of computational biomedicine, accelerating the development and management of new tools, services, and processes.

11.8.3 Improved Interoperability of Biomedical Information and Knowledge

Through fostering collaborations between disparate multidisciplinary research groups from many countries, the VPH initiative has substantially improved the interoperability of biomedical information and knowledge. In order to ensure that the resulting products are practically usable, potential users, most importantly clinicians, are directly involved in developing and evaluating the tools. The *European Clinical Research Infrastructure* is helping to facilitate the effective integration of research and to guarantee best practices in data management.

Establishing federated storage services for large data objects will enable this interoperability and address one of the more difficult barriers to data reuse today: the compartmentalization of data and information within providers and systems. The VPH and allied initiatives support a semantic basis for data integration to improve the interoperability of data and knowledge. This, clearly, involves the development and use of quality-assured ontologies that are appropriate for the domains of biomedical knowledge required for the models.

Hospital information systems, and even repositories for clinical trial data, differ greatly in their use of semantic resources. Several projects funded within the VPH initiative are annotating this stored data using widely accepted standards (such as HL7, ICD, and SNOMED CT), which provides the basis for better data integration that will assist research in computational biomedicine and other collaborative projects well after the end of the VPH programme.

11.8.4 Increased Acceptance and Use of Realistic and Validated Models that Allow Researchers from Different Disciplines to Exploit, Share Resources, and Develop New Knowledge

All large clinical trials are conducted internationally, and therefore the researchers involved in setting them up need to consider many different national and international laws, regulations, and guidelines. VPH projects, and similar ones across the world, have been set up using legal frameworks that protect the safety of all participants and guarantee the quality and security of clinical trial results and other individual patient data. This is establishing a solid legal framework that can help build the trust of clinicians, patients and the general public and thus increase the acceptance of large-scale, ICT-intensive clinical trials.

Tools in computational biomedicine can only be used if they are registered, and this means that they will need to be tested in clinical trials: eventually, in large, international clinical trials. These will evaluate not only the efficacy of the tools but how easy they are for physicians, data managers, and occasionally even patients or patient groups to use. Trial results will be published, which should lead to the tools being exploited much more widely throughout Europe and beyond. In return, the tools will be further modified and optimized to make clinical trials faster and data easier to interpret, and to ensure that the tools are used only within the internationally accepted standards of Good Clinical Practice (GCP).

11.8.5 Industrial Leadership and Strengthened Multidisciplinary Research Excellence in Supporting Innovative Medical Care

One important aspect of the VPH initiative that has not been mentioned previously is the attention it gives to the role of industry. Both large companies and small and medium-sized enterprises (SMEs) are expected to play a central role in most projects that are funded through the initiative. The role of these companies in promoting innovation is vital if new computational models are to be transformed into commercially successful clinical tools.

At present, the academic community, regulatory authorities such as the EMA [15] and FDA [16], and the pharmaceutical and biotech industries all agree that the existing tools are much better at analysing data 'after the event', such as adverse events that occur during clinical trials, than at predicting outcomes in advance. Improving the predictive accuracy of models and using these in (for example) personalized clinical trial design and biomarker identification can be expected to have a significant impact on the future development of the healthcare industries within Europe and potentially beyond.

These impacts are summarized in Table 11.3.

11.9 Sustainability

All scientific projects have a finite lifetime, and computational biomedicine projects are no exception to this. If the momentum provided by the VPH and other similar initiatives is not to be lost, researchers will need to take steps to ensure that the project results are managed, disseminated, and exploited over a longer term than the project itself allows. Here, setting out a coherent strategy for the sustainability of the project over the long term in good time is of crucial importance. It is also important that the different interests of all potential stakeholders are adequately considered in this process. Within the VPH initiative, for example, a great deal of attention has been and is being given to managing and transferring knowledge between the various partners in the initiative and to communicating it more widely within the scientific community and to the general public.

The role of coordinating activities related to ensuring the sustainability of the tools, knowledge, and good practice developed as part of the VPH initiative has been taken by the VPH NoE [12]. This network provided a coordinating infrastructure for all the projects funded through the initiative. It is involved in supporting training activities for researchers and other stakeholders and in disseminating and exploiting the results of funded projects. This should enable the results of the projects to be adopted and exploited by the whole biomedical community.

The various strategies adopted by the VPH NoE in ensuring the sustainability of the project are discussed in more detail below. You will find that similar strategies will have been adopted by all similar successful initiatives; in fact, these strategies will be important in ensuring the long-term success of large-scale scientific initiatives in disciplines far removed from computational biomedicine.

11.9.1 Dissemination

Definition of Relevant Target Groups and the Corresponding Stakeholders

Before deciding on a strategy for disseminating the results and other outcomes from a scientific endeavour it is important to work out exactly who your

Table 11.3 Summary of the expected impacts of VPH-funded projects in European biomedicine, and the steps that are being taken to bring about these impacts

Expected impact	Solutions provided by P-Medicine[3] and other VPH-funded projects
More predictive, individualized, effective, and safer healthcare	Developing *in silico* treatment-support systems Providing data-mining tools and solutions for the effective analysis of biomedical data Developing clinical decision-support tools for the translation of an integrative research approach into healthcare practice Building a collaborative environment for patient empowerment and thus involving patients more closely in the healthcare process
Accelerated development of medical knowledge discovery and management, development of devices and procedures using *in silico* environments	Developing integrated *in silico* models as clinical research tools Providing data-mining patterns to improve the sharing, reuse, and collaboration on data mining solutions Developing and implementing a flexible reference architecture to support a whole spectrum of possible biomedical research tools and services
Improved interoperability of biomedical information and knowledge	Developing a middle layer ontology to provide a basis for future ontology development in healthcare Developing and enhancing tools to annotate and integrate external data Developing data warehouses for the storage and retrieval of large, heterogeneous medical data types Developing a semantic framework for access to biobanks
Increased acceptance and use of realistic and validated models that allow researchers from different disciplines to exploit, share resources, and develop new knowledge	Involving end users early in the modelling process through cooperation with clinical research infrastructures (e.g. ECRIN) Establishing a legal and ethical framework (and the corresponding security infrastructure) as a trust-building measure for clinicians, patients, and the general public Developing models in a modular way to facilitate the uptake of *in silico* methods in different medical disciplines Providing data-mining patterns to standardize this process Continuously developing and disseminating modules and tools via (e.g.) the VPH Toolkit[4] to facilitate the exploitation and sharing of resources
Reinforced leadership of European industry and strengthened multidisciplinary research excellence in supporting innovative medical care	Involving industrial partners in promoting innovation and in evaluating and in exploiting the results Cooperating with other research projects or organizations. These include, for example, European Advanced Translational Research Infrastructure (EATRIS) [17], Center for the Development of a Virtual Tumor (CViT), Massachusetts General Hospital, Boston, MA, USA [18], European Clinical Research Infrastructures Network (ECRIN) [19], and European Organisation for Research and Treatment of Cancer (EORTC) [20].

audience is. The VPH NoE established a thorough definition of its target groups of stakeholders at the beginning of the project and developed it further over time. Each of the many projects funded through the initiative also developed its own strategy and defined its own target groups.

The target groups established by the NoE, and the specific fields of research with which each will be concerned, are set out in Table 11.4.

---

[3] http://www.p-medicine.eu/

[4] http://toolkit.vph-noe.eu/

Table 11.4 Specific fields of dissemination of VPH and similar initiatives, as related to their target groups

Target group	Specific field
Scientific communities	Technology to support computational biomedicine including data sharing, ontologies, and semantic interoperability
	Healthcare ICT
	System biology
	Clinical research infrastructures
	Specific medical domains and disciplines, e.g. cardiology, musculoskeletal medicine, oncology, immunology, etc.
	Legal and ethical aspects of ICT for healthcare, including data protection
	Biobanking
	Usability and validation
	Certification
Clinicians	Clinical trials, education of medical students, and continuing professional development
Patients	Patient empowerment
Political stakeholders	Impact of computational biomedicine tools and techniques on healthcare on a global level; funding, sustainability, and maintenance
General public/citizens	Impact of computational tools on healthcare on an individual level; personalized medicine

Activities and Methods of Dissemination

As we mentioned above, each of the 30 projects or more that have been funded by the EU within the VPH initiative defines its own strategy for dissemination. Each project is involved in a wide range of different dissemination activities, including, but not necessarily restricted to, websites, newsletters, flyers, posters, press releases and conferences, publications, scientific conferences, meetings, workshops, and summer schools. In particular, each project is expected to maintain its own website as a focus for dissemination of its activities to a variety of different target groups. A typical project website will include:

- an introduction to the project and its main objectives, and a short profile of each of the project partners;
- a frequently updated summary of the project's main achievements;
- news and publications;
- details of conferences and training events;
- online training materials;

- an option to subscribe to a newsletter;
- an option to subscribe to an RSS feed that keeps the subscriber continuously updated with a brief news feed, perhaps including news from similar projects and networks;
- links to related projects and initiatives.

A major part of the dissemination strategy for a VPH project or, indeed, any specific project within computational biomedicine, will be teaching and training. Different groups of stakeholders—basic scientists, clinicians, information technologists, lawyers, and patient groups, to name but a selection—will have different levels of background knowledge and different concerns. The VPH community has prioritized organizing workshops and summer schools to involve researchers; engaging clinicians and IT experts in conferences led by researchers; and developing e-learning tools to educate and train end users (sometimes including patient groups). It is, of course, absolutely crucial to gain the support and understanding of clinicians if they are to be motivated to use the tools that are being developed in their routine practice. It is clear that encouraging the widespread dissemination of both research and good practice in computational biomedicine will benefit not only the project developers but its end users and, through economic and social improvements, the wider community.

Acceptance of computational biomedicine tools and models will also be improved by enrolling patients into the wider initiative and, as we said earlier, in empowering them. Such empowered patients will encourage the spread and usage of computational models, which should help to improve diagnosis or treatment, or even contribute to the prevention of disease. In the field of cancer, for example, the ecancer website [21] is becoming the leading European channel for cancer communication and education. It provides peer-reviewed papers, news and reviews, and a wide range of education and training resources (including videos) aimed at both oncologists and patients. Its aim is to optimize both care and outcomes for patients with cancer in Europe and beyond.

Exploitation

Initiatives funded under the VPH umbrella can be expected to fuel a growth in the diagnostics industry, which has been traditionally considered a poor relative of its pharmaceutical cousin. Computational biomedicine models are able to contribute to improved diagnostics via associated growth in the use of disparate types of individual patient data. This will feed into improved patient stratification and clinical trial design through the use of predictive tools that can, for example, identify biomarkers or predict which patients are more likely to suffer an adverse event from a particular therapy.

Business models are always, to some extent, dependent on the capabilities offered by their underlying technological platforms. The computational infrastructure provided within the VPH initiative, and made available to stakeholders including companies, will allow these commercial users to develop scientific models into sustainable business models that preserve their intellectual property and enable commercial development.

Management of Intellectual Property

The promotion of the use in the clinic of models and tools developed by initiatives such as the VPH cannot be achieved unless those tools are exploited commercially. In the case of the VPH, the whole computational biomedicine research community, within and beyond Europe, is already taking steps to ensure this exploitation happens; for example, by identifying and documenting those results that have the greatest potential for commercial exploitation. Once they have ensured that there are no significant practical drawbacks (in terms of, for example, the cost of any product or the need to conform to established standards) they will identify the best way to develop each model into a product. This will undoubtedly involve the protection of innovative results through patenting: this is always a complex and expensive business that requires the assistance of lawyers and other skilled professionals.

11.10 Conclusions

Computational biomedicine is becoming an established scientific discipline in its own right. There are now probably over 10 000 researchers worldwide working on this type of project, with over 2000 in Europe alone. This total includes the clinicians who are ensuring that the tools, models, and services developed are clinically relevant, and the industrial researchers who are working towards developing them into commercial applications.

In this, the final chapter of the book, we have described some of the concerns that need to be addressed before any *in silico* medical technology can be used in clinical practice: the need for the HTA approach in assessing computational biomedicine technologies; regulatory issues for medical devices (the most appropriate category for computational models) including the important issues of verification, sensitivity, and validation; and how such technologies can be assessed in the clinic, drawing parallels with the well-established procedure of clinical trials for the assessment of medicines. We have also considered the implications of these technologies for each group of stakeholders, including the important issues of dissemination and commercialization.

We hope this book will have given you a sense of just how ambitious the research agenda in computational biomedicine is, and how its undoubted promise is still far from being realized. The development of a set of comprehensive, integrated computational models of the human body in health and disease that are accurate, trustworthy, and user-friendly enough for busy clinicians to use is one of the 'grand challenges' of modern medicine [22]. Yet there is much that still needs to be done before this goal can be achieved. Once these methods have been fully established, however, computational biomedicine will come to play an important—perhaps uniquely important—role in the creation of a P4 medicine for the twenty-first century: truly predictive, preventative, personalized, and participatory [23].

And there are already some 'low-hanging fruits': computational models that are now, at least, ready for a full clinical assessment. Once this has been carried out, and some computational models and tools are entering routine clinical practice and showing their efficacy, we will begin to see how computer simulation has the potential to transform the practice of healthcare as radically as it has already done for many other sectors.

And we hope that you, the reader of this book, will come to play your own part in these developments, and help forge the next revolution in medical practice.

Recommended Reading

Brody. T. (2012) *Clinical Trials: Study Design, End-points and Biomarkers, Drug Safety, FDA and ICH Guidelines.* Oxford: Elsevier.

Deisboeck, T.S. and Stamatakos, G.S. (eds) (2011) *Multiscale Cancer Modelling.* Chapman and Hall/CRC Mathematical and Computational Biology Series, vol. 34. Boca Raton, FL: CC Press Taylor & Francis.

Taylor, R. and Taylor, R. (2009) What is health technology assessment? April 2009. **http://www.whatisseries.co.uk/whatis/pdfs/What_is_health_tech.pdf**.

References

1. Taylor, R.S., Drummond, M.F. et al. (2004) Inclusion of cost effectiveness in licensing requirements of new drugs: the fourth hurdle. *British Medical Journal* 329: 972–975.
2. VPH Network of Excellence (2008): *D3.2 VPH Requirements and Technology Assessment Exercise,* p.2. http://www.vph-noe.eu/vph-repository/doc_details/347-d32-vph-requirements-and-technology-assessment-exercise.pdf.
3. Hutton. J., McGrath, C. et al. (2006) Framework for describing and classifying decision-making systems using technology assessment to determine the reimbursement of health technology assessment? Health technologies (fourth hurdle systems). *International Journal of Technology Assessment in Health Care* 22: 10–18.
4. http://eur-lex.europa.eu/LexUriServ/LexUriServ.do?uri=OJ:L:2007:247:0021:0055:en:PDF
5. http://en.wikipedia.org/wiki/ISO_13485
6. http://www.iso.org/iso/iso_catalogue/catalogue_tc/catalogue_detail.htm?csnumber=36786
7. Sargent, R.G. (2010) Verification and validation of simulation models. In *Proceedings of the 2010 Winter Simulation Conference*, Johansson, S.J.B., Montoya-Torres, J. et al. (eds), pp. 166–183.
8. Roache, P.J. (2002) Code verification by the method of manufactured solutions. *Journal of Fluids Engineering* 124: 4–10.
9. Grassi, L., Schileo, E. et al. (2012) Accuracy of finite element predictions in sideways load configurations for the proximal human femur. *Journal of Biomechanics* 45(2): 394–399.
10. Viceconti, M. (2011) A tentative taxonomy for predictive models in relation to their falsifiability. *Philosophical Transactions. Series A, Mathematical, Physical, and Engineering Sciences* 369(1954): 4149–4161.
11. Meinert, C.L. and Tonascia, S. (1986) *Clinical Trials: Design, Conduct, and Analysis.* New York: Oxford University Press.
12. http://www.vph-noe.eu/
13. http://p-medicine.eu
14. NESSI (2006) *Framing the Future of the Service Oriented Economy.* NESSI Strategic Research Agenda, vol. 1. Public Draft 1—Version 2006-2-13-Revision 3.1. http://www.nessi-europe.com/files/ResearchPapers/NESSI_SRA_VOL_1.pdf.
15. European Medicines Agency, http://www.ema.europa.eu
16. US Food and Drug Administration, http://www.fda.gov/
17. http://www.eatris.eu/
18. https://www.cvit.org/
19. http://www.ecrin.org/
20. http://www.eortc.be/
21. http://ecancer.org/
22. Hunter, P., Coveney, P.V. et al. (2010) A vision and strategy for the virtual physiological human in 2010 and beyond. *Philosophical Transactions. Series A, Mathematical, Physical, and Engineering Sciences* 368(1920): 2595–2614.
23. Hood, L. (2013) Systems biology and P4 medicine: past, present and future. *Rambam Malmonides Medical Journal* 4(2): 1–15.

Chapter written by Marco Viceconti, University of Sheffield, Sheffield, UK and Norbert Graf, Saarland University, Saarbrücken, Germany.

APPENDIX
Markup Languages, Standards, and Model Repositories

Having reached this stage of the book you may well be starting to try some modelling for yourself. If so, you will need to get to grips with some of the tools, databases, and markup languages that we have discussed. As computational biomedicine is an international, multidisciplinary endeavour, it is likely that, sooner or later, you will need to collaborate with others beyond your own institution and discipline. It will therefore be important to use the same standard languages and terminologies so that the work you do is easily shared amongst a potentially large group of collaborators. This Appendix outlines some of the tools, model databases, and markup languages that have been developed in the computational physiology community in recent years.

The resources that will be discussed here include three primary standards for encoding models (CellML, SBML, and FieldML), a simulation description standard (SED-ML), and a data format standard (BioSignalML). The last of these is the equivalent, for the modelling community, of the Digital Imaging and Communications in Medicine (DICOM) standard for biomedical and clinical imaging (see Chapter 4). It is important that all tools and models that are developed adhere to these standards so that models from different groups can be curated, annotated, reused, and combined. Minimum standards for data and models, and ontology-based metadata standards (such as RICORDO, which has been discussed extensively in this book) are also necessary, and these should of course be complementary to the markup language standards.

We also describe a repository for models written in CellML and FieldML, the Physiome Model Repository (PMR), and the application programming interfaces (APIs) that facilitate access to the models from visualization, computational, or other software. You should note that, at least for now, data standards are not as well developed as the equivalent standards for encoding models.

A.1 Introduction

As you have read, an enormous volume of reductive knowledge about the human body has been generated over the last half century through initiatives such as the Human Genome Project. As yet, however, there have been relatively few efforts to integrate this back into physiological understanding and evidence-based medicine [1]. Bioengineering has of course had a substantial impact on medical practice through medical devices and diagnostic imaging equipment, but other influences of the engineering disciplines on medicine have so far been relatively slight [2]. However, this is about to change. This book has described how the quantitative and predictive power of the mathematical sciences is entering the biological sciences, and how both an engineering approach to biological materials and a more systematic approach to the knowledge management of multi-scale physiological data are beginning to impact the clinical sciences.

Many of the initiatives and projects that we have described in this book derive from a series of international initiatives that began in 1997 when the International Union of Physiological Sciences (IUPS) initiated its Physiome Project. A related US initiative by the Interagency Modelling and Analysis Group (IMAG) was begun in 2003. The European Commission initiated the Virtual Physiological Human (VPH) project[1] in 2007. These projects and similar initiatives in Japan[2] and Korea are now coordinated to some extent and are sometimes collectively referred to here as the VPH-Physiome Project.

The primary goal of all these projects is to develop, promote, and facilitate the use of computational models, data, and software tools (including web services) for understanding the human body as an integrated system and to facilitate predictive, preventive, personalized, and participatory medicine. Medical practice will benefit from technologies in which digital data enable predictable outcomes through quantitative models that integrate physical processes across multiple spatial scales. Computational tools are also being developed to link individual patient data with virtual population databases using mathematical models of biological processes.

[1] See www.vph-noe.eu for details on projects funded under the VPH calls.
[2] See www.physiome.jp for details on the Japanese Physiome Project.

The standards, knowledge-representation strategies, repositories, and tools that we describe here have all been developed in projects under the VPH-Physiome umbrella. We present them in a practical way that we hope will be useful for novice modellers, illustrated with examples where appropriate.

A.2 Infrastructure for Computational Biomedicine

The overall aim of integrated computational biomedicine is to develop a systematic framework for understanding physiological properties in terms of human anatomy and biophysical mechanisms on different spatial and temporal scales. The model and data standards that we describe in the rest of this Appendix form a solid foundation for projects in computational physiology, making it easier to reproduce, reuse, and combine models.

We now have a framework that includes several Extensible Markup Language (XML)-based markup languages (see Section A.3 below). These all specify well-designed sets of tags for representing data and models in a way that is both semantically rich and easy to interpret automatically. They are mathematically rigorous; some, such as CellML and Systems Biology Markup Language (SBML; Sections A.4.1 and A.4.3), are designed to encode systems of ordinary differential equations, whereas others, such as FieldML (Section A.4.2), are designed to specify spatially varying fields using systems of partial differential equations. In all these cases the parameters and variables in the mathematical models are annotated with metadata that provide them with biological or physical meaning. These languages encourage modularization and have import mechanisms so complex models can easily be constructed from simpler, modular components. Model repositories and software tools have been created to create, visualize, and execute these models; almost invariably, these have been made freely available through open-source licensing.

We will take cell-level modelling (see Chapter 5) as an example. Many models of biological processes at the subcellular level ignore the detailed three-dimensional structure of a cell, and model cell parameters that may include excitability, mechanics, transient currents, motility, signalling, metabolism, or gene regulation (depending on the cell type) in a 'lumped parameter' system of differential equations. These models generally either make the spatial variation over the model discrete or assume that the whole model will remain spatially homogeneous; the latter approach is known as the well-stirred reactor assumption. These models sometimes include non-linear algebraic equations or need to solve constrained optimization problems but they do not require the solution of partial differential equations.

The markup language CellML (Section A.4.1) has been developed to provide an unambiguous definition for these models. This language is designed to support the definition and sharing of models of biological processes by including information about the following features of each model:

1. the model structure (how the parts of a model are organizationally related to one another);
2. mathematics (equations describing the underlying biological processes); and
3. metadata (additional information about the model that allows scientists to search for specific models or model components in a database or other repository, and which indicates the biological meaning associated with a model and its parts, allowing the model to be manipulated or interpreted correctly; see Chapter 2 for more detail).

The development of a biophysically based mathematical model is, at present, a creative endeavour, often requiring a great deal of insight into the physical processes being modelled and personal judgement about the approximations needed. Once created, however, all models should be independently reproducible and testable. The model and data files that can demonstrate the reproducibility of a model on an automated basis are termed the *reference description* of the model. These files include protocols for running models with appropriate parameter sets and comparing simulation results against suitably encoded experimental data (possibly multiple times to generate typical phenotypic outputs).

The issues of robustness and reproducibility are particularly important when models are to be combined, and this is especially true when the combined models are to be used in a clinical setting. To be worthy of reuse in this fashion, all components (perhaps developed independently) should be demonstrably 'correct' in that they should have both *biological validity* (i.e. should match some aspect of biological reality) and *mathematical validity* (i.e. they should be consistent with a range of mathematical and physical principles, such as using consistent and appropriate units).

The standards we describe here have been developed using the following basic framework:

1. develop markup languages for encoding models (including both data and metadata);
2. develop APIs which specify how specific software components interact with each other: here, to read

and write models in markup languages and to run simulations using them;

3. develop libraries of open-source tools that can read and write markup language-encoded files;

4. develop data and model repositories based on the markup languages;

5. develop reference descriptions to demonstrate model reproducibility;

6. develop services and libraries to automate or facilitate a variety of tasks including running models, comparing results against experimental data, optimizing and testing new parameters.

Figure A1 provides an overview of how these standards, tools, and databases can work together. Models at the cell, tissue, or organ level are encoded in the CellML, SBML, or FieldML standards (using open-source tools) and deposited in model repositories. These models are curated to check their validity and their variables and parameters are annotated with relevant ontology terms. A model can then be loaded from a database into a relevant simulation environment. Population data are stored in a separate atlas database along with medical images in DICOM format.

Another XML standard called SED-ML (see Section A.4.3) is being developed to specify the numerical simulation itself, including the choice of numerical algorithm. These software environments also allow access to time-varying signal data that are encoded and stored in a separate database. Finally, relevant genomic and proteomic data can be accessed from bioinformatics databases via a separate API.

These standards are discussed in more detail in [1].

A.3 Syntax, Semantics, and Annotation of Models

In order for a model to be checked and implemented completely automatically, it is necessary to set standards for the minimum set of information (including annotation) that is needed to specify a model. These standards are also useful, if not necessary, for improving the robustness and reproducibility of models in the published literature.

There are already minimum information specifications for experimental measurements in different biological disciplines. An early set of standards to be defined was the Minimum Information about a Microarray Experiment (MIAME) standards for transcriptomics. This establishes the following minimum standard for reporting a microarray experiment:

- the raw data for each experiment;
- the final processed data, in the form of a gene-expression data matrix;
- sample annotation;
- the experimental design;
- array annotation;
- the protocols used for processing the data.

Examples of minimum standards that are particularly relevant to computational biomedicine include Minimal Information Required In the Annotation of Models (or MIRIAM; **http://biomodels.net/miriam**) and Minimum Information About a Simulation Experiment (MIASE; **http://biomodels.net/miase**). The website **http://mibbi.org** has been set up to provide a one-stop portal to a number of similar sets of standards.

The markup languages that we describe in this Appendix are examples of languages based on the general Extensible Markup Language, XML. This has been developed by the World Wide Web (W3C) Consortium (**http://www.w3.org/XML/**) to define sets of tags that are related to specific domains of knowledge. These tags are used to associate meaning to entities in models in a computer-readable form, so the models can be checked and interpreted automatically. MathML and its subset Content MathML are examples of XML-based markup languages used for encoding mathematics.

As a very simple example of markup language syntax, take the expression x + 2. In Content MathML this is represented in the following way:

```
<apply> <plus/> <ci> x </ci> <cn> 2 </cn>
</apply>
```

Here the tags `<ci>` and `</ci>` indicate that x is an identifier, `<cn>` and `</cn>` indicate that 2 is a number, and `<apply>` and `</apply>` indicate that the `<plus/>` operator is applied to x and 2.

In computational biomedicine, CellML has been designed to encode lumped parameter systems of ordinary differential equations (ODEs) and linear and non-linear algebraic equations (together known as differential algebraic or DAE systems) that are based on the biophysical parameters of cells [3]. Examples of CellML code are shown in Figure A2. FieldML has been designed to encode field information that varies in space and time, including anatomical structure, the spatial distribution of protein density or the computed variation of electrical potential or oxygen concentration throughout a tissue [4]. The SBML (**http://sbml.org**) that has been developed

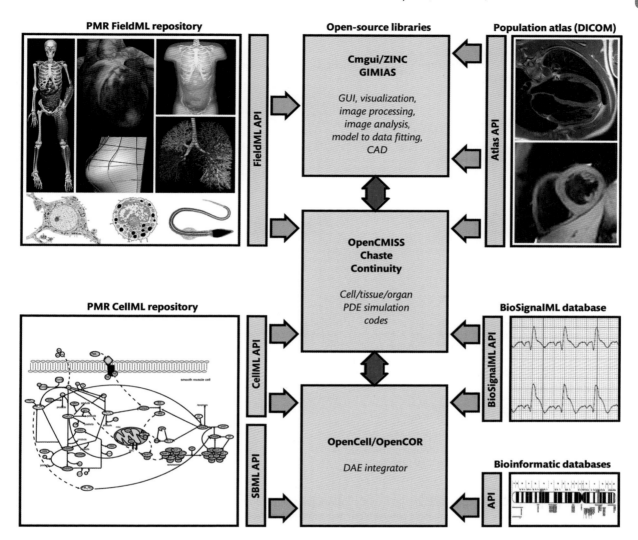

PMR FieldML repository

Open-source libraries

Population atlas (DICOM)

FieldML API

**Cmgui/ZINC
GIMIAS**

*GUI, visualization,
image processing,
image analysis,
model to data fitting,
CAD*

Atlas API

**OpenCMISS
Chaste
Continuity**

*Cell/tissue/organ
PDE simulation
codes*

PMR CellML repository

CellML API

smooth muscle cell

SBML API

OpenCell/OpenCOR

DAE integrator

BioSignalML database

BioSignalML API

Bioinformatic databases

API

Figure A1 An overall framework for the use of the software tools, models, and data repositories discussed in this Appendix.

by the systems biology community is similar in format to CellML but is used more specifically for representing models of biochemical reactions. All these examples are taken from **http://www.cellml.org/getting-started/cellml-primer**.

A useful way of viewing the development of standards is shown in Table A1. The standards presented here include specifications for the minimal requirements for data, models, and simulation experiments; standards for the syntax of data, models, and simulation experiments; and ontologies for annotating the semantic meaning of terms used in modelling and simulation.

We have mentioned already that if models are to be combined and reused effectively it is essential for them to be annotated with standard terms taken from one or more relevant ontologies. The RICORDO methodology that is described in this book (see Chapter 3 in particular) provides a coherent framework and infrastructure for annotating and adding semantic meaning to data and model resources that are used in computational physiology. In practice, RICORDO uses a strategy of interoperability that is based on the use of standard ontologies that have been set up within the molecular biology and systems biology communities. Ontology terms (such as those in the Gene Ontology) are easily embedded in the metadata that are associated with data and model resources.

RICORDO provides the computational biomedicine community with the ontology and metadata standards that will be necessary if these models are to be adopted by the industrial and clinical communities. Annotation-bearing

Markup Languages, Standards, and Model Repositories

(a) **Component**

```
<component name="environment">
    <variable name="time" public_interface="out" units="second"/>
</component>
```

(b) **Equation**

```
<math id="1" xmlns="http://www.w3.org/1998/Math/MathML">
 <apply><eq />
   <ci> C </ci>
   <apply><plus />
     <ci> A </ci>
     <ci> B </ci>
     <cn cellml:units="second"> 20.0 </cn>
   </apply>
 </apply>
</math>
```

(c) **Connection**

```
<connection>
  <map_components component_1="membrane" component_2="sodium_current"/>
  <map_variables variable_1="V" variable_2="V"/>
  <map_variables variable_1="i_Na" variable_2="i_Na"/>
</connection>
```

(d) **Units**

```
<units name="nM">
    <unit prefix="nano" units="mole" />
    <unit units="litre" exponent="-1" />
</units>
```

Figure A2 Snippets of CellML code defining: (a) a model component ('environment'), (b), the equation C = A + B + 20 s, represented using MathML, (c) a connection between the two components 'sodium_current' and 'membrane', with two variables mapped between them, and (d) some units built up from basic SI units.

metadata in RICORDO is encoded using the Resource Description Framework (RDF), which has a serialization in XML. This enables models to be combined more easily with the existing minimum information standards discussed above.

A.4 Markup Languages

This section describes in detail some of the most important markup languages that are used to develop models in computational biomedicine.

A.4.1 CellML

The markup language CellML (**http://www.cellml. org**) has been developed to store, exchange, and reuse mathematical models at the level of the single cell and its

components. Like many markup languages, it separates the structure and mathematics of a model from the biological and physical meanings of its components and parameters. These are defined within the model metadata by reference to relevant ontologies, and mathematical equations are encoded in Content MathML (see **http://www.w3.org/TR/MathML2/chapter4**). All CellML models can be broken down into components, so complex models can be built through importing modular components from libraries. In contrast, SBML is more closely tied to the concepts of biochemical and genetic networks, and does not maintain such a clear separation between the mathematical representation and the biological semantics. FieldML encodes anatomical features at multiple spatial scales by allowing hierarchies of material coordinate systems that preserve anatomical relationships (e.g. coronary arteries embedded in a deforming myocardial tissue that is itself part of a heart contained within a torso). These three standards

Table A1 Minimum information standards for data, models, and simulations

	Data	Models	Simulation
Minimal requirements	MIBBI, http://www.mibbi.org	MIRIAM, http://biomodels.net/miriam	MIASE, http://biomodels.net/miase
Standard formats	PDB, http://www.rcsb.org/pdb DICOM, http://medical.nema.org BioSignalML, http://www.embs.org/techcomm/tc-cbap/biosignal.html HDF5, http://www.hdfgroup.org/HDF5/	SBML, http://sbml.org CellML, http://www.cellml.org FieldML, http://www.fieldml.org	SED-ML, http://www.sed-ml.org
Ontologies	GO, http://www.geneontology.org Biopax, http://www.biopax.org FMA, http://sig.biostr.washington.edu/projects/fm/AboutFM.html SBO, http://www.ebi.ac.uk/sbo OPB, http://sig.biostr.washington.edu/projects/biosim/opb-intro.html	GO Biopax FMA SBO OPB	KiSAO, http://www.ebi.ac.uk/compneur-srv/kisao

are recommended both by the US National Institutes of Health (NIH) and by the European Commission for the VPH projects.

The structure of CellML is simple and is based on connected components. The connections between components allow users to share information by associating variables that are visible in the interface of one component with those in the interface of another component. Appropriate and dimensionally compatible physical units must be assigned whenever variables are connected.

The interfaces between components may be public or private, which provides another mechanism for hiding or abstracting information. These general features of CellML allow new models to be constructed by combining existing hierarchies of components into model hierarchies.

An overview of the structure of CellML is shown in Figure A3; the CellML 1.1 standards are available at **http://www.cellml.org/specifications/cellml_1.1** and an overview of metadata standards is available in [5]. The language is still being developed[3], with the intention of simplifying and extending the capabilities of the language, and providing a clear separation between normative and informative descriptions. This makes it easier to evaluate and approve secondary specifications based on CellML. One example of this is a proposal to support descriptions of uncertain parameters by enabling Content MathML to reference operators that associate variables with probability distributions.

The software that is used to read and write CellML models is known as the CellML API. This API also provides a number of services that are associated with the use of CellML models; these are listed below. It can be thought of as a collection of interfaces that describe objects from CellML models, operations and attributes on each object. It is formally described in an interface description language, or IDL[4]. *Bindings* describe how this IDL maps on to particular programming languages and *bridges* describe how users can access the objects, operations, and attributes of one binding from another.

A reference implementation is available with the API. This is a full open-source implementation of all API interfaces, written in C++, and it is described as the Physiome C++ Mapping or PCM. Different bindings have been written between PCM and Java, Javascript (via XPCOM), Python, and CORBA. The goal is to provide a common interface between the implementations (which are CellML libraries) and the applications, so implementations are interchangeable and applications can efficiently communicate parsed models. This makes it easier for applications to process CellML properly, and aims to increase the support for, and the correctness of, CellML processing. It also avoids unnecessary duplication of effort spent on CellML processing software.

The API is divided up into a 'core' part, which provides the basic data model, and its services. The core provides an object for every element type. MathML is accessed using the W3C MathML Document Object Model (DOM)

---

[3] http://www.cellml.org/specifications/specifications-under-development
[4] See http://www.omg.org/cgi-bin/doc?formal/02-06-39.

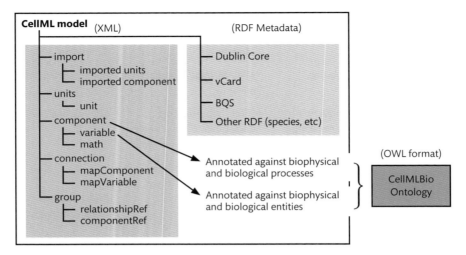

Figure A3 Entities in a CellML model. The base CellML model file is shown with its imports, units, components, connections, and groups described in XML format, and the metadata in RDF format. Metadata illustrate the biological and biophysical meaning of the data, specify the simulation parameters using the SED-ML standard, and provide a graphical output standard.

specification, and extension elements are accessed via the W3C DOM specification. Access to this can be provided optionally.

The *services* offered by the CellML API are listed below:

- the CyVariable Association Service (CeVAS) facilitates dealing with variables connected to each other in multiple components;
- the MathML Language Expression Service (MaLaES) allows MathML fragments to be converted into text-based expressions. This is used as part of the CCGS service (see below) and the language is described in a format called MAL;
- the CellML Units Simplification and Expansion Service (CUSES) allows CellML units to be resolved into the base units that comprise them, thus facilitating comparison between units;
- the CellML Code Generation Service (CCGS) allows models to be translated into imperative code. The MaLaES transformer can be provided so that a wide range of languages is supported (although some, like FORTRAN77, require substantial post-processing);
- the CellML Language Export Definition Service (CeLEDS and CeLEDSExporter) provides an XML wrapper around MAL. This provides an XML description of all configurable parts of CCGS code generation, so that a single XML file can define an imperative language to generate code in. The CellML Integration Service (CIS) is used to run numerical simulations of models.
- the Validation Against CellML Specification Service (VACSS) allows models to be validated; the errors

and warnings may be either representational (invalid XML or elements in the wrong place) or semantic. For example, dimensional inconsistency on connections triggers an error; dimensional or units inconsistency in equations triggers warnings.

- the SED-ML Processing Service (SProS) provides a way to access SED-ML descriptions of simulation experiments (see Section A.4.3) from the CellML API. Currently this is just the object model (SED-OM), but more services for fully processing SED-ML descriptions referencing CellML models are planned.

Planned future developments in CellML include better support for interpreted languages, better support for SED-ML, and more language bindings. Many other planned items can be viewed at **https://tracker.physiomeproject.org** and the current draft of the CellML metadata specification is at **http://www.cellml.org/specifications/metadata/mcdraft**.

A.4.2 FieldML

FieldML (**http://www.fieldml.org**) has been developed to formulate models that include spatial information. Files written in FieldML include all the parameters and mathematical expressions that are needed to define fields that change over space, time, and other domains. The variation in parameters over space or time is defined using a finite element mesh with a given topology, with values assigned to parameters at each node. Dense data formats (which are used, for example, to represent images embedded inside models) and arbitrary functions of existing fields can also be represented.

This language is designed as a standard for communicating computational field descriptions and data [4]. It defines the spaces that make up a body of interest termed a domain using basic primitives, and maps positions on those spaces to parameter values (the fields) using explicit mathematical expressions. Domains may be composed of aggregates of subdomains, or there may be multilevel subdomains within domains which will allow meshes to be described hierarchically and using multiple scales. They may also be embedded within other domains, enabling models to represent fields that depend on one another. One example of this is structures that are materially fixed to another, such as blood vessels embedded in a deformable tissue.

Important principles in the design of FieldML include:

- the use of a minimal set of interoperable concepts;
- no limits on the generality of a domain, which may be discrete, continuous, or a combination and may represent, for example, space, time, parameter-space, population, or any combination of these so that all quantities are 'field-like';
- no 'low-level objects', only domains and sets that have attributes mapped to them;
- extensibility to support arbitrarily complex field functions;
- support for the reuse of domain descriptions and functional expressions;
- efficiency in storage and performance;
- a preference for homogeneous data.

The approach taken in FieldML contrasts with typical 'finite element' field interchange formats which build fields out of fixed data structures such as 'nodes' and 'elements' with fixed relationships between them. This limits the description of more general fields. FieldML can represent these legacy formats, but only if the definition of sets of such 'objects' and the relationships between them are explicitly described. Therefore, FieldML is essentially a meta-language for field interchange.

Reading and writing files in the FieldML format is supported by the companion FieldML API, written in C. This provides an interface for accessing and working with FieldML documents and objects plus support for bulk data input and output.

A.4.3 SED-ML

The Simulation Experiment Description Markup Language (SED-ML; **http://www.sed-ml.org**) is an XML-based format for encoding simulation experiments that follows the requirements defined in the MIASE guidelines. It specifies the experimental tasks to run, the models to use, and the

results to produce. It is built from six main classes of description: the Model class, the Change class, the Simulation class, the Task class, the DataGenerator class, and the Output class.

The *Model class* is used to reference the models used in simulation experiments. SED-ML itself is independent of the encoding underlying the models. The only requirement is that all models must be referenced using unambiguous identifiers so they can be easily located [for example, by using a MIRIAM uniform resource identifier (URI)]. A set of predefined language resource names is provided to specify the language in which the model is encoded.

The SED-ML *Change class* allows changes to be applied to the models referenced. These may include changes to the XML attributes (e.g. changing the value of an observable), computing the change of a value using mathematics, or general changes to the XML elements such as substituting a piece of XML for an updated one.

The *Simulation class* defines the simulation settings and the steps taken during a simulation. These include the particular type of simulation and the algorithm used for its execution. Here it is preferred that references to simulation algorithms should be unambiguous, and this can be achieved by using a controlled vocabulary or ontology. The Kinetic Simulation Algorithm Ontology (KiSAO) is an example of an ontology of simulation algorithms. The Simulation class is also used to encode information that is dependent on the simulation type such as the step size and simulation duration.

SED-ML makes use of the *Task class* to combine a defined model (from the Model class) and a defined simulation setting (from the Simulation class). Each task always holds one reference.

Raw results from a simulation often do not correspond to the output that is desired; a modeller may, for example, wish to normalize a plot or apply a post-processing algorithm before the results are used. The *DataGenerator class* allows this type of post-processing to be applied to simulation results before output. This post-processing may use any addressable variable or parameter of any model defined using the Model class, and new entities can be specified using Content MathML definitions.

The *Output class* defines the form of the output of the simulation; that is, what the results look like. An output type such as a two- or three-dimensional plot or a data table is defined and each corresponding axis or column is assigned to a specified instance of the DataGenerator class.

A.4.4 BioSignalML

Many experiments in molecular biology, physiology, and biomedicine produce signals that change over time (biosignal data). There are more than a hundred different file formats that have been developed to store and exchange these data [6]. It is likely that several of these will be used in any given project. This is not by design—in fact it is almost bound to prove inconvenient—but the manufacturers, vendors, and in-house developers of the software used in a complex project are very unlikely to have chosen the same format. The use of archived data and collaborations between teams from different institutions are likely to compound this problem.

Although there are a number of tools available for converting between data formats, such as those catalogued in the BioSig project [6], these do not provide a standard view of the metadata included in a biosignal file. The lack of any standard, domain-neutral format for working with signal data has hampered the integration of this data into physiological models.

The BioSignalML project has been set up to address problems that arise from this multitude of formats and lack of standard metadata. This is providing a framework for the storage and exchange of physiological time series or biosignal data (**http://www.embs.org/techcomm/tc-cbap/biosignal.html**).

The framework provided by BioSignalML:

- encapsulates common features of biosignal file formats in an abstraction layer;
- uses ontologies to define biosignal terms and attributes;
- can be extended to incorporate domain-specific concepts and terms;
- provides software tools and libraries that allow the use of disparate signal file formats;
- includes a repository component that allows recordings to be accessed using standard web software as a Linked Data resource.

We will discuss the role of BioSignalML using a project initiated by the research team that is developing this language as an example. These researchers are working with biosignal recordings that are associated with sleep research, known as polysomnograms. Polysomnograms can contain a comprehensive range of physiological signals measuring brain, heart, breathing, eye, and muscle activity during sleep, often at different sampling rates. The recordings are taken usually over several hours and there may be several recordings for a single patient. A variety of signal-storage formats are used, including both proprietary formats and standard ones such as the European Data Format (EDF and EDF+; [7]). BioSignalML can facilitate the use of biosignal data with physiological modelling languages such as CellML, with an HDF5-based file format that has been specified for biosignal exchange and storage (**http://www.hdfgroup.org/HDF5/**), and a streaming format used for real-time connections between simulation components.

BioSignal Metadata

Metadata attached to biosignal data files can be used to address questions such as: What are the general attributes of biosignals? What does a particular signal represent? When was it recorded, how, and by whom? What processing has been applied? What is the purpose of a recording besides being a collection of signals? And, finally, do other people know the meaning of the terms that have been used to describe properties? Current proprietary and standard biosignal file formats usually have a limited number of fields available for metadata and these are usually only pertinent to the domain that the format was designed for. These metadata fields often contain free-format text with no controlled vocabulary or ontology to specify content, leading to possible future ambiguity.

BioSignalML addresses this problem by specifying a core set of terms and relationships for metadata. Some of these are defined in the BioSignalML ontology; others are taken from other relevant standard ontologies, and additional domain-specific concepts and terms can easily be added. Metadata statements about biosignals are made using a RDF, with the Web Ontology Language (OWL) used to specify ontologies. This use of both RDF and OWL allows biosignal data to be integrated into the broader context of the Semantic Web without restricting its future applications and extensions. Some of the ontologies that are applicable to biosignal annotation are already well established as international standards, and there are more that are at different stages of development. Both the OBO Foundry (**http://www.obofoundry.org**) and the NCBO BioPortal (**http://bioportal.bioontology.org**) provide repositories of publicly available biological and biomedical ontologies. These include:

- Dublin Core Terms (**http://dublincore.org/documents/dcmi-terms**),
- Timeline Ontology (**http://purl.org/NET/c4dm/timeline.owl**),
- Foundational Model of Anatomy (**http://sig.biostr.washington.edu/projects/fm**),
- Relation Ontology (**http://www.obofoundry.org/ro**),
- Cardiovascular Research Grid ECG Ontology (**http://www.cvrgrid.org/?q=Ontologies**),
- Sleep Domain Ontology (**http://www.bioontology.org/sleep-ontology**).

Concepts from the abstract model have been set up as objects and methods in an API and software library. This library allows signal files and their metadata to be created and accessed in a format-independent way, and they form the basis of a web-accessible biosignal repository.

The BioSignalML Repository

Biosignal recordings are usually stored in files in the original format that they were recorded in. Exchanging recordings with colleagues will often involve format conversion. The emphasis in the laboratory is generally on electronic transfer of the signal data without its accompanying metadata, with the metadata kept and exchanged in the form of paper laboratory notes.

 As an alternative to simply working with flat computer files, BioSignalML provides a repository application in which metadata are treated as an important component of the biosignal data. This repository is an extensible, cross-platform resource that integrates with existing signal processing workflows and has been designed to be easy to use. It stores BioSignalML resources in the form of *Recordings*, *Signals*, *Annotations*, and *Events*. Each resource has its own URI—a unique string of characters that can represent it—and a request for a resource returns a representation of it. This might be actual signal data in a particular format, an HTML web page describing the object, or an RDF description of the resource complete with metadata links to other resources. The resource representation returned is determined by the original request. The repository provides a HTTP interface and may be accessed using standard web browsers.

Signal recordings submitted to the repository are kept in their original format, with metadata about BioSignalML objects extracted into an RDF. Each recording has a Named Graph [8] to hold its metadata, so that all metadata from the recording can be referenced by a single resource. Domain-specific metadata is mapped to ontological terms in a user-extensible way, allowing future refinement as further ontologies are developed.

Signal recordings are exchanged with the repository either as files or as a stream of continuous signal data (a telemetry stream). The repository supports a number of standard and proprietary file formats, and it is straightforward to add new ones using a software module for conversion into the BioSignalML abstract model and mapping statements for domain-specific metadata.

The use of URIs to identify biosignals is fundamental to the repository and its interfaces. In line with Linked Data guidelines the repository uses the 'http' scheme for all URIs other than local file-system resources. As full 'http' URIs can be unwieldy for users, however, the API and user tools allow shorter relative URLs to be used. Time segments of a recording or signal can be requested by using a comma separated list of intervals as the query component

of the object's URI, and this interval can be in the form of either 'start:end' or 'start:duration'. A request of this type will always return signal data spanning the requested interval(s) and may, depending on the format of the request, also return other data.

There are several other markup languages that are relevant to the computational bioscience community. These include NeuroML (**http://neuroml.org**), for models that are relevant to the neurosciences, and InSilicoML (**http://physiome.jp/insilicoml**), which has been developed by Physiome Project groups in Japan to describe multilevel hierarchical models of physiological processes.

A.5 Model Repositories

Simply encoding a model in a widely used and extensible markup language is not enough to ensure that it can be easily reused; it must also be made easily accessible. Storing models in repositories that are free and easy to access is a useful way of ensuring this. An example of a public model repository is the Physiome Model Repository (PMR), which has been used to store many of the models that have been encoded in the markup languages described here. This repository has been designed primarily, but not exclusively, for CellML and FieldML models; SBML models are mainly available through **http://biomodels.net**.

In this section we will discuss the features of model repositories by using PMR as an example. Its key features include:

- facilitating exchange of models directly between modellers, without reliance on the central repository;
- a detailed revision history for each model;
- user access workflows, which are used to control privacy when this is required;
- embedded workspaces that allow models to be reused and promote modular development.

A few specific features of the PMR are described in the next sections.

A.5.1 Workspaces and Model Revision History

Within the PMR all the files that are related to a particular model are stored together within a defined *workspace*, which is analogous to a folder on a desktop computer. The implementation of workspaces uses a system called the Distributed Version Control System (DVCS), which provides version tracking to ensure that users in a group cannot accidentally overwrite or purge changes made by other users. Furthermore, each change made to a model or its associated files is recorded as a single changeset: this is a time-stamped, informative comment from an identifiable user, which describes the changes they have made. As model files are progressively altered, the changesets preserve the history of development of each model. Finally, even if two users simultaneously change the same file, distinct changesets allow their work to be merged later in a controlled manner.

The DVCS also has the advantage of allowing users working on the same model to collaborate directly, independent of a centralized online repository. Each collaborator working on a particular model will own and have full access to a complete clone of the workspace with the model and related files they are working on. This allows each individual collaborator to work and commit changes to their local copy, creating new changesets which may then be shared between themselves or pushed into the centralized repository when their work is ready to be reviewed or released.

A.5.2 Embedded Workspaces

It is possible to store composite models encoded using CellML 1.1 in the PMR. These models use the import feature of the repository to reference components and units from other models. PMR provides support for embedded workspaces; that is, workspaces into which one or more other workspaces may be included by reference. Embedded workspaces are intended to manage the separation of a core model from its subcomponents, thereby facilitating the sharing and reuse of the components of a model independently from its source model. This allows each subcomponent to be developed independently of the main model, with the versions of the embedded workspaces also tracked.

Changes made to a workspace that has been embedded in another will not affect the embedding workspace until the author explicitly chooses to update the available version. This gives authors the opportunity to review changes to embedded submodels until they are happy that these changes will not adversely affect the models into which these submodels are embedded. These also allow components to be imported into models via relative URIs, thus promoting modular development.

A.5.3 Model Exposures

Another important feature of the PMR is the way that the models and their associated data are presented to their users. It is possible to select a single changeset—a 'snapshot' of the model and data at a particular time—and create an exposure from it. At the moment, when an exposure is created the contents of the selected revision will be presented and the URL associated with that exposure will be included in the category listings for the main repository. Curators may annotate their assessment of the quality of that version of the coded model using curation stars.

The repository has been designed so it can be extended to support a range of presentation styles to suit a growing range of different file types. This requires a system where plug-ins for different file formats can be installed with ease, and this has been implemented using the Zope Component Architecture. This registers and activates plug-ins based on their names. Therefore, software developers may construct specific plug-ins to generate the presentation styles that are required for any model type, and these may be installed and enabled on the developers' local instance of the repository. It is also possible to develop and activate specific browser plug-ins to create a rich web interface for viewing models.

The access control and presentation layer of PMR is managed by a content-management system known as Plone [9]. The access control features of Plone allow authorized users to manage permissions for other users, so, for example, model developers may choose to allow their collaborators to view a private workspace, push changesets, create exposures, and update workflow states such as expiring exposures.

The CellML component of the PMR repository currently lists over 600 models, and these are categorized under the following headings:

- calcium dynamics,
- cardiovascular circulation,
- cell cycle,
- cell migration,
- circadian rhythms,
- electrophysiology,
- endocrine,
- excitation–contraction coupling,
- gene regulation,
- hepatology,
- immunology,
- ion transport,
- mechanical constitutive laws,
- metabolism,
- myofilament mechanics,
- neurobiology,
- pH regulation,
- pharmacokinetics/pharmacodynamics,
- signal transduction,
- synthetic biology.

These categories change as new classes of models are developed. The pharmacokinetics/pharmacodynamics classification, for example, has been added recently to accommodate models that are now being encoded in CellML by this research community.

A.6 Conclusions

Currently, the main goal of research into computational biomedicine is to improve our quantitative and integrative understanding of the processes of human physiology through developing models that are based on human anatomy and biophysics. The application of these models to medical practice is a further goal that is secondary in terms of maturity of development but not in terms of importance. Eventually we believe that these models will be applicable to many medical disciplines and endeavours including diagnostics, therapy, device design, clinical decision making (including planning surgery), and drug discovery.

Achieving these ambitious goals, and particularly the clinical ones, requires collaboration between modellers in different disciplines and countries, and therefore necessarily requires the development of standards for encoding data and models; the establishment of web-accessible databases for accessing models and freely available open-source software libraries based on these standards to facilitate developing them. International initiatives include the Physiome Project, which has been under way since 1997, and the European VPH project, which has been fully established since 2007. Through these projects and other similar ones there is a now a robust framework for these standards in place. This includes standards for markup languages such as CellML, SBML, and FieldML and their associated databases, and open-source codes incorporating the APIs for model and data repositories. However, much remains to be done. In this Appendix we outline some of the remaining gaps.

The standards that are built into markup languages such as CellML and FieldML allow users to verify that models are properly curated (i.e. they have consistent units, obey appropriate physical laws, etc.) and that they are annotated with metadata that assigns terms from appropriate ontologies to the model's variables and parameters. This ensures, for example, that appropriate models can easily be retrieved using electronic searches and incorporated into more complex composite models.

Completed and validated models should be published in the peer-reviewed literature if at all possible. One of the many advantages of this is that models should always be checked for validity when they are reviewed for publication. However, much more still needs to be done to clarify the provenance of parameters that are used in models and the limitations of those models for application to new settings. The types of questions that should be addressed include the species, tissue, and cell types that are used in the experimental determination of parameters, and the experimental conditions under which this data was obtained. Other important considerations to be addressed include the robustness of the models, the sensitivity of the models to small changes in parameters, and the extent to which the model reflects the robustness in the biological system. This involves, for instance, the extent to which measured parameter values may differ from one cell to a neighbouring one.

The standards for CellML have been developed through a project known as VPH-Share to include the stochastic variability of parameters and to provide for multiple instances of the same model in adjacent cell types using stochastically generated parameter values that are consistent with the specified distribution function for the parameters. The codes used in simulation need themselves to be verified and databases of benchmark problems are being developed to help establish this. The STACOM challenge (**http://cardiacatlas.org/web/guest/stacom2011**), run in 2011, is an example of the type of competition that has been run to encourage the modelling community to participate in this.

Another markup language, ModelML, is being planned and developed to encode the workflows that are needed to set up the physical equations that are needed for computational physiology modelling, and their boundary and initial conditions. This is currently done mainly using a scripting language such as Perl or Python, but this is not ideal. Using a workflow that has been marked up using formal descriptors will make it much easier to deal with complex problems that involve many different types of equation. It is possible that this language will eventually merge with SED-ML.

The early chapters of this book described some of the genomics, proteomics, and clinical imaging data, and associated models of tissue and organ structure, that can now be obtained from patients. One of the major goals of the VPH-Physiome projects is to link this information from individuals into subject-specific models that can be of practical use in, for example, clinical decision making. To achieve this goal the projects will need to integrate much more with experimental data and models at the level of molecular systems biology. Good links have been set up between the computational physiology community and systems biologists who are studying biochemical networks and who have generally adopted the

SBML standard. However, making connections between our community and molecular biophysicists who are engaged in analysing data and setting up models that take three-dimensional spatial information into account will be major goals for the next few years.

We have already noted that the goals of computational biomedicine include both the generation of verified, validated, annotated, and reproducible models and their application in a clinical setting. The second goal brings a completely new set of challenges in making the models accessible to and accepted by the clinical community. Several of the European projects that have been described elsewhere in this book, however, can be thought of as particularly helpful in addressing these challenges.

One of these is the RICORDO project, led from the European Bioinformatics Institute (EBI) near Cambridge, UK, and discussed extensively in this book, particularly in Chapter 3. This is developing a multi-scale ontological framework for computational biomedicine to improve interoperability between data and modelling resources. As long as ontologies that are relevant to clinical medicine are incorporated into this framework this interoperability should allow clinicians to understand and accept the models more easily.

The other project that will arguably be particularly applicable to the clinic is VPH-Share. This provides essential services, building on the computational infrastructure set up within VPH-Physiome projects for sharing of clinical and research data and tools. It facilitates the construction and operation of new workflows and collaborations between the members of the computational biomedicine community, and, therefore, is well placed to provide tools for sharing data and models with clinicians.

We have stressed throughout this book how advances in reductionist science during the last 50 years have enabled us to understand genotype–phenotype relationships and create rich genomic and proteomic databases. The biomedical and bioengineering communities now have the opportunity to assemble these molecular pieces by linking the data stored in these databases to anatomy and function at the cell, tissue, and organ level through biophysically based computational modelling of the human body. Once this is robust enough to be applied to the diagnosis and treatment of disease it is likely that this will revolutionize the practice of medicine.

This book has described in detail the different forms of modelling that are being developed within the ambit of computational biomedicine and the infrastructure and tools that have been set up to enable their use. The success of these exciting and important developments is highly dependent on the development, adoption, and integration of many associated markup languages, tools, model repositories, and workflows. In this Appendix we have described some of those that are being developed within various international VPH and Physiome projects. We hope that this book will have inspired you to begin modelling yourself, and that you will find here detailed information about some of the computational tools that will prove practically useful in your endeavours.

References

1. Hunter, P.J. and Viceconti, M. (2009) The VPH-Physiome Project: standards and tools for multi-scale modelling in clinical applications. *IEEE Reviews in Biomedical Engineering* 2: 40–53.

2. Hunter, P.J. (2004) The IUPS Physiome Project: a framework for computational physiology. *Progress in Biophysics and Molecular Biology* 85(2-3): 551–569.

3. Cuellar, A.A., Lloyd, C.M. et al. (2003) An overview of CellML 1.1, a biological model description language. *SIMULATION: Transactions of the Society for Modelling and Simulation International* 79(12): 740–747.

4. Christie, G.R., Nielsen, P.M.F. et al. (2000) FieldML: Standards, tools and repositories. *Philosophical Transactions of the Royal Society A* 367: 1869–1884.

5. Beard, D.A. (2005) A biophysical model of the mitochondrial respiratory system and oxidative phosphorylation. *PLoS Computational Biology* 1(4): e 36.

6. Schlögl, A. (2009) An overview on data formats for biomedical signals. In Image Processing, Biosignal Processing, Modelling and Simulation, Biomechanics, O. Dössel and A. Schlegel (eds), World Congress on Medical Physics and Biomedical Engineering. *IFMBE Proceedings* 25(4): 1557–1560.

7. Kemp B., and Olivan, J. (2003) European data format 'plus' (EDF+), an EDF alike standard format for the exchange of physiological data. *Clinical Neurophysiology* 114(9): 1755–1761.

8. Carroll, J.J., Bizer, C. et al. (2005) Named graphs, provenance and trust. In *Proceedings of the 14th International Conference on World Wide Web*, WWW '05, pp. 613–622. New York: ACM.

9. Aspell, M. (2007) *Professional Plone Development*. Birmingham: Packt Publishing.

Appendix written by
Poul Nielsen, Auckland Bioengineering Institute (ABI), University of Auckland, New Zealand;
Bernard De Bono, Auckland Bioengineering Institute (ABI), University of Auckland, New Zealand; Centre for Health Informatics & Multiprofessional Education (CHIME), University College London, 3rd Floor, Wolfson House, Stephenson Way, London NW1 3HE, UK;
Peter Hunter, Auckland Bioengineering Institute (ABI), University of Auckland, New Zealand and University of Oxford, Oxford, UK.

GLOSSARY

abstraction The process in computer science in which generic concepts are separated from specific instances in which they are implemented, removing all details of the hardware and software used.

action potential A process in which the electrical membrane potential of a cell rapidly rises and falls, leading to a short-lived change in polarity that activates a specific intracellular process. Action potentials arise in 'excitable cells' that, in humans, include neurons, muscle cells, and pancreatic beta cells.

active transport The process in which molecules move across cell membranes against a concentration gradient (i.e. from low to high concentrations) with the expenditure of energy. In primary active transport this energy is chemical energy in the form of adenosine triphosphate (ATP); in secondary active transport the energy is provided by a second molecule moving across the same membrane along its concentration gradient.

acyclic (of a graph) In graph theory, a graph—that is, a collection of points (vertices) and connecting edges—that contains no cycles. This means that it is impossible to start at one vertex and return to the same one by following a sequence of edges.

allele Any one of two or more different forms of the same gene that occur at the same position on the same chromosome; alleles have different bases at one or more places in their DNA sequence.

anonymization A process through which a set of data taken from an individual is stripped of any information that might lead to the identification of that person.

antiporter A protein that is embedded in the membrane of a cell or organelle and that is involved in active transport. Two different compounds move in opposite directions across the membrane through an antiporter, one with and the other against its concentration gradient.

apoptosis The process of programmed cell death, in which cells in multicellular organisms die in a controlled, regulated fashion in response to a complex series of biochemical signals.

application programming interface (API) A set of computational functions that specifies how different components of software interact with each other and with hardware devices. An API is often presented in human-readable (text) form as a library of specifications for data structures, routines, etc.

authentication The generally automatic process of checking the credentials of an individual requiring access to computers and data in order for access can be granted; a common authentication method involves the use of a username/password pair.

autophagy The metabolic process in which cellular components that are no longer needed are broken down or digested into smaller molecules by lysosomes.

avatar A computational representation of an individual as a virtual human, an alter ego, or character. In computational physiology, an avatar is the computational representation of an individual's genetics, metabolism, physiology, and disease that can be used for diagnosis or clinical decision making.

binding affinity The strength of the interaction between a small molecule (ligand) bound to a protein or other macromolecule experimentally by the thermodynamic free energy of the binding process. A large binding affinity derives from multiple strong interactions between atoms in the ligand and in the binding site.

biomarker A measured biological characteristic (such as the expression of a particular gene) that can be used as an indicator of a less easily measured condition (such as the likely presence or absence of a disease).

boundary conditions A set of conditions required to fully solve a set of mathematical equations, most often ordinary or partial differential equations in space and time. Such boundary conditions are often imposed at the edges of the system or spatial domain to which the differential equations apply, and at the initial time when the model commences. A molecular dynamics simulation of a protein in a box of water molecules is a typical example of a simulation involving boundary conditions imposed on Newton's equations of motion; in this case, periodic boundary conditions apply at the edges of the water box, with initial temporal boundary conditions that specify the positions and velocities of all the atoms within the simulation.

Branscomb pyramid A qualitative diagram representing the prevalence and availability of computing systems as a function of their speed. The commonest and least powerful systems, typically home and workplace desktop computers, are found at the base of the pyramid and the most powerful supercomputers at the apex.

causally cohesive genotype-to-phenotype (cGP) (modelling) A mathematical representation of the relationship between a genotype and the resulting phenotypes, showing how that genotype is manifested at all relevant levels from the molecular and cellular to the whole organism.

cellular automata A modelling technique that is based on specifying spatially and temporally discrete variables at points on a lattice and defining rules that determine how sites interact based on the state of neighbouring sites at each discrete timestep. Generally contrasted with modelling continuous variables using differential equations.

cluster computing A method of computing such that a single computer is assembled by connecting smaller computer components together so that they can be used as a single larger machine. Computer clusters are able to execute multiple serial and parallel codes, the performance of the latter being determined by the speed of the network interconnects between cores, processors, and nodes.

codon A triplet of three consecutive bases in a sequence of messenger RNA that codes for either a particular amino acid or a START or STOP signal.

cofactor A molecule (generally a small organic molecule) that must bind to an enzyme in order for that enzyme to catalyse a reaction but that is not changed during the reaction. Nucleotides and nucleotide phosphates are common cofactors.

complex automata A framework for computational modelling in which a multi-scale system is broken down into a number of interacting single-scale systems, primarily based on cellular automata, the complete model being built up from simpler models of each of those systems. Agent-based models are examples of complex automata.

computational fluid dynamics A branch of applied mathematics and engineering (but today of much broader concern) aimed at solving fluid flow problems (usually based on the Navier–Stokes equations) using numerical (computational) methods. Common biomedical problems that can be solved using computational fluid dynamics include those involving blood flow.

compute node A term for a component of a sizeable computer of a parallel architecture which contains its own set of cores, processors, and associated memory. Modern supercomputers contain many nodes and may have many thousands of computer cores, hundreds of nodes, and a complicated memory hierarchy.

computed tomography A medical imaging technique (also known as X-ray computed tomography or CT scanning) in which X-rays are transmitted through a human body (or other object) and the attenuated X-rays are measured at many angles. These projections are combined to produce images of slices through the body.

confidentiality (legal) The legal requirement for patient data to be kept confidential, meaning that only authorized individuals (such as the doctor treating that patient) are granted access to that data unless access is explicitly granted by the patient concerned.

coupling The process through which different submodels of a multi-physics model are connected together, or the degree to which they share data and rely on each other to function. The submodels in a tightly coupled system are more interdependent and share more data than those in a loosely coupled one.

data audit trail A secure set of chronological records providing the evidence of a sequence of manipulations of one or more sets of data.

data parallelism A form of parallel computing in which a dataset is split into a large number of pieces, each being distributed to a different processor or node, with the same algorithm being run on each packet.

Data Protection Directive (Directive 95/46/EC) A European Union directive that provides legal protection to individuals in the European Union concerning the use and processing of their personal data.

data warehouse A repository for data from a variety of sources that enables the data to be accessed and used by a project. The hardware components of the warehouse may, but need not, be in a single physical location.

DICOM (Digital Imaging and Communications in Medicine) A widely used standard format for imaging data, defining how those data should be stored, transmitted between computer systems, and displayed.

differentiation (cellular) The process in cell biology through which an immature cell type matures into a more specialized one. The least differentiated cells are stem cells, which can create progressively more differentiated cells through the process of cell division.

diploid cell A eukaryotic cell that contains two paired copies of each chromosome. All normal human cells apart from gametes (sex cells) are diploid. These cells are also known as somatic cells.

directive (in European Union law) A set of requirements laid out in European Union law which each member state is free to implement in its own way using its own legislation. One example is the Data Protection Directive.

domain (protein) A subsequence of amino acids within a protein that can fold independently into a stable structure with a similar structure to that part of the intact protein. One domain of a multidomain enzyme will contain all the amino acids that make up that enzyme's active site.

efficacy In pharmacology and medicine, the efficacy of a drug or other intervention is defined as its ability to exert an effect (such as alleviating pain or curing disease). By extension, the efficacy of a computational model is defined as its effectiveness as a tool to, for example, diagnose a disease or select an intervention.

e-health A generic term for all aspects of health research and healthcare that are supported by electronic (i.e. digital) communication and/or electronic processing tools.

e-infrastructure A generic term for computational hardware, software, and data processing, generally distributed across the global Internet.

electrophysiology The subdiscipline of physiology that is concerned with the electrical properties of cells and tissues, ranging in scale from the study of ion channels in cell membranes to the electrical properties of whole organs such as the heart.

epistasis A phenomenon in genetics in which the effect of a mutation in one gene depends on the allele or alleles present at one or more other genes.

epigenetics Any alteration in gene expression that is not related to the sequence of DNA bases along the gene. Epigenetic modifications to DNA include base methylation and changes to the way in which the DNA and its associated proteins are packaged into the cell nucleus.

epithelium One or more densely packed layers of cells that line the cavities and surfaces of many organs and tissues. The epithelium forms one of the four basic types of animal tissue; the others are connective tissue, nervous tissue, and muscles.

ex vivo Literally, Latin for 'out of the living'. In drug and medical device development, an *ex vivo* experiment is one that involves tissue taken from a living organism and used under controlled conditions.

finite element mesh In finite element modelling methods, three-dimensional objects are presented as a collection of finite volume elements. The properties of the medium are assigned at each point and may be used to calculate, for example, deformation under stress. These elements can be connected using a finite element mesh, and this defines the shape of the object.

Fourier transform A reversible mathematical transformation that can be used to transform a signal between the domains of real space and real time to reciprocal space and frequency. In magnetic resonance imaging (MRI) the spatial image is obtained from a pattern of magnetic resonance frequencies by inverse Fourier transformation while the absorption spectrum of frequencies is derived from the free induction decay of the signal by Fourier transformation.

gene action The gene action at a particular locus on the genome is a numerical formulation of the relationship between the phenotype of a heterozygote at the locus compared to those of homozygotes for the different alleles.

gene expression The process through which the information in a gene sequence is expressed in a cell or tissue as a functional RNA or protein.

genomics The study of the sequence, structure, and function of whole genomes: that is, of the complete complement of DNA in an organism.

genotype The genetic makeup of an individual, defined in relation to the sequence of each allele of one or more particular genes. Most commonly, an individual's genotype at a locus is defined as a combination of alleles, each of which may but need not be dominant or recessive.

Good Clinical Practice An international quality standard developed by the International Conference on Harmonization that provides a set of guidelines for the ethical conduct of clinical research, including clinical trials.

grid computing A distributed computational e-infrastructure set up to give authenticated, authorized users access to a wide range of distributed resources that are owned and operated by different organizations that are sometimes located in different countries. Cloud computing is a form of grid computing in which the resources are provided by a third party that seeks to make money from the enterprise.

GUI (graphical user interface) Any interface that allows users to interact with computers and other electronic devices through graphical icons rather than text in a command-line interface.

half-life Generically, the time in which any decaying quantity will fall to half its initial value, which is constant in exponential decay. The half-lives of radioactive isotopes (radionuclides) that are used in imaging techniques such as positron emission tomography determine whether and how they can be used effectively.

haploid cell A eukaryotic cell containing only a single copy of each chromosome. Thus, normal human gametes (egg and sperm cells) are described as haploid.

haplotype The specific pattern and order of alleles on a chromosome, or on a gene, or series of genes within a chromosome, that are inherited together, is termed its haplotype.

health technology assessment The evaluation of a health technology (drug, device, or other intervention) in terms of its efficacy, safety, cost-effectiveness, and any ethical or social implications of its implementation. This multidisciplinary process aims to sum up the value that the technology adds to established clinical practice.

homeostasis The regulation of a system by altering variables so that its overall condition remains constant despite changes in its environment. An important concept in physiology applying to, for example, the regulation of body temperature and the pH of body fluids.

homology Organisms or species are said to show homology (be homologous) if they share a common ancestor; that is, if they are related despite divergent evolution. In bioinformatics, nucleic acid or protein sequences are said to be homologous if they share a high degree of sequence similarity. Homology modelling techniques require considerable sequence similarity or their predictions will be unreliable.

hormone A biochemical that can alter cell metabolism by binding to a target protein, thus affecting one of many physiological and behavioural processes. Hormones are generally synthesized away from their target sites and transported there in the bloodstream.

hybridization The process of combining two complementary or near-complementary strands of nucleic acid to form a double-helical duplex. This occurs naturally but can also be induced; hybridization between messenger RNA and immobilized, complementary DNA sequences forms the basis of the ubiquitous microarray technology for the measurement of gene expression as messenger RNA.

hypothesis testing A procedure for determining whether there is enough experimental evidence to prove that a theory or hypothesis is true to a statistically significant level. In the physical sciences, computational modelling is regularly used in hypothesis testing for the formulation of predictions to be tested by experiment, and it is beginning to become established in such a mode in medical science, as this book attests.

informed consent The ability of an individual, most often a patient, to understand the implications of a test or procedure in order to give consent for it to be carried out. The need for informed consent raises particular problems for researchers involved with some groups of patients, including minors and those with mental impairments.

Infrastructure as a Service (IaaS) A model of grid or cloud computing in which the resource providers offer users remote access to the components of one or more 'virtual machines'—central processing unit (CPU), memory, and storage—for a fee. The users control and run programs on the remote resources as if they were local machines.

in silico A test of a drug or other procedure that is run as a simulation on a computer is referred to as an in silico test, by analogy with the more familiar terms in vitro (for chemical or cell-culture experiments) and in vivo (for experiments using living systems).

instance The single execution of a workflow with a particular set of specified inputs, parameters, and conditions.

intellectual property A legal concept through which authors and those who develop ideas and products have exclusive rights to their creations that can, but need not, lead to others being charged fees for their use or exploitation. The management of the intellectual property in a discovery forms a necessary part of its commercial exploitation.

interactome A term used to refer to the complete set of interactions between genes, proteins, and other molecules in a cell. The term has been coined by analogy with genome, proteome, and other 'omes'.

interoperability If two systems (such as computer programs or web services) are able to exchange data and to consume the data that have been exchanged, then those systems are said to be interoperable. Semantic interoperability requires each system to interpret the data in the same way.

ion channel A protein that is embedded in a cell membrane and that controls the passage of ions into or out of the cell. Ion channels have many important functions in physiology, controlling processes that involve rapid change at a cellular level, including heartbeat and nerve responses.

ischaemia A restriction in the blood supply to a tissue, generally caused by damage to or restriction in one or more blood vessels.

isotropy In general terms, a substance or object is isotropic if it is uniform in all directions. Conversely, anisotropy is used to describe a substance that has different properties in separate directions (such as human bone).

kinematics A branch of classical mechanics that is used to study objects and systems of objects in motion. In computational physiology, it can be used to model moving organs and smaller structures (e.g. muscles, the heart).

linkage disequilibrium If two or more alleles occur together in the same individual at a higher frequency than would be predicted from their frequencies in the whole population, then those alleles are said to be in linkage disequilibrium.

lumen Most commonly, the space within a tubular anatomical structure, such as a blood vessel or intestine. The term can also be used by analogy at the subcellular level to refer to the space within a tubular organelle such as the endoplasmic reticulum.

magnetic resonance imaging (MRI) A medical imaging technique in which the image is produced from a radiofrequency signal usually generated by the nuclei of hydrogen atoms in the body when a patient is placed in a strong, external magnetic field.

Markov model A type of mathematical model or simulation that depends on the assumption that the probability of a particular future state arising in a process depends only on its present state; that is, it has no dependence on its past states. Hidden Markov models, in which the current state is only partially observable but the output is known, are common in inference-based modelling.

markup language A system for annotating or marking up text with codes that give additional meaning to that text. The Extensible Markup Language (XML) is a generic markup language that has been designed to be both human- and machine-readable. XML format is used in numerous specialist markup languages, some of which (e.g. CellML, Systems Biology Markup Language or SBML) are useful in computational biomedicine.

medical device Any piece of apparatus or equipment, including computer hardware and software, that has been designed to be used specifically for detecting, diagnosing, or treating human disease.

Message Passing Interface (MPI) The standard application programming interface (API) for parallel programming, which manages the passing of messages between computational processes running in parallel on separate nodes, processors, or cores on parallel machines.

metabolite Any small or medium-sized molecule that forms a substrate, intermediate, or product for one or more metabolic reactions within a living cell. The term metabolome refers to the sum total of all the metabolites in a single cell or cell type, and metabolomics is the study of the metabolome.

metadata Digital information that is accessory to or describes the main data held in a data file; for example, the metadata associated with a patient data record might include the date and time the sample was taken and demographic data about the patient.

microarray An ordered two-dimensional array of fragments of nucleic acids or proteins [most often complementary DNA (cDNA)] that is used to detect or measure molecules that bind to those fragments in biological samples in a high-throughput manner.

microstructure The structural properties of a natural or synthetic material that can be visualized or measured at a scale that is appropriate for a light microscope (i.e. at the micron level). The microstructure of bone and some other human tissues can be useful as a biomarker (*which see*) of disease presence or progression.

middleware A layer of software that sits between a computer operating system and specific tools and applications in distributed computer systems that provides generic services such as input/output, file transfers, resource allocation, and communication between resources.

molecular dynamics The simulation, according to Newton's laws of motion, of a collection of atoms and molecules (e.g. a protein and its ligands in a solvent 'bath') at a particular temperature and pressure, over a timescale that today is generally recorded in tens to hundreds of nanoseconds, but with the advent of increasingly powerful computers is now reaching the millisecond domain.

molecular machine A complex of proteins and sometimes nucleic acids or other macromolecules that is large on the molecular scale and that carries out a biological function involving a conformational change. Examples include ribosomes and ATP synthase, which catalyse the synthesis of proteins and ATP respectively.

Monte Carlo simulation A stochastic simulation technique that uses random-number sampling to produce a range of numerical values and a probability distribution for an outcome. Monte Carlo techniques involve multiple runs of a simulation with different random or near-random starting values and have many applications in computational biomedicine.

morphology The physical properties (appearance, shape, colour, etc.) of an organism, an organ, or a smaller part of a living organism. The term morphology is often compared to and distinguished from physiology, which is more concerned with function.

multi-scale modelling Any computational modelling procedure that can be broken down into submodels involving distinct spatial and/or temporal scales. For example, a model of cardiac dynamics might involve separate models involving ion channels, heart muscle cells, and the gross anatomy of the heart, comprising electrophysiology and mechanical motion.

Navier–Stokes equations The mass, momentum, and energy-conservation equations that describe the motion of fluids in mathematical terms, derived by applying Newton's laws of motion to fluids viewed as comprised of continuous matter. Computational fluid dynamics (*which see*) is based on the numerical solution of these equations. Applications of the Navier–Stokes equations to physiology include modelling the flow of blood through the human arterial tree.

ontology In computer and information science, a representation of a body of knowledge as a structured, hierarchical vocabulary of terms and definitions which enable one domain to be related to another. Many ontologies have been introduced as controlled vocabularies in different subdisciplines of the biomedical sciences; the best known of these is perhaps the Gene Ontology.

organelle A subcellular component that is specialized for a particular function; most biologists now only use the term to refer to those subcellular components that are surrounded by membranes. Examples of organelles in animal cells that match this definition include the nucleus, mitochondria, vacuoles, and endoplasmic reticulum.

osmotic gradient The difference between the concentration of a solute on one side of a semi-permeable membrane and the concentration of the same solute on the other side of the membrane. The process of osmosis involves the passage of solvent molecules across the membrane into the more concentrated solution, tending to equalize the concentrations.

P4 medicine A vision of medicine as predictive, preventive, personalized, and participatory (P4) formalized by Leroy Hood, one of the pioneers of systems biology. Computational biomedicine is an essential component of P4 medicine, which it is hoped will improve healthcare while decreasing cost.

parallelism In this book, the term parallelism is used to refer to parallel computing, that type of computation in which a problem is divided into many subcomponents which are solved simultaneously on multiple compute nodes or processors, using the Message Passing Interface (*which see*).

parameter A measured or calculated numerical quantity that defines part of a system and that can be incorporated into a model as a constant value (as opposed to a variable). Parameters used in computational biomedicine may be universal but often differ between individual patients.

patient identity management system (PIMS) A system for controlling and managing the assignation of pseudonyms and other identifiers to patient data records to ensure that data can always be traced to the correct patient if necessary, which is an important factor in maintaining data quality.

perfusion The process through which blood is delivered to the network of capillaries within a tissue or organ and thence to the tissue or organ concerned.

petaflops The term flops (floating point operations per second) is an important measure of the power of a computer system. Currently (2014) the most powerful computers available globally are petaflops machines, capable of executing over 10^{15} flops (peta = 10^{15}).

Petri net A mathematical concept in which a distributed system including events and conditions is defined and represented graphically, and progress through the system represented by the consumption of tokens. Petri nets can be useful formulations in the development and analysis of workflows.

phase (of clinical trials) Any medicine or medical device must be tested on patients before it can be licensed, and these tests (clinical trials) are divided into four phases, from phase I (toxicity tests on a few patients or volunteers) to phase IV (post-registration monitoring involving thousands of patients).

phenotype The observable characteristics of an organism, comprising its appearance, physiology, and (where appropriate) behaviour. The phenotype of an organism is derived from both its genetic composition (genotype) and its environment.

pipeline A set of computer programs, processes, or simulations that are connected in series so that the output of one program becomes the input of the next. The steps in a pipeline do not need to be executed on the same system or even in the same country. A pipeline is a particular linear form of a more general workflow (*which see*).

pleiotropy A phenomenon in genetics in which the function of a single gene affects several unrelated physiological systems. A deleterious mutation in a gene of this type may therefore cause apparently unrelated symptoms in several organs or systems.

positron A fundamental particle with an electric charge of +1 electron unit (+1e) and the same mass as an electron. A positron is the anti-matter equivalent of an electron (charge −1e); collision between a positron and an electron results in the annihilation of both particles and the release of two photons of high-energy electromagnetic radiation (gamma rays).

positron emission tomography A nuclear imaging technique in which radioactive particles that emit positrons (*which see*) are injected into the patient. Each positron travels a short distance before colliding with an electron and annihilating to produce two gamma ray photons. These photons are collected by a scanner to produce an image of the uptake of the original radionuclides.

post-translational modification Any chemical modification to a protein that is made after that protein has been synthesized at the ribosome; these include the addition of phosphate groups and simple sugar molecules to amino acids and the cross-linking of the protein chain via disulphide bonds.

proteome The complete complement of protein molecules expressed by the genome in a particular cell (or cell type) at a particular time. As various sets of proteins are expressed in different cells under diverse conditions, a multicellular organism has a single genome but many proteomes.

provenance Provenance in computer science refers to the origin, source, and thus history of a process or a data object. The provenance of a database item, for example, might include when, why, how, and by whom that object was created; the provenance of a measured parameter will include how it was obtained.

pseudocode A piece of structured text that describes the operation of a computer program or routine step by step. Pseudocode is not written in any recognizable computer language and is easier for humans to read than such languages.

pseudonymization The process through which data obtained from individual patients are stripped of information that links them to the patients concerned, while retaining a code that allows the link to be restored if required (e.g. if the data suggest that a patient would benefit from a treatment under test) by means of a trusted third party (*which see*).

quantitative genetics The genetics of continuously varying and measureable traits, such as body height in humans, caused by the effects of multiple genes and the environment, so treated using statistics, as distinct from traits that take a distinct value in each individual such as eye colour or the presence or absence of a single-gene (Mendelian) disorder.

regulatory network A network representing the interactions between genes in which a pair of genes is connected if one regulates the other; that is, if the expression level of one gene controls that of the other. The connections between nodes (genes) in a regulatory network are directional, one direction representing the relationship 'regulates' and the other 'is regulated by'.

repository In general terms, a repository is a location where a collection of objects are stored. In this book, the term is used to refer to a storage location on the Internet from where programs, models, or database entries may be obtained and installed on a potential user's computer. Most if not all the repositories referred to in this book are open access; that is, their contents may be accessed free of charge.

reproducibility An experiment or simulation can be described as reproducible if it is capable of being re-run by someone other than and independent from the original investigator and produces the same results. It is a fundamental tenet of the scientific method that its findings are reproducible.

resource broker The software component of a computational grid system that is responsible for distributing jobs between the resources available on the grid, taking into account the demands that each job will make on storage, memory, and central processing unit (CPU) and the current load on the grid.

root mean square error (RMSE) A statistical parameter that is often used to measure the deviation between a group of experimentally determined values and the values predicted for those measurements. It is obtained by measuring the difference between each pair of actual and predicted values, squaring each, and taking the square root of the mean of the squared values.

scale separation map A schematic diagram with axes representing space and time, on which each submodel within a multimodel system can be represented by a non-overlapping domain that describes its spatial and temporal scales.

sensitivity analysis An assessment of a predictive model that describes the extent to which variations in the model outcome can be attributed to variations in the input data and/or its parameters.

sequential workflow A workflow is defined as sequential if each place in the workflow is limited to one input and one output; it may then be referred to as a pipeline (*which see*). There is therefore no possible conflict in the logic flow through the workflow, although concurrent processes are allowed. An alternative workflow formulation is the state machine (*which see*); the difference between sequential workflows and state machines is best understood when visualized using Petri nets (*which see*).

signal transduction The process through which a molecule (the ligand) binds to part of a target receptor protein, causing a conformational change in the target that leads to one or more defined biochemical reactions within the target cell. In many cases the receptor is bound to a membrane, with the ligand binding occurring on the outside of the cell and the signal response in its interior.

single nucleotide polymorphism (SNP) A polymorphism in a DNA sequence in which a change (e.g. from a nucleotide base A to C) occurs at one base position only, leaving the neighbouring bases unchanged. SNPs occur in coding and non-coding regions of DNA, and even if they occur in coding regions they may be silent (have no effect on the

protein sequence) if the base change has no effect on the amino acid coded for.

solute Any substance that is dissolved in a solvent to form a solution.

splicing In eukaryotes, the post-translational modification of newly translated messenger RNA sequences to remove regions that do not code for protein sequences (the introns) and join the coding sequences (exons) together. Often exons can be skipped or joined in different orders to form splice variants, enabling several different protein sequences to be coded by the same gene.

state machine A workflow is defined as a state machine if each transition in the workflow is limited to one input and one output. This allows for conflict but not concurrent processes in the workflow. An alternative formulation is a sequential workflow (*which see*); the difference between sequential workflows and state machines is best understood when visualized using Petri nets (*which see*).

state transaction A transition between two states within a workflow. Each state encapsulates an action or a set of actions.

stem cells Immature, undifferentiated cells that have the capacity to both differentiate into specialized cell types and divide to produce more stem cells. Mammals including humans have two distinct types: embryonic stem cells from blastocysts, which can differentiate into any cell type, and adult stem cells, which can normally only differentiate into cells in a particular lineage (e.g. blood cells).

stirred tank An approximation often used in chemical engineering that states that the liquid contents of a vessel can be modelled as homogeneous. In computational biomedicine the stirred-tank model is frequently applied to modelling cytoplasm within cells.

substrate A molecule that binds to the active site of an enzyme and is chemically changed by the reaction that the enzyme catalyses into the product(s) of the reaction, which are released from the active site. An enzyme may have one or more substrates.

system dynamics (or non-linear system dynamics) A mathematical approach, based on differential equations (or their spatially and/or temporally discrete equivalents), to the study of the dynamical behaviour of complex systems over time, incorporating feedback loops and time delays.

systems biology and systems medicine A multidisciplinary approach to biology and medicine that combines experimental observations and models at multiple levels, which may range from the molecular to the whole organism or population. Often contrasted with the reductionist approach that assumes that a biological system will eventually be understood from its component parts, starting with the genes.

task parallelism A form of parallel computing in which the execution processes that operate on a single dataset are split and distributed to different nodes, with a different task run on each node. The term inherently parallel is synonymous with task parallel.

topology A mathematical description of the shape of an object, which may equally be applied to the graphical representation of a sequence or flowchart. Topology may be used to describe the way that submodels are joined together to form a single multi-scale or multilevel systems medicine model.

transcription The process through which DNA is copied by the enzyme RNA polymerase into a molecule of RNA with a complementary sequence. RNA transcription is the first step of the process of gene expression, in which genetic information is encoded into proteins.

translation (protein) The process through which a protein molecule is synthesized from its constituent amino acids by ribosomes, which translate each three-base codon of the RNA sequence into its corresponding amino acid via the genetic code.

translational medicine The activity of translating discoveries in basic biomedical science into medicines, diagnostic tools, procedures, and other advances of practical use in clinical medicine.

trusted third party A data-storage facility that is separate from but trusted by all the organizations that are involved in the generation and analysis of patient data. When data is pseudonymized (*which see*), the keys that connect data records to the patients concerned will generally be stored with a trusted third party.

ultrasound Sound waves at frequencies higher than the upper limit of normal human hearing are termed ultrasound. The term can also be applied (as it is in this book) to refer to a medical imaging technique based on the detection of ultrasound waves that have been reflected from structures within the body.

validation The process of testing the inherent accuracy of a computational model by comparing results obtained by the model with those of measurements obtained from comparable controlled experiments.

verification The assessment of the error arising in solving a mathematical model of a process. Verification can be split into code verification, which investigates the accuracy to which the mathematical problem can be solved using a particular algorithm, and model verification, which investigates the accuracy to which that model reflects the biological process concerned.

virtualization The abstraction of computational resources so that the details of a hardware or software system are hidden from its users. Virtualization is often used in cloud and grid computing.

workflow A set of connected steps or operations in which each step follows on automatically from the preceding one. Workflows have many applications including within complex multi-scale modelling when the output from one submodel becomes the input to the next, including the possibility of concurrent information exchange.

workflow engine A computational tool that controls the way that data passes through the components of a workflow in individual instances, calling workflow components or submodels as they are required, for example keeping a job in a queue if the next component it requires is busy.

Young's modulus (or Young modulus) A numerical parameter that describes the stiffness of an elastic material, and that is defined as the ratio of the stress along an axis of the material to the strain along that axis resulting from the stress. It therefore describes the amount by which a material distorts if stretched or shortens if compressed.

zygote In sexual reproduction, the single cell that is formed when the two gametes fuse (i.e. when an egg is fertilized by a sperm).

INDEX